Initial-Boundary Value Problems and the Navier-Stokes Equations

This is Volume 136 in
PURE AND APPLIED MATHEMATICS

A list of titles in this series appears at the end of this volume.

Initial-Boundary Value Problems and the Navier-Stokes Equations

Heinz-Otto Kreiss
Jens Lorenz
Applied Mathematics
California Institute of Technology
Pasadena, California

ACADEMIC PRESS, INC.
Harcourt Brace Jovanovich, Publishers
Boston San Diego New York
Berkeley London Sydney
Tokyo Toronto

ACADEMIC PRESS, INC.
1250 Sixth Avenue, San Diego, CA 92101

United Kingdom Edition published by
ACADEMIC PRESS, INC. (LONDON) LTD.
24-28 Oval Road, London NW1 7DX

Library of Congress Cataloging-in-Publication Data

Kreiss, H. (Heinz)
Initial-boundary value problems and the Navier-Stokes equations / Heinz-Otto Kreiss, Jens Lorenz.
p. cm. — (Pure and applied mathematics ; v. 136)
Bibliography: p.
Includes index.
ISBN 0-12-426125-6
1. Initial value problems. 2. Boundary value problems. 3. Navier-Stokes equations. I. Lorenz, Jens. Date— . II. Title.
III. Series: Pure and applied mathematics (Academic Press) : 136.
QA3.P8 vol. 136
[QA378]
510 s—dc 19 88-7618
[515.3'5] CIP

Printed in the United States of America
89 90 91 92 9 8 7 6 5 4 3 2 1

Contents

Introduction

The aim of this book is to develop a theory of initial-boundary value problems for linear and nonlinear partial differential equations. There are already many books available, and we shall list some of them after the introduction. However, the area is vast, and any one book can only treat certain aspects of the theory. Our choice of material is very much influenced by the availability of fast computers. They have made it possible to solve rather complex problems for which the classical theory of second-order equations is not adequate. Existence and regularity questions play a fundamental rôle in computations because the resolution required depends on the smoothness of the solution, and there is always the danger that one tries to compute things which do not exist. Another fundamental question concerns admissible boundary conditions, which we shall discuss in great detail. In computations the boundary conditions cause most of the problems.

We believe that this book can fill a gap between elementary and rather abstract books. To illustrate our theory, we have chosen the compressible and incompressible Navier-Stokes (N-S) equations, which describe fluid flows ranging from large scale atmospheric motions to the lubrication of ball bearings. The choice was dictated by the desire to find a system which is so rich in phenomena that the whole power of the mathematical theory is needed to discuss existence, smoothness and boundary conditions. We hasten to add, however, that we only scratch the surface of the diversity in which its solutions can behave. For example, turbulent flow is described by the N-S equations, and at present no adequate mathematical theory is available.

There are different ways to develop the theory. One way is to start with weak solutions and then to discuss their smoothness. This approach can lead to difficulties when it comes to boundary conditions and to nonlinear equations. For numerical calculations one is very much interested in knowing the exact smoothness behavior "up to" the boundary. Also, for nonlinear problems it is often difficult to show that the weak solutions have sufficient regularity (smoothness) in order to make them unique.

We proceed instead in the following way: First we show, by using difference approximations for linear problems and linearization for nonlinear problems, that there is a set of C^∞-smooth data, dense in L_2, for which the equations we discuss have C^∞-smooth solutions. These solutions and their derivatives can be estimated in terms of the data. Then we use the usual closure argument to define weak solutions if the data are less smooth. This process is much closer to computing than the previous one: If one wants to compute solutions with discontinuous data then one obtains better results if one approximates the discontinuous data by smoother data.

Also, in computations of solutions of nonlinear problems, one adds terms in the equation (numerical dissipation) so that the solutions do not develop discontinuities. We proceed analogously in the analytic theory of hyperbolic equations and treat these as a limit of parabolic ones. The latter have C^∞-smooth solutions for C^∞-smooth data. For the Euler equations (which are identical with the N-S equations without viscous term) this corresponds to replacing the inviscid equations by the viscid equations. In general, this process of adding a *formally small* higher-order derivative term is doubtful: There is no assurance that one obtains anything meaningful in the limit as the term goes to zero. However, for problems coming from applications, the extra terms are often present in the full equations and have only been neglected – without mathematical justification – to formally simplify the equation.

We would like to thank our students who partly read the manuscript and suggested valuable improvements. Thanks is due also to Linda Soha, who expertly typed a large part of the manuscript. We wish to acknowledge the support of our research by the National Science Foundation (contract number DMS-8312264), by the Office of Naval Research (contract number N-00014-83-K-0422), and by the Department of Energy (contract number DE-AS03-76ER72012).

Text and Reference Books

Courant, R., and Hilbert, D. (1962). "Methods of mathematical physics", Vol II, Interscience.

Friedman, A. (1964). "Partial differential equations of parabolic type", Reprint, Krieger, 1983.

Garabedian, P.R. (1964). "Partial differential equations", Wiley.

Hellwig, G. (1964). "Partial differential equations, an introduction", Blaisdael.

Henry, D. (1981). "Geometric theory of semilinear parabolic equations", Lecture Notes in Math. 840, Springer.

John, F. (1971). "Partial differential equations", Springer.

Mizohata, S. (1973). "The theory of partial differential equations", Cambridge University Press.

Petrovskii, I.G. (1954). "Lectures on partial differential equations", Interscience.

Treves, F. (1975). "Basic linear differential equations", Princeton University Press.

Weinberger, H.F. (1965). "A first course in partial differential equations with complex variables and transform methods", Wiley.

1 The Navier-Stokes Equations

In this preliminary chapter we first outline some questions which will be treated in this book. Then we derive the Navier-Stokes equations. Though the derivation will not be used later, it is of interest to understand the underlying logical and physical assumptions, because the mathematical theory of the equations is not complete. There is no existence proof except for small time intervals. Thus it has been questioned whether the N-S equations really describe general flows. If one changes the stress tensor such that diffusion increases when the velocities become large, then existence can be shown. This change of the equations does not seem to be justified physically, however. For example, certain similarity laws – valid for the Navier-Stokes equations – are well-established experimentally, but the modified equations do not allow the corresponding similarity transformations. Possibly a lack of mathematical ingenuity is the reason for the missing existence proof, and the N-S equations are physically correct.

The N-S equations form a quasilinear differential system, and much of our understanding of such systems is gained through the study of *linearized* equations. These will, in general, have *variable* coefficients. By *freezing* the coefficients in such a problem, one obtains systems with *constant* coefficients. It is much easier to analyse the latter, as will be later shown in Chapter 2. However, the relation between variable-coefficient and constant-coefficient equations is not trivial. The fundamental ideas of linearization and localization are discussed in Section 1.3.

1.1. Some Aspects of Our Approach

1.1.1. The Equations, Initial and Boundary Conditions

Neglecting effects due to thermodynamics, i.e., assuming constant entropy, the full N-S equations consist of the three momentum equations

$$\rho \frac{D\mathbf{u}}{Dt} + \text{grad } p = \mathbf{P} + \rho \mathbf{F},$$

the continuity equation

$$\rho_t + \text{div}\,(\rho \mathbf{u}) = 0,$$

and an equation of state

$$p = r(\rho).$$

Here the velocity $\mathbf{u} = (u, v, w)$, the density ρ, and the pressure p are the unknowns, $\mathbf{F}$ is a given forcing, and $r(\rho)$ with $dr/d\rho > 0$ is assumed to be a known function. Furthermore, we have used the common notations

$$\text{grad}\, p = \begin{pmatrix} p_x \\ p_y \\ p_z \end{pmatrix} = \begin{pmatrix} \partial p/\partial x \\ \partial p/\partial y \\ \partial p/\partial z \end{pmatrix},$$

$$\text{div}\,(\rho \mathbf{u}) = (\rho u)_x + (\rho v)_y + (\rho w)_z,$$

$$\frac{D}{Dt} = \frac{\partial}{\partial t} + u\frac{\partial}{\partial x} + v\frac{\partial}{\partial y} + w\frac{\partial}{\partial z}.$$

The term $\mathbf{P}$ is defined by

$$\mathbf{P} = (\mu + \mu')\text{grad}\,(\text{div}\,\mathbf{u}) + \mu\Delta\mathbf{u} = (\mu + \mu')\begin{pmatrix} (\text{div } \mathbf{u})_x \\ (\text{div } \mathbf{u})_y \\ (\text{div } \mathbf{u})_z \end{pmatrix} + \mu \begin{pmatrix} \Delta u \\ \Delta v \\ \Delta w \end{pmatrix}$$

where Δ denotes the Laplacian, and μ and μ' are nonnegative material constants, assumed to be known.

In most applications one deals with simplified equations:

Case 1. *Incompressible* flow, $\rho = \rho_0$ a known constant, the equation of state is dropped. Most of the mathematical theory is done for this case. After choosing suitable units, one can assume $\rho = 1$, and the equations read

$$\frac{D\mathbf{u}}{Dt} + \text{grad } p = \nu\Delta\mathbf{u} + \mathbf{F}, \quad \text{div } \mathbf{u} = 0 \quad (\nu = \mu/\rho_0).$$

Note that the momentum equation and the equation div $\mathbf{u} = 0$ are of different type. This makes the existence theory more difficult. Indeed, the small-time existence theory for the *compressible* equations is somewhat easier than the theory for the incompressible case. However, as we will show, the vorticity formulation allows a systematic treatment of the incompressible equations. Using the idea of initialization for problems with different time scales, one can also treat the incompressible case as a limit of compressible ones:

Case 2. Compressible but *inviscid* flow, i.e., $\mu' = \mu = 0$. One obtains the full equations with $\mathbf{P} \equiv 0$. The resulting system – possibly augmented by an energy equation – is fundamental in gas dynamics;

Case 3. Inviscid, incompressible flow. The equations are known as **Euler equations**:

$$\frac{D\mathbf{u}}{Dt} + \text{grad } p = \mathbf{F}, \quad \text{div } \mathbf{u} = 0.$$

The differential equations have to be supplemented by initial and boundary conditions. The mathematically easiest case is the **Cauchy problem** with periodic initial data, where one seeks for solutions periodic in space. Loosely formulated, the following results are known for the viscous equations:

1. If the initial data are sufficiently smooth, then there is a time $T > 0$ such that the N-S equations have a unique smooth solution for $0 \le t \le T$. The time T depends on the initial data.
2. In the incompressible case one can say more: The solution is analytic for $0 < t \le T$. Furthermore, if the velocity $\mathbf{u}$ satisfies a bound in maximum norm* like

$$\max_{x,y,z} |\mathbf{u}(x, y, z, T)| = |\mathbf{u}(\cdot, T)|_\infty \le C,$$

then the solution can be continued for a time $\Delta t > 0$, which only depends on C. This latter result guarantees – for the viscous incompressible case – that the solution can be continued as long as the velocities stay uniformly bounded. In other words, it is *not* possible that the solution ceases to exist because just derivatives of $\mathbf{u}$ become large; $\mathbf{u}$ has to become large itself. In experiments extremely large velocities are not observed. The maximal velocity hardly ever exceeds twice the mean velocity. This nourishes the hope that it might be possible to derive uniform bounds for $\mathbf{u}$ and thus

*By $|\mathbf{u}| = (u^2 + v^2 + w^2)^{1/2}$ we denote the Euclidean length of a vector $\mathbf{u}$. The notion *maximum norm* refers to maximization w.r.t. to the arguments x, y, z.

to prove existence at all times. Thus far, however, this has not been achieved.

In practical applications one studies flows in finite domains and most of the interesting phenomena happen near the boundary. Therefore, it is of great interest to study *initial-boundary* value problems. Indeed, most of the results formulated for the Cauchy problem can be carried over to suitably posed initial-boundary value problems using the general theory of partial differential equations. It is one of the main aims of the second part of these notes (starting in Chapter 7) to describe techniques for choosing correct boundary conditions, i.e., boundary conditions which lead to a mathematically well-posed problem.

1.1.2. Shocks and Weak Solutions

For the *inviscid* compressible case of the N-S equations the solution can cease to exist though **u** stays uniformly bounded. This can be illustrated by the inviscid **Burgers' equation**:

$$u_t + \frac{1}{2}(u^2)_x = 0. \tag{1.1.1}$$

Discontinuities (shocks) for $u(x,t)$ can develop in finite time though u stays uniformly bounded. We will show this by the method of characteristics. Thus a classical solution can cease to exist in a finite time. To obtain existence of a solution for *all* time, one can broaden the solution concept and allow for *weak* solutions. Let us illustrate this for equation (1.1.1) together with an initial condition

$$u(x,0) = f(x), \tag{1.1.2}$$

where $f(x)$ is a smooth* function with compact support. Later we shall show that the (regularized) problems

$$u_t + \frac{1}{2}(u^2)_x = \epsilon u_{xx}, \qquad \epsilon > 0, \tag{1.1.3}$$

together with the initial condition (1.1.2) have a unique smooth solution $u_\epsilon(x,t)$ existing for all time. Furthermore,

$$|u_\epsilon(\cdot,t)|_\infty \le |u_\epsilon(\cdot,0)|_\infty, \qquad \int_{-\infty}^{\infty} |u_{\epsilon x}(x,t)|\, dx \le \int_{-\infty}^{\infty} |u_{\epsilon x}(x,0)|\, dx.$$

*A function is called *smooth* if it has sufficiently many continuous derivatives. We use this convenient term whenever more precise smoothness specifications are unimportant for our arguments.

This allows us to show that there is a unique function $u_0(x,t)$, satisfying the above inequalities, such that for every smooth test function $\phi(x,t)$ with compact support

(1.1.4)

$$\lim_{\epsilon \to 0} \int_0^\infty \int_{-\infty}^\infty u_\epsilon^j(x,t)\phi(x,t)dx\,dt = \int_0^\infty \int_{-\infty}^\infty u_0^j(x,t)\phi(x,t)dx\,dt, \quad j = 1,2.$$

The function u_0 is called a weak solution of the problem (1.1.1), (1.1.2).

It is desirable to introduce and characterize weak solutions in a more direct way, without studying a regularization. To this end, let $\phi = \phi(x,t)$ denote a smooth function with compact support. If we multiply (1.1.3) by ϕ and integrate over

$$-\infty < x < \infty, \qquad 0 \le t < \infty$$

we find after integration by parts

$$\begin{aligned} &\epsilon \int_0^\infty \int_{-\infty}^\infty u_\epsilon \phi_{xx}\,dx\,dt \\ &\quad = -\int_0^\infty \int_{-\infty}^\infty (u_\epsilon \phi_t + \frac{1}{2}u_\epsilon^2 \phi_x)\,dx\,dt - \int_{-\infty}^\infty u_\epsilon(x,0)\phi(x,0)\,dx. \end{aligned}$$

Using the initial condition and sending $\epsilon \to 0$, one finds from (1.1.4) that u_0 satisfies

$$\int_0^\infty \int_{-\infty}^\infty (u\phi_t + \frac{1}{2}u^2\phi_x)\,dx\,dt + \int_{-\infty}^\infty f(x)\phi(x,0)\,dx = 0. \tag{1.1.5}$$

This relation does not involve the ϵ-dependent equation (1.1.3), and thus it motivates the

Definition. A function $u(x,t)$ is called a weak solution of (1.1.1), (1.1.2) if (1.1.5) holds for all test functions ϕ.

Our derivation shows that $u_0(x,t)$ is *a* weak solution. It turns out, however, that the given definition does not characterize a *unique* function. There is, in general, more than one weak solution if we accept the definition given above.

From our point of view (and for physical reasons) it is natural to define weak solutions of the inviscid equations as the limit of solutions of the viscous equations. But unfortunately, for more general equations or systems of equations, one generally does not know whether this limit exists. Therefore, it seems to be advantageous to define weak solutions *directly*, without going through such a limit process. This is usually done by requesting an integral relation for certain test functions, as in (1.1.5), for example. In this way one can make the concept of a solution so weak that solutions do exist. However, often *uniqueness* becomes a problem. For the N-S equations there is no solution concept known which leads to both existence and uniqueness.

1.1.3. Scaling Arguments and Limiting Processes

One of the most powerful tools for discussing properties of the N-S equations are scaling arguments. As an example we shall discuss some aspects of these arguments for the 2D inviscid compressible equations, which we write in the form

(1.1.6)

$$\rho(u_t + uu_x + vu_y) + p_x = 0, \quad \rho(v_t + uv_x + vv_y) + p_y = 0,$$

$$\rho_t + u\rho_x + v\rho_y + \rho(u_x + v_y) = 0, \qquad p = r(\rho).$$

Assume that for a given flow we can scale the dependent and independent variables by

$$u = Uu', \quad v = Uv', \quad \rho = Q\rho', \quad p = Pp',$$

$$x = Lx', \quad y = Ly', \quad t = Tt',$$

so that u', v', ρ', p' and their first derivatives are of order $O(1)$. Here U, Q, P, L, T are positive constants and Q, P are assumed to be compatible with the equation of state, i.e., $P = r(Q)$. If, for example, $p = \rho^\gamma$ then $P = Q^\gamma$, and thus $p' = \rho'^\gamma$; i.e., the equation of state remains unchanged by the transformation. Introducing the new variables into the equations above, we find

$$\rho'\left(u'_{t'} + \frac{UT}{L}(u'u'_{x'} + v'u'_{y'})\right) + \frac{UT}{L}\frac{1}{M^2}p'_{x'} = 0, \quad M^2 = \frac{U^2Q}{P},$$

$$\rho'\left(v'_{t'} + \frac{UT}{L}(u'v'_{x'} + v'v'_{y'})\right) + \frac{UT}{L}\frac{1}{M^2}p'_{y'} = 0,$$

$$\rho'_{t'} + \frac{UT}{L}\left(u'\rho'_{x'} + v'\rho'_{y'} + \rho'(u'_{x'} + v'_{y'})\right) = 0.$$

Assume now that $UT/L = 1$ and that the equation of state remains unchanged by the transformation. If we drop the $'$ sign in the notation, the equations become

$$\rho(u_t + uu_x + vu_y) + \frac{1}{M^2}p_x = 0,$$

$$\rho(v_t + uv_x + vv_y) + \frac{1}{M^2}p_y = 0,$$

$$\rho_t + u\rho_x + v\rho_y + \rho(u_x + v_y) = 0,$$
$$p = r(\rho).$$

It follows from these equations – as we shall discuss later – that the type of the flow depends crucially on the so-called Mach number M introduced above. If $M < 1$, $M = 1$, $M > 1$, then the flow defined by the state $\big(U, P = r(Q)\big)$ is called *subsonic, transsonic*, or *supersonic*, respectively.

Consider, for example, the case $M \to \infty$. Formally we obtain in the limit the equations

$$u_t + uu_x + vu_y = 0,$$
$$v_t + uv_x + vv_y = 0,$$
$$\rho_t + u\rho_x + v\rho_y + \rho(u_x + v_y) = 0.$$

Here the pressure has been eliminated, and the velocity equations are uncoupled from the density equation.

We can also obtain the incompressible equations by a limit process from the equations governing compressible flows. To this end, assume that

$$\rho = \rho_0 + Q\rho', \quad \rho_0 = \text{const}, \quad |Q| << 1, \quad \rho' = O(1).$$

Thus the density is almost constant. For the pressure we obtain that

$$p = r(\rho) = r(\rho_0 + Q\rho') \approx r(\rho_0) + Q\frac{dr}{d\rho}(\rho_0)\rho',$$

and therefore the pressure is correctly scaled if we write

$$p = p_0 + Pp', \quad p_0 = r(\rho_0), \quad P = Q\frac{dr}{d\rho}(\rho_0).$$

Introducing $\rho = \rho_0 + Q\rho'$, $p = p_0 + Pp'$ and using the same scalings of u, v, x, y, t as above, we obtain from (1.1.6) the following system, where again $'$ is dropped in the notation:

$$(\rho_0 + Q\rho)\big(u_t + \frac{UT}{L}(uu_x + vu_y)\big) + \frac{UT}{L}\frac{Q}{M^2}p_x = 0,$$
$$(\rho_0 + Q\rho)\big(v_t + \frac{UT}{L}(uv_x + vv_y)\big) + \frac{UT}{L}\frac{Q}{M^2}p_y = 0,$$

$$\rho_t + \frac{UT}{L}(u\rho_x + v\rho_y) + \frac{(\rho_0 + Q\rho)UT}{QL}(u_x + v_y) = 0,$$

$$M^2 = \frac{U^2 Q}{P} = \frac{U^2}{dr(\rho_0)/d\rho},$$

$$p = \rho + \cdots.$$

Considering the limit process

$$\frac{UT}{L} = 1, \quad Q \to 0, \quad M \to 0 \quad \text{but} \quad \frac{Q}{M^2} = 1,$$

we obtain formally the Euler equations, where the density equation above goes over in $u_x + v_y = 0$. We shall discuss this limit process in Chapter 10.

Another limit is obtained for

$$Q \to 0, \quad \frac{UT}{L} \to 0 \quad \text{but} \quad \frac{UTQ}{LM^2} \to \alpha \neq 0, \quad \frac{UT}{QL} \to \beta \neq 0.$$

The resulting equations are

$$\rho_0 u_t + \alpha p_x = 0, \quad \rho_0 v_t + \alpha p_y = 0,$$

$$\rho_t + \rho_0 \beta(u_x + v_y) = 0, \qquad \rho = p.$$

Combining these, we obtain the wave equation for ρ:

$$\rho_{tt} = c^2(\rho_{xx} + \rho_{yy}), \quad c^2 = \frac{T^2}{L^2} dr(\rho_0)/d\rho.$$

(In the original coordinate system where $L = T = 1$ the quantity $(dr(\rho_0)/d\rho)^{1/2}$ denotes the sound speed.)

The above scaling arguments show that the N-S equations describe vastly different types of problems; these appear when certain parameters approach extreme values. One could believe that these extreme cases are practically uninteresting. However, the opposite is true: they represent the usual behavior. The reason is that the parameters do not need to take extreme values for the flow to have qualitatively the same behavior as suggested by the limiting equations. For example, a flow with $M = 2$ is in certain aspects not much different from the limiting flow corresponding to $M = \infty$. Therefore, the analysis of the limiting equations is important.

1.1.4. Remarks on Solution Estimates and Computations

From a purely mathematical point of view the theory of the viscous N-S equations is in rather good shape, except for the question of long-time existence and the related question of blow-up. Once we assume a bound for the maximum norm of the solution, we can estimate all its derivatives and we know what boundary conditions to prescribe. Also, one understands various limit processes as long as ν is fixed.

In the incompressible case one can show that the j-th derivatives of the flow-variables are bounded in terms of

$$(|D\mathbf{u}(\cdot, t)|_\infty \nu^{-1})^{j/2},$$

where $|D\mathbf{u}|_\infty$ denotes the maximum norm of the velocity gradient. Let us discuss the implications in two space dimensions first. In 2D, the values $|D\mathbf{u}(\cdot, t)|$ are essentially bounded by the initial data $|D\mathbf{u}(\cdot, 0)|$. If we scale the variables such that $|D\mathbf{u}(\cdot, 0)| = 1$, then the j-th derivatives are bounded in terms of $\nu^{-j/2}$. In applications a value $\nu = 10^{-6}$ is not uncommon. If the derivatives are of the order $O(10^3)$ everywhere, then we need $O(10^6)$ computational points in a two-dimensional region. This stresses the capacities of today's machines, but it is computationally feasible. In three space dimensions the velocity gradients are not bounded independently of ν, but one can make a plausibility argument that $|D\mathbf{u}(\cdot, t)| = O(\nu^{-1/2})$. For $\nu = 10^{-6}$ one would need $O(\nu^{-9/4}) = O(10^{27/2})$ computational points to resolve the flow; this is clearly beyond today's (and tomorrow's) machines.

Fortunately, the situation is not always this bad. There are many flows for which the derivatives are bounded independently of ν. Other flows, particularly those in (compressible) gas dynamics, have derivatives bounded independently of ν except near isolated sharp fronts (shocks). In this case one can replace a very small ν by a larger value, without changing the structure of the flow away from the fronts. Burgers' equation is a good example, and we shall discuss it in detail.

To treat situations where derivatives are large in extended regions, we need a mathematical theory to identify functionals which depend smoothly on ν as $\nu \to 0$. These could be evaluated using larger ν-values for computations, and could be extrapolated. These questions are closely related to turbulence models and so-called subgrid modeling.

1.2. Derivation of the Navier-Stokes Equations

In this section we give a brief derivation of the N-S equations. Our presentation relies heavily on Meyer (1971) and Chorin and Marsden (1979). For additional information, see also Serrin (1959). The general logical assumptions underlying continuum mechanics – in particular smoothness assumptions – are not easily formulated and we do not discuss them here. Instead we give a rather uncritical intuitive description of the ideas of particle paths and velocity fields. Both concepts can serve to describe flows. The relation between both descriptions is discussed first. Then the laws of conservation of mass and change of momentum are formulated; these laws – in their differential form – constitute the fundamental differential equations discussed in this book. Specific assumptions about the stress tensor appearing in the momentum equations lead to the N-S system. The equations obey an important similarity law, which is strongly confirmed experimentally. This justifies indirectly the assumptions made about the stress tensor.

1.2.1. Eulerian and Lagrangian Descriptions

Consider a body of fluid which – at time $t = 0$ – occupies some open bounded domain Ω_0 with boundary $\partial\Omega_0$. We think of every point $\mathbf{a} \in \Omega_0$ as a fluid particle whose motion we can follow for $t \geq 0$. Its position at a later time t is denoted by $\Phi(\mathbf{a}, t)$. At this time the fluid elements occupy the domain

$$\Omega_t = \{\Phi(\mathbf{a}, t) : \mathbf{a} \in \Omega_0\}.$$

The curve $t \to \Phi(\mathbf{a}, t)$ describes the trajectory of the fluid particle which at time $t = 0$ has the initial coordinate $\mathbf{a} = \Phi(\mathbf{a}, 0)$. It is assumed that $\Phi(\mathbf{a}, t) \neq \Phi(\mathbf{b}, t)$ for $t \geq 0$ if $\mathbf{a} \neq \mathbf{b}$; i.e., *the fluid particles keep their identity.* We formalize the basic assumption underlying the description of the flow in

Assumption 1.2.1. There exists a smooth function $\Phi(\mathbf{a}, t)$ with the following properties:

1. $\Phi(\mathbf{a}, 0) = \mathbf{a}$.
2. If $\mathbf{a} \neq \mathbf{b}$ then $\Phi(\mathbf{a}, t) \neq \Phi(\mathbf{b}, t)$ for $t \geq 0$.
3. The mapping $\mathbf{a} \to \Phi(\mathbf{a}, t)$ has a smooth inverse.

These conditions are assumed to hold in $\Omega_0 \cup \partial\Omega_0$, i.e., the region including the boundary. Then, by general topological arguments, at any later time t the boundary $\partial\Omega_t$ of Ω_t is the image of the original boundary:

$$\partial\Omega_t = \{\Phi(\mathbf{a}, t) : \mathbf{a} \in \partial\Omega_0\}.$$

This will be used below.

Let $\mathbf{u}(\mathbf{x}, t)$ denote the fluid-velocity at $\mathbf{x} \in \Omega_t$ and time t. Then

$$\mathbf{u}(\mathbf{x}, t) = \frac{\partial}{\partial t}\Phi(\mathbf{a}, t) \quad \text{if} \quad \mathbf{x} = \Phi(\mathbf{a}, t),$$

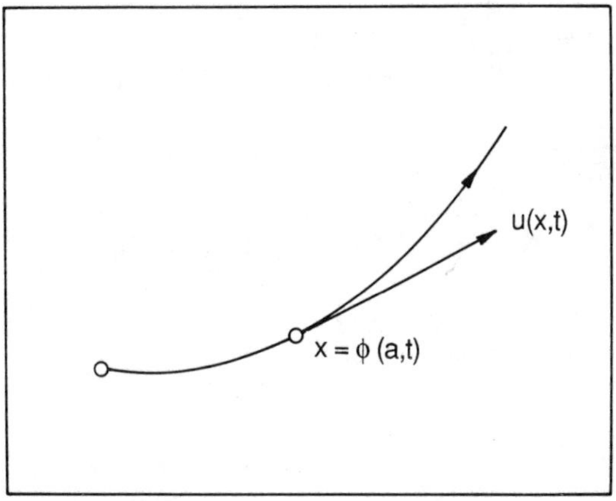

FIGURE 1.2.1. Trajectory with velocity vector.

and thus we have

$$\text{(1.2.1)} \qquad \mathbf{u}(\Phi(\mathbf{a}, t), t) = \frac{\partial}{\partial t}\Phi(\mathbf{a}, t), \quad \mathbf{a} \in \Omega_0.$$

In principle, if the mapping $\Phi = \Phi(\mathbf{a}, t)$ describing the trajectories is known, then the velocity field $\mathbf{u} = \mathbf{u}(\mathbf{x}, t)$ can be obtained from (1.2.1). Conversely, if we know the velocity field $\mathbf{u}$, then we can reconstruct the trajectories $t \to \Phi(\mathbf{a}, t)$ by solving – for each fixed $\mathbf{a} \in \Omega_0$ – the (ordinary) initial value problems (1.2.1) with initial condition $\Phi(\mathbf{a}, 0) = \mathbf{a}$. Thus one can describe the fluid motion in terms of t and the initial position $\mathbf{a} \in \Omega_0$ (the so-called material coordinate) or one can use t and the spatial coordinate $\mathbf{x} \in \Omega_t$. The first way is known as **Lagrangian** description whereas the second is the **Eulerian** point of view.

One can ask which description one should use in actual calculations. If discontinuity surfaces (e.g., density discontinuities) are present which move with the flow, then the Lagrangian way is advantageous: The location of the discontinuities in terms of the material coordinates remains fixed in time, and one can work with a description which is smooth with respect to t. On the other hand, watching the distortion which a blob of ink undergoes in a turbulent flow makes us wonder if the Lagrangian description is always useful. Here the initial coordinates lose their meaning. One prefers the Eulerian description if large distortions are present. Also, in numerical calculations it is quite common to use a combination of both descriptions.

Let $f = f(\mathbf{x}, t)$ denote a scalar quantity depending smoothly on the Eulerian coordinates $\mathbf{x} \in \Omega_t$ and t. For example, f can denote the local fluid-density or the temperature. One can ask how the quantity f changes in time if one follows a trajectory $\mathbf{x}(t) = \Phi(\mathbf{a}, t)$. The chain rule and (1.2.1) imply that

$$\begin{aligned} \frac{d}{dt} f(\mathbf{x}(t), t) &= \{\frac{\partial}{\partial t} f + u\frac{\partial}{\partial x} f + v\frac{\partial}{\partial y} f + w\frac{\partial}{\partial z} f\}(\mathbf{x}(t), t) \\ &= \{\frac{\partial}{\partial t} f + \mathbf{u} \cdot \operatorname{grad} f\}(\mathbf{x}(t), t). \end{aligned} \tag{1.2.2}$$

(Here $\mathbf{u} = (u, v, w)$ and $\mathbf{x} = (x, y, z)$.)

This motivates us to introduce the operator D/Dt acting on functions $f = f(\mathbf{x}, t)$ in Eulerian coordinates:

$$\frac{D}{Dt} f(\mathbf{x}, t) = \frac{\partial}{\partial t} f(\mathbf{x}, t) + \mathbf{u}(\mathbf{x}, t) \cdot \operatorname{grad} f(\mathbf{x}, t).$$

The function $(D/Dt)f(\mathbf{x}, t)$ is known as the material derivative of f. One easily shows the product rule:

$$\frac{D}{Dt}(fg) = f\frac{D}{Dt}g + g\frac{D}{Dt}f. \tag{1.2.3}$$

We consider now the time-change of integrals of the quantity f. Using the mapping Φ to transform all integrals to the fixed domain Ω_0, one can show with the rules of calculus the

Transport Theorem. *Under appropriate smoothness assumptions the following holds:*

$$\frac{d}{dt}\int_{\Omega_t} f(\mathbf{x}, t)\, d\mathbf{x} = \int_{\Omega_t} \{\frac{\partial}{\partial t} f + \operatorname{div}(f\mathbf{u})\}(\mathbf{x}, t)\, d\mathbf{x} = \int_{\Omega_t} \{\frac{D}{Dt} f + f \operatorname{div} \mathbf{u}\}(\mathbf{x}, t)\, d\mathbf{x}$$

where

$$\operatorname{div}(f\mathbf{u}) = (fu)_x + (fv)_y + (fw)_z = \mathbf{u} \cdot \operatorname{grad} f + f \operatorname{div} \mathbf{u}.$$

A proof of the Transport Theorem is given, for example, in Hughes and Marsden (1976). With Gauss' Theorem

$$\int_\Omega \operatorname{div} \mathbf{w}\, d\mathbf{x} = \int_{\partial\Omega} \mathbf{w} \cdot \mathbf{n}\, dS$$

one obtains further that

$$\frac{d}{dt}\int_{\Omega_t} f(\mathbf{x}, t)\, d\mathbf{x} = \int_{\Omega_t} \frac{\partial}{\partial t} f(\mathbf{x}, t)\, d\mathbf{x} + \int_{\partial\Omega_t} (f\mathbf{u} \cdot \mathbf{n})(\mathbf{x}, t)\, dS,$$

where $\mathbf{n}(\mathbf{x}, t)$ denotes the unit outward normal of $\partial\Omega_t$ and dS is the surface element. *The amount of the quantity f in Ω changes due to direct changes of f and due to the flow of f across the boundary.*

If we apply the Transport Theorem with $f \equiv 1$, we obtain that

$$\frac{d}{dt}\int_{\Omega_t} 1\,d\mathbf{x} = \int_{\Omega_t} \operatorname{div}\mathbf{u}\,d\mathbf{x},$$

and thus each volume

$$\textbf{volume}\,(\Omega_t) = \int_{\Omega_t} 1\,d\mathbf{x}$$

is preserved in time if and only if $\operatorname{div}\mathbf{u} \equiv 0$.

1.2.2. Conservation of Mass and Balance of Momentum

We start with the conservation of mass, which can be formalized as follows:

Assumption 1.2.2. There exists a smooth function $\rho = \rho(\mathbf{x}, t) \geq 0$ defined on each fluid domain Ω_t such that for all $t \geq 0$,

$$\int_{\Omega_t} \rho(\mathbf{x}, t)\,d\mathbf{x} = \int_{\Omega_0} \rho(\mathbf{x}, 0)\,d\mathbf{x}. \tag{1.2.4}$$

The function ρ is the fluid-density, and the integral (1.2.4) is the mass contained initially in Ω_0 and conserved in time. Using the Transport Theorem, we find that

$$\int_{\Omega_t} \{\frac{\partial}{\partial t}\rho + \operatorname{div}(\rho\mathbf{u})\}\,d\mathbf{x} = 0$$

holds for any domain Ω_t; therefore, the integrand must vanish, and we obtain for ρ the so-called **continuity equation**:

$$\frac{\partial}{\partial t}\rho + \operatorname{div}(\rho\mathbf{u}) = 0. \tag{1.2.5}$$

This equation often takes a simpler form: A fluid is called *incompressible* if

$$\rho(\mathbf{x}, t) = \rho_0 = \text{const.}$$

For an incompressible fluid the continuity equation reads $\operatorname{div}\mathbf{u} = 0$, thus volumes are preserved. Conversely, if the flow is volume preserving, then $\operatorname{div}\mathbf{u} = 0$ and the continuity equation becomes $D\rho/Dt = 0$. Under this assumption the Transport Theorem implies that ρ is constant w.r.t. $\mathbf{x}$ and t if the density is homogeneous at $t = 0$, i.e., $\rho(\mathbf{x}, 0) = \text{const.}$

Let us now discuss the **balance of momentum**. The momentum of the fluid in the domain Ω_t is

$$\int_{\Omega_t} \rho(\mathbf{x}, t)\mathbf{u}(\mathbf{x}, t)\,d\mathbf{x}.$$

By **Newton's law**, its time derivative equals the force which acts on the mass in Ω_t. We assume that this force is the sum of an *external* body force (e.g., gravity, the Coriolis force, or an electromagnetic force) and an *internal* force, which acts on the boundary of the fluid domain. The internal force is due to friction between fluid elements. We assume that it can be written in terms of a stress tensor field:

Assumption 1.2.3. There exists a force field $\mathbf{F} = \mathbf{F}(\mathbf{x}, t)$ and a stress tensor field

$$\mathbf{S} = \mathbf{S}(\mathbf{x}, t) = \begin{pmatrix} \mathbf{s}_1 \\ \mathbf{s}_2 \\ \mathbf{s}_3 \end{pmatrix} = \begin{pmatrix} s_{11} & s_{12} & s_{13} \\ s_{21} & s_{22} & s_{23} \\ s_{31} & s_{32} & s_{33} \end{pmatrix}$$

such that, for any fluid domain Ω_t,

$$\frac{d}{dt}\int_{\Omega_t} \rho(\mathbf{x}, t)\mathbf{u}(\mathbf{x}, t)\, d\mathbf{x} = \int_{\Omega_t} \rho(\mathbf{x}, t)\mathbf{F}(\mathbf{x}, t)\, d\mathbf{x} + \int_{\partial\Omega_t} \mathbf{S}(\mathbf{x}, t)\mathbf{n}(\mathbf{x}, t)\, d\mathbf{x}. \tag{1.2.6}$$

Note that this is a vector equation. Using Gauss' Theorem, we can write

$$\int_{\partial\Omega_t} \mathbf{S}\mathbf{n}\, d\mathbf{x} = \int_{\Omega_t} \operatorname{div} \mathbf{S}\, d\mathbf{x}, \qquad \operatorname{div} \mathbf{S} = \begin{pmatrix} \operatorname{div} \mathbf{s}_1 \\ \operatorname{div} \mathbf{s}_2 \\ \operatorname{div} \mathbf{s}_3 \end{pmatrix}.$$

If we apply the Transport Theorem to each component of (1.2.6) we obtain that

$$\int_{\Omega_t} \{\frac{D}{Dt}(\rho\mathbf{u}) + \rho\mathbf{u} \operatorname{div} \mathbf{u} - \rho\mathbf{F} - \operatorname{div} \mathbf{S}\}\, d\mathbf{x} = 0.$$

Under appropriate smoothness assumptions the integrand must vanish, and, using the continuity equation, we find that

$$\rho\frac{D}{Dt}\mathbf{u} = \rho\mathbf{F} + \operatorname{div} \mathbf{S}. \tag{1.2.7}$$

We refer to (1.2.7) as the *momentum equation.*

1.2.3. Forms of the Stress Tensor

Any further analysis of the fluid motion depends on knowledge about the stress tensor **S**. We must relate **S** to the other flow variables. In many situations – in particular in gas dynamics – one can use a very simple form for **S**:

Definition. A flow is called *inviscid* if the stress tensor **S** has the following form:

(1.2.8)
$$\mathbf{S}(\mathbf{x},t) = -p(\mathbf{x},t)\begin{pmatrix} 1 & 0 & 0 \\ 0 & 1 & 0 \\ 0 & 0 & 1 \end{pmatrix} = -p(\mathbf{x},t)I.$$

The scalar quantity $p = p(\mathbf{x}, t)$ is the pressure. For an inviscid flow the momentum equation (1.2.7) becomes

(1.2.9)
$$\rho \frac{D}{Dt}\mathbf{u} + \operatorname{grad} p = \rho \mathbf{F}.$$

Together with the continuity equation (1.2.5), the vector equation (1.2.9) represents a first-order system of four equations in the five scalar unknowns ρ, p, u, v, w. There are two important different ways to complete the system. First, if the fluid is incompressible, i.e., if $\rho(\mathbf{x}, t) = \rho_0$ is a known constant, we have the Euler equations:

(1.2.10)
$$\frac{D}{Dt}\mathbf{u} + \frac{1}{\rho_0} \operatorname{grad} p = \mathbf{F}, \quad \operatorname{div} \mathbf{u} = 0.$$

A flow governed by these equations, i.e., an inviscid incompressible flow, is said to be *ideal.*

Another way to complete the set of equations (1.2.5), (1.2.9) is to add an equation of state. In the absence of thermodynamic effects it takes the form

$$p = r(\rho)$$

where r is a known function. The resulting system describes inviscid compressible flow.

Let us consider again the viscous case, which requires a more general form of the stress tensor $\mathbf{S}$ than (1.2.8). The appropriate dependence of $\mathbf{S}$ on the other flow variables can only be inferred from observations. If the motion is uniform, i.e., if all velocity gradients are zero, then experiments show that the simple form (1.2.8) for $\mathbf{S}$ is still appropriate. Thus it is reasonable to assume – at least as a first approximation – that $\mathbf{S} + pI$ depends *linearly* on the matrix of velocity gradients:

(1.2.11)
$$\mathbf{S} = -pI + \tilde{\mathbf{S}}(T), \quad \tilde{\mathbf{S}} \text{ linear in } T, \quad T = T(\mathbf{x},t) = \begin{pmatrix} u_x & u_y & u_z \\ v_x & v_y & v_z \\ w_x & w_y & w_z \end{pmatrix}.$$

Heuristically one can think of $\tilde{\mathbf{S}}(T)$ as the stresses due to viscous phenomena. The assumption (1.2.11) is made in most of classical fluid mechnics. We want to argue now that $\tilde{\mathbf{S}}$ only depends on the *symmetric* part of T, i.e., on the *deformation tensor*

$$D = \frac{1}{2}(T + T^*),$$

and not on the local *rotation* of the fluid, described by the antisymmetric part $\frac{1}{2}(T - T^*)$ of T. Consider the

Couette experiment. Assume that the space between two long concentric cylinders is filled with fluid. Let the inner cylinder be forced to rotate slowly and let the outer cylinder be free. Then the fluid and the outer cylinder also begin to rotate. This can only be due to tangential stresses. After some time the outer cylinder and the fluid move with the same velocity as the inner cylinder. Thus, the motion has become like the rotation of a solid body; there is no momentum transfer anymore, and we have $\mathbf{S} = -pI$. The matrix T corresponding to a solid rotation is antisymmetric. Therefore, it is reasonable to assume that $\tilde{\mathbf{S}}$ is zero for an antisymmetric argument and depends on the symmetric part $D = \frac{1}{2}(T + T^*)$ of T only.

One can prove (see Gurtin and Martins (1976))

Theorem 1.2.1. *Assume that $\tilde{\mathbf{S}} = \tilde{\mathbf{S}}(D)$ is a linear function of D which is invariant under all rotations of the coordinate system, i.e.,*

$$\tilde{\mathbf{S}}(UDU^*) = U\tilde{\mathbf{S}}(D)U^*$$

for all orthogonal 3 by 3 matrices U and all symmetric 3 by 3 matrices D. Then $\tilde{\mathbf{S}}$ can be written as

$$\tilde{\mathbf{S}}(D) = \mu'(d_{11} + d_{22} + d_{33})I + 2\mu D,$$

where μ and μ' are constants independent of D.

An application of this result gives us for the stress tensor:

$$\mathbf{S} = -pI + \mu' \operatorname{div} \mathbf{u}\, I + \mu(T + T^*).$$

If we substitute this expression into the momentum equation (1.2.7), we obtain – together with the continuity equation – the so-called Navier-Stokes system. The material constants μ and μ' depend on the temperature and the chemical properties of the fluid. The coefficient μ is known as the shear viscosity and $\zeta = \mu' + \frac{2}{3}\mu$ is the bulk viscosity.

Again, one distinguishes incompressible and compressible flow. In the first case one has $\rho(\mathbf{x}, t) = \rho_0$, $\operatorname{div} \mathbf{u} = 0$; for compressible flow one uses an equation of state $p = r(\rho)$ to complete the system. The resulting equations are listed in Section 1.1.

1.2.4. The Similarity of Flows and the Reynolds Number

One of the most important tools in hydrodynamics is to scale the variables. This enables one both to simplify the equations by neglecting terms and to use one calculation or experiment to obtain results for another similar problem. We shall explain the procedure for incompressible flow without external forces. To this end, assume we have a fluid with density ρ and viscosity μ. Its motion satisfies

$$\frac{\partial}{\partial t}\mathbf{u} + (\mathbf{u} \cdot \operatorname{grad})\mathbf{u} + \frac{1}{\rho}\operatorname{grad} p = \frac{\mu}{\rho}\Delta\mathbf{u}, \qquad \operatorname{div}\mathbf{u} = 0. \tag{1.2.12}$$

We consider (1.2.12) in a time-independent domain Ω with initial conditions in Ω and boundary conditions on $\partial\Omega$.

Suppose now that we want information about another fluid with density $\tilde{\rho}$ and viscosity $\tilde{\mu}$ in a geometrically similar domain $\tilde{\Omega} = L\Omega$. The functions $\tilde{\mathbf{u}} = \tilde{\mathbf{u}}(\tilde{\mathbf{x}}, \tilde{t})$, $\tilde{p} = \tilde{p}(\tilde{\mathbf{x}}, \tilde{t})$ satisfy

$$\frac{\partial}{\partial \tilde{t}}\tilde{\mathbf{u}} + (\tilde{\mathbf{u}} \cdot \operatorname{grad})\tilde{\mathbf{u}} + \frac{1}{\tilde{\rho}}\operatorname{grad} \tilde{p} = \frac{\tilde{\mu}}{\tilde{\rho}}\Delta\tilde{\mathbf{u}}, \qquad \operatorname{div}\tilde{\mathbf{u}} = 0,$$

and are subject to initial conditions in $\tilde{\Omega}$ and boundary conditions on $\partial\tilde{\Omega}$. Suppose further that we can find scaling transformations

$$\tilde{\mathbf{u}} = U\mathbf{u},\ \tilde{\mathbf{x}} = L\mathbf{x},\ \tilde{p} = Pp,\ \tilde{t} = Tt, \quad U, L, P, T = \text{const},$$

such that the transformed functions

$$\mathbf{u}'(\tilde{\mathbf{x}}, \tilde{t}) := U\mathbf{u}(\mathbf{x}, t) = U\mathbf{u}(\tilde{\mathbf{x}}/L, \tilde{t}/T), \quad p'(\tilde{\mathbf{x}}, \tilde{t}) := Pp(\mathbf{x}, t) = Pp(\tilde{\mathbf{x}}/L, \tilde{t}/T)$$

satisfy the same initial and boundary conditions as $\tilde{\mathbf{u}}$ and $\tilde{p}$. From

$$\frac{\partial}{\partial \tilde{t}}\mathbf{u}' = \frac{U}{T}\frac{\partial}{\partial t}\mathbf{u}, \qquad \operatorname{grad}\mathbf{u}' = \frac{U}{L}\operatorname{grad}\mathbf{u},$$
$$\Delta\mathbf{u}' = \frac{U}{L^2}\Delta\mathbf{u}, \qquad \operatorname{grad} p' = \frac{P}{L}\operatorname{grad} p$$

and (1.2.12) we find that

$$\frac{\partial}{\partial \tilde{t}}\mathbf{u}' + \frac{L}{UT}(\mathbf{u}' \cdot \operatorname{grad})\mathbf{u}' + \frac{UL}{\rho TP}\operatorname{grad} p' = \frac{\mu L^2}{\rho T}\Delta\mathbf{u}', \qquad \operatorname{div}\mathbf{u}' = 0.$$

Thus the functions $\mathbf{u}', p'$ satisfy the same equations as $\tilde{\mathbf{u}}, \tilde{p}$ if

$$\frac{L}{UT} = 1, \quad \frac{UL}{TP} = \frac{\rho}{\tilde{\rho}}, \quad \frac{L^2}{T} = \frac{\tilde{\mu}\rho}{\mu\tilde{\rho}}. \tag{1.2.13}$$

If these relations for the scaling parameters are satisfied and if the initial-boundary value problem for $\tilde{\mathbf{u}}, \tilde{p}$ has a *unique* solution, then we can conclude that

$$\tilde{\mathbf{u}}(\tilde{\mathbf{x}}, \tilde{t}) = U\mathbf{u}(\mathbf{x}, t) = U\mathbf{u}(\tilde{\mathbf{x}}/L, \tilde{t}/T), \quad \tilde{p}(\tilde{\mathbf{x}}, \tilde{t}) = Pp(\mathbf{x}, t) = Pp(\tilde{\mathbf{x}}/L, \tilde{t}/T).$$

The requirements simplify if the coefficients in the boundary conditions do not depend on time, because we can choose T arbitrarily. If we take $T = L/U$, then (1.2.13) requires that

$$\frac{U^2\tilde{\rho}}{P} = \rho, \qquad \frac{UL\tilde{\rho}}{\tilde{\mu}} = \frac{\rho}{\mu}.$$

Roughly speaking, flows are similar if their respective values for

$$\frac{U^2\tilde{\rho}}{P} \quad \text{and for} \quad \frac{UL\tilde{\rho}}{\tilde{\mu}} = R$$

coincide. The number R is called the **Reynolds number** of the flow.

As mentioned above, the similarity law has been confirmed by experiments over and over again. Indirectly, this experimental fact supports and justifies the basic assumption underlying the Navier-Stokes equations, namely that the stress tensor depends linearly on the velocity gradients and not on other fluid variables.

1.3. Linearization and Localization

In this section we outline some basic mathematical questions which will be treated more thoroughly in the subsequent chapters. We want to solve the so-called **Cauchy problem** for systems of quasilinear partial differential equations

(1.3.1) $$u_t = P(x, t, u, \partial/\partial x)u + F(x, t), \qquad x \in \mathbf{R}^s, \quad 0 \le t \le T,$$

with initial conditions

(1.3.2) $$u(x, 0) = f(x), \qquad x \in \mathbf{R}^s.$$

Here $x = (x_1, \cdots, x_s) \in \mathbf{R}^s$ is the vector of independent space variables and $u = u(x, t)$ denotes the solution to be determined, which takes values in $\mathbf{C}^n$.

The operator $P = P(x, t, u, \partial/\partial x)$ is a quasilinear differential operator of order m of the general form

$$P(x, t, u, \partial/\partial x) = \sum_{|\nu| \le m} A_\nu(x, t, u) \frac{\partial^{|\nu|}}{\partial x_1^{\nu_1} \cdots \partial x_s^{\nu_s}}.$$

Here $|\nu| = \nu_1 + \cdots + \nu_s$ for any multi-index $\nu = (\nu_1, \cdots, \nu_s)$. The coefficients

$$A_\nu = A_\nu(x, t, u) \in \mathbf{C}^{n,n}$$

are given smooth matrix functions of the arguments

$$x \in \mathbf{R}^s, \quad 0 \leq t \leq T, \quad u \in \mathbf{C}^n.$$

Also, the forcing function F in (1.3.1) and the initial function f in (1.3.2) are assumed to be given. More specific assumptions and boundary conditions will be discussed later. From a purely mathematical point of view there are two fundamental questions:

1. When can one guarantee that the problem (1.3.1), (1.3.2) has a unique solution, at least for a small time interval?
2. Assuming that (1.3.1), (1.3.2) has a unique solution u for some fixed F, f, what is the influence on the solution of small perturbations added to F and f?

The second question can be rephrased more precisely as follows: Replace F by $F + \delta F$ in (1.3.1) and f by $f + \delta f$ in (1.3.2), and thus consider a perturbed problem

(1.3.3) $$v_t = P(x, t, v, \partial/\partial x)v + F + \delta F, \quad x \in \mathbf{R}^s, \quad 0 \leq t \leq T,$$

(1.3.4) $$v(x, 0) = f(x) + \delta f(x), \quad x \in \mathbf{R}^s.$$

If (1.3.1), (1.3.2) is "physically reasonable", one can hope that the perturbed problems (1.3.3), (1.3.4) are also uniquely solvable, at least for sufficiently small perturbations, and that their solutions can be written as

$$v = u + \delta u,$$

where δu satisfies an estimate

(1.3.5) $$\|\delta u\|_{(1)} \leq K(\|\delta F\|_{(2)} + \|\delta f\|_{(3)}).$$

Here $\|\delta u\|_{(1)}, \|\delta F\|_{(2)}, \|\delta f\|_{(3)}$ are certain norms which still have to be specified. The constant K in (1.3.5) should be independent of δF and δf as long as the perturbations are kept sufficiently small. This leads to the preliminary

Definition. Assuming that (1.3.1), (1.3.2) has a unique solution u for some fixed $F, f,$ we say that the nonlinear problem (1.3.1), (1.3.2) is *well-posed* at u if there is an $\epsilon > 0$ s.t. for all smooth functions δF and δf with

$$\|\delta F\|_{(2)} + \|\delta f\|_{(3)} \leq \epsilon$$

the perturbed problem (1.3.3), (1.3.4) is also uniquely solvable, and $\delta u := v - u$ satisfies an estimate (1.3.5) with K independent of δF and δf.

Note that we still have to specify the norms. Indeed, different norms lead to different concepts of well-posedness, as we will discuss below.

We will try to show the existence of a solution of the nonlinear problem (1.3.1), (1.3.2) in a sufficiently small time interval by considering the sequence $u^k = u^k(x,t)$ of functions defined iteratively through the *linear* equations

$$\begin{aligned} u_t^{k+1} &= P(x,t,u^k(x,t),\partial/\partial x)u^{k+1} + F(x,t), \\ u^{k+1}(x,0) &= f(x), \quad k = 0,1,\ldots, \\ u^0(x,t) &\equiv f(x). \end{aligned}$$

For this reason, we first have to study the Cauchy problem for linear equations

(1.3.6) $$u_t = P(x,t,\partial/\partial x)u + F(x,t).$$

Here

(1.3.7) $$P(x,t,\partial/\partial x) = \sum_{|\nu|\le m} A_\nu(x,t)\frac{\partial^{|\nu|}}{\partial x_1^{\nu_1}\cdots\partial x_s^{\nu_s}},$$

and the coefficients $A_\nu = A_\nu(x,t)$ depend smoothly on (x,t). Also, the question of well-posedness at u, which we formulated above for the nonlinear problem (1.3.1), (1.3.2), is closely related to well-posedness of linearized equations. Roughly speaking, the following *linearization principle* holds: *A nonlinear problem is well-posed at u if the linear problems which are obtained by linearizing at all functions near u are well-posed.*

Thus we will study linear equations first. In order to understand the basic questions of existence, uniqueness, and well-posedness for them, it suffices to treat just the homogeneous case, i.e., one can take $F \equiv 0$ in (1.3.6). Inhomogeneous equations can then be treated by **Duhamel's principle**. More importantly, it is desirable to relate the well-posedness of the Cauchy problem for a *variable*-coefficient equation

(1.3.8) $$u_t = P(x,t,\partial/\partial x)u$$

to the well-posedness for the *constant*-coefficient equations

(1.3.9) $$u_t = P(x_0,t_0,\partial/\partial x)u$$

which are obtained by freezing the coefficients

$$A_\nu = A_\nu(x_0,t_0)$$

at arbitrary but fixed points (x_0, t_0). This idea is called *localization.* By Fourier transformation, the discussion of constant-coefficient problems can be reduced to purely algebraic questions; this will be carried out in Chapter 2.

The *localization principle* one would like to have, can be formulated as follows: *If all frozen-coefficient problems are well-posed then the corresponding variable-coefficient problem is also well-posed.* It should be noted in advance that this principle is *not* valid for *general* linear variable-coefficient operators. However, the differential operators which appear in the context with the N-S equations are of parabolic, of hyperbolic, or of mixed hyperbolic-parabolic type, and for these classes of operators the above localization principle turns out to be applicable.

Formal linearization. To explain the process of linearization, we consider Burgers' equation as an example:

$$u_t = uu_x + \epsilon u_{xx}, \quad \epsilon > 0. \tag{1.3.10}$$

Suppose $U = U(x,t)$ is a smooth function. We may think of U as a known approximate solution of (1.3.10), but this is not essential. If we substitute

$$u = U + u', \quad u' = u'(x,t)$$

into (1.3.10), then we obtain

$$u'_t = U_x u' + U u'_x + u' u'_x + \epsilon u'_{xx} + F, \quad F = UU_x + \epsilon U_{xx} - U_t.$$

Thus far, no terms have been neglected. Since we consider u' as a *small* correction to U, we neglect the quadratic term $u'u'_x$ and write v instead of u'. Thus we obtain the *linear* equation

$$v_t = U_x v + U v_x + \epsilon v_{xx} + F.$$

This equation is called the linearization of (1.3.10) at U, or the *linearized equation.* Note that $F \equiv 0$ if and only if U solves (1.3.10) exactly. The linearized equation governs the influence of small perturbations on U. This is another important reason to study linear equations.

2 Constant-Coefficient Cauchy Problems

The main tool to discuss constant-coefficient problems (with initial data given on the whole space) is the Fourier transformation. It allows us to "decompose" general initial data into pure exponentials. One easily observes that – in case of constant coefficients – the time evolution of each pure (spatial) exponential can be treated separately. We introduce the symbol $P(i\omega)$ of a differential operator $P(\partial/\partial x)$ and obtain

$$u(x,t) = e^{i\langle\omega,x\rangle} e^{P(i\omega)t} \hat{f}(\omega)$$

as the solution for initial data $u(x,0) = e^{i\langle\omega,x\rangle}\hat{f}(\omega)$. Well-posedness of the Cauchy problem can be characterized in terms of estimates for the symbol.

We will start out with some special cases, namely hyperbolic and parabolic systems in one space dimension. For these, the conditions for the symbol are easily checked, and one can solve the Cauchy problem. In Section 2.3 we *characterize* families of matrices A for which the exponential e^{At}, $t \geq 0$, is uniformly bounded. This result, which is central in a general theory of well-posedness, will be applied in Section 2.4 to characterize those first-order systems (in any number of space dimensions) which lead to well-posed Cauchy problems; the corresponding systems are called *strongly hyperbolic*. An important example is given, the compressible Euler equations linearized about a constant flow. Similarly, linearization about a constant flow of the viscous compressible Navier-Stokes equations leads to a mixed hyperbolic-parabolic system; applying the same general principles, we obtain well-posedness of the Cauchy problem and can solve these linearized problems by Fourier transformation.

Though constant-coefficient equations are of some interest by themselves, they are too restrictive for most applications. If one freezes variable coefficients at arbitrary points, then obviously constant-coefficient equations are obtained. Unfortunately however, it is possible that *all* frozen-coefficient problems are well-posed and the given variable-coefficient equation is ill-posed nevertheless. Roughly speaking, switching between variable and constant coefficients introduces lower-order terms. Naturally, this leads to the question: Which constant-coefficient problems can be perturbed by (arbitrary) lower-order terms without losing well-posedness? As we will show in Section 2.6, the answer is that only strongly hyperbolic and parabolic problems have this property. To obtain this result, we use a restrictive but simple concept of well-posedness: It must be possible to estimate the L_2-norm of the solution at later times in terms of the L_2-norm of the initial function; derivative terms of the initial function are not permitted in the estimate. This idea guides our discussion of well-posedness.

At the end of this chapter, in Section 2.7, we will outline generalizations to variable-coefficient problems. The details will be presented in Chapters 3 and 6.

2.1. Pure Exponentials as Initial Data

The solution of constant-coefficient Cauchy problems is particularly simple if the initial function is a pure exponential; it can be written down in terms of the symbol of the differential operator. The behaviour of the symbol for large wave-numbers will determine whether or not the Cauchy problem is well-posed.

2.1.1. Introductory Examples

We start with some simple examples.

Example 1. (The simplest equation with wave-solutions.) The differential equation

$$u_t(x,t) + a u_x(x,t) = 0, \quad x \in \mathbf{R}, \quad t \geq 0, \tag{2.1.1}$$

(with a given constant $a \in \mathbf{R}$) is the simplest hyperbolic equation. Suppose that an initial condition

$$u(x,0) = e^{i\omega x}\hat{f}(\omega), \quad x \in \mathbf{R}, \tag{2.1.2}$$

is prescribed at $t = 0$. Here $\omega \in \mathbf{R}$ and $\hat{f}(\omega) \in \mathbf{C}$ are constants; the factor $\hat{f}(\omega)$ is added for later purposes. To solve the equation, we use an ansatz in separated variables,

$$u(x,t) = e^{i\omega x}\hat{u}(\omega,t), \tag{2.1.3}$$

and obtain

$$\hat{u}_t(\omega,t) + i\omega a\hat{u}(\omega,t) = 0, \quad \hat{u}(\omega,0) = \hat{f}(\omega).$$

The solution of this ordinary initial value problem is

$$\hat{u}(\omega,t) = e^{-i\omega at}\hat{f}(\omega).$$

Thus, if the problem (2.1.1), (2.1.2) has a solution of the form (2.1.3), it is necessarily given by

$$u(x,t) = e^{i\omega(x-at)}\hat{f}(\omega).$$

One easily checks that the above function indeed solves the problem; the solution represents a wave of constant amplitude traveling with speed a; for $a > 0$ or $a < 0$ the wave travels to the right or left, respectively.

Example 2. (The heat equation.) Another simple equation is the so-called heat equation

$$u_t = u_{xx}, \quad x \in \mathbf{R}, \quad t \geq 0.$$

Assuming the same initial condition (2.1.2), we use the same ansatz (2.1.3); the ordinary initial value problem for $\hat{u}(\omega,\cdot)$ reads

$$\hat{u}_t(\omega,t) = -\omega^2\hat{u}(\omega,t), \quad \hat{u}(\omega,0) = \hat{f}(\omega).$$

Hence the problem is solved by

$$u(x,t) = e^{i\omega x - \omega^2 t}\hat{f}(\omega). \tag{2.1.4}$$

Example 3. (The heat equation in backward time.) If we apply the transformation $t \to -t$ to the equation $u_t = u_{xx}$, $t \leq 0$, we obtain

$$u_t = -u_{xx}, \quad x \in \mathbf{R}, \quad t \geq 0.$$

Proceeding in the same way as above, we find

$$u(x,t) = e^{i\omega x + \omega^2 t}\hat{f}(\omega) \tag{2.1.5}$$

as a solution for the initial data (2.1.2).

If $|\omega|$ is large, the solutions (2.1.4) and (2.1.5) behave completely differently for increasing t: Whereas (2.1.4) decays rapidly, the solution (2.1.5) grows exponentially. For increasing $|\omega|$, there is no bound on the exponential growth rate in time. As stated more precisely below, the Cauchy problem for the heat equation in backward time is ill-posed.

2.1.2. The Symbol of a Constant-Coefficient Operator

The previous examples can be generalized to systems of constant-coefficient equations in any number of space dimensions.

Notations. With $x = (x_1, ..., x_s) \in \mathbf{R}^s$ we denote the space variable. A multi-index ν is a vector with nonnegative integers as components:

$$\nu = (\nu_1, \ldots, \nu_s), \quad \nu_j \in \{0, 1, 2, \ldots\}.$$

Its order is

$$|\nu| = \nu_1 + \ldots + \nu_s.$$

Each multi-index ν determines a differential expression

$$D^\nu = \frac{\partial^{|\nu|}}{\partial x_1^{\nu_1} \cdots \partial x_s^{\nu_s}}.$$

Furthermore, we use the following basic notations:
for $\omega, x \in \mathbf{R}^s$,

$$\langle \omega, x \rangle = \sum_{j=1}^{s} \omega_j x_j;$$

for $u, v \in \mathbf{C}^n$,

$$\langle u, v \rangle = \sum_{j=1}^{n} \overline{u}_j v_j, \qquad |u| = \langle u, u \rangle^{1/2}.$$

If $A \in \mathbf{C}^{n,n}$ is a matrix, then its norm is defined by

$$|A| = \max\{|Au|, |u| = 1\}.$$

If $H_1, H_2 \in \mathbf{C}^{n,n}$ are Hermitian matrices, then

$$H_1 \le H_2 \qquad \text{if and only if} \qquad \langle u, H_1 u \rangle \le \langle u, H_2 u \rangle \qquad \text{for all } u \in \mathbf{C}^n.$$

Consider a differential equation

$$u_t = P(\partial/\partial x)u \equiv \sum_{|\nu| \le m} A_\nu D^\nu u, \quad x \in \mathbf{R}^s, \quad t \ge 0, \tag{2.1.6}$$

with constant matrices $A_\nu \in \mathbf{C}^{n,n}$ and assume an initial condition

$$u(x,0) = e^{i\langle\omega,x\rangle}\hat{f}(\omega), \quad x \in \mathbf{R}^s, \tag{2.1.7}$$

where $\omega \in \mathbf{R}^s$ and $\hat{f}(\omega) \in \mathbf{C}^n$ are fixed. The unknown solution

$$u(x,t) = \begin{pmatrix} u_1(x,t) \\ \vdots \\ u_n(x,t) \end{pmatrix}$$

takes values in $\mathbf{C}^n$. A basic observation is

$$\frac{\partial}{\partial x_j} e^{i\langle\omega,x\rangle} = i\omega_j e^{i\langle\omega,x\rangle},$$

$$D^\nu e^{i\langle\omega,x\rangle} = (i\omega_1)^{\nu_1}\dots(i\omega_s)^{\nu_s} e^{i\langle\omega,x\rangle},$$

and therefore

$$P(\partial/\partial x)e^{i\langle\omega,x\rangle}\hat{f}(\omega) = \sum_{|\nu|\le m} A_\nu (i\omega_1)^{\nu_1}\dots(i\omega_s)^{\nu_s} e^{i\langle\omega,x\rangle}\hat{f}(\omega).$$

Thus, application of the differential operator $P(\partial/\partial x)$ to $e^{i\langle\omega,x\rangle}\hat{f}(\omega)$ results in multiplication by the matrix

$$P(i\omega) := \sum_{|\nu|\le m} A_\nu (i\omega_1)^{\nu_1}\cdots(i\omega_s)^{\nu_s}.$$

This matrix $P(i\omega)$, which is formally obtained by substitution of $i\omega_j$ for $\partial/\partial x_j$, is called the *symbol* of the differential operator $P(\partial/\partial x)$. We show

Lemma 2.1.1. *The initial value problem* (2.1.6), (2.1.7) *has the solution*

$$u(x,t) = e^{i\langle\omega,x\rangle}e^{P(i\omega)t}\hat{f}(\omega). \tag{2.1.8}$$

Proof. The ansatz

$$u(x,t) = e^{i\langle\omega,x\rangle}\hat{u}(\omega,t)$$

leads to

$$\hat{u}_t(\omega,t) = P(i\omega)\hat{u}(\omega,t), \quad \hat{u}(\omega,0) = \hat{f}(\omega).$$

This ordinary initial value problem has the solution

$$\hat{u}(\omega,t) = e^{P(i\omega)t}\hat{f}(\omega),$$

and the formula (2.1.8) is obtained.

2.1.3. Well-Posedness in Terms of the Symbol

Thus far we have considered only initial data of the form

$$u(x,0) = e^{i\langle \omega, x\rangle} \hat{f}(\omega);$$

i.e., the spatial behaviour of the initial function is essentially determined by a single wave-vector $\omega \in \mathbf{R}^s$; the constant vector $\hat{f}(\omega) \in \mathbf{C}^n$ only allows us to multiply $e^{i\langle \omega, x\rangle}$ by different constants in different components. If more general initial data $f(x)$ are given, we can try to write these as

$$f(x) = \frac{1}{(2\pi)^{s/2}} \int_{\mathbf{R}^s} e^{i\langle \omega, x\rangle} \hat{f}(\omega)\, d\omega, \quad x \in \mathbf{R}^s, \tag{2.1.9}$$

where

$$\hat{f}(\omega) = \frac{1}{(2\pi)^{s/2}} \int_{\mathbf{R}^s} e^{-i\langle \omega, x\rangle} f(x)\, dx, \quad \omega \in \mathbf{R}^s, \tag{2.1.10}$$

is the Fourier transform of $f(x)$. (See Section 2.2 for more details.) According to Lemma 2.1.1, the evolution of each individual term of the integrand of (2.1.9),

$$e^{i\langle \omega, x\rangle} \hat{f}(\omega), \quad \omega \text{ fixed},$$

is known, and it is tempting to believe that

$$u(x,t) = \frac{1}{(2\pi)^{s/2}} \int_{\mathbf{R}^s} e^{i\langle \omega, x\rangle} e^{P(i\omega)t} \hat{f}(\omega)\, d\omega$$

describes the evolution to general data $f(x)$. This formula expresses the *principle of superposition.* Indeed, as we will discuss in Section 2.2, this process is often justified. However, it is also clear that there might be serious convergence problems if

$$|e^{P(i\omega)t}|$$

is unbounded for $|\omega| \to \infty$. We want to give here an *operational* definition of the well-posedness of the Cauchy problem

$$\begin{aligned} u_t &= P(\partial/\partial x)u, \quad && x \in \mathbf{R}^s, \quad t \geq 0 \\ u(x,0) &= f(x), && x \in \mathbf{R}^s \end{aligned} \tag{2.1.11}$$

in terms of the symbols $P(i\omega)$. The definition is operational in the sense of providing conditions which can be checked in applications. In Section 2.2 we will show that this definition of well-posedness is equivalent to another one which might be more familiar, namely that

(i) for all initial data f in a certain class there is a unique solution in a certain class;
(ii) the solution depends continuously on f with respect to certain norms.

The operational definition is

Definition 1. The Cauchy problem (2.1.11) is called well-posed if there are constants α, K such that

$$|e^{P(i\omega)t}| \leq Ke^{\alpha t}$$

for all $t \geq 0$ and all $\omega \in \mathbf{R}^s$.

2.1.4. Examples of Well-Posed and of Ill-Posed Problems

Let us illustrate the previous definition by some examples.

Example 4. (The wave equation as a hyperbolic system*.) The wave equation $y_{tt} = y_{xx}$ leads to the system

$$u_t = Au_x, \quad A = \begin{pmatrix} 0 & 1 \\ 1 & 0 \end{pmatrix}, \quad x \in \mathbf{R}, \quad t \geq 0, \tag{2.1.12}$$

if one introduces the variables $u_1 = y_t$, $u_2 = y_x$. There is a unitary transformation U such that

$$UAU^* = \begin{pmatrix} 1 & 0 \\ 0 & -1 \end{pmatrix}.$$

Therefore,

$$P(i\omega) = i\omega A = U^* \begin{pmatrix} i\omega & 0 \\ 0 & -i\omega \end{pmatrix} U,$$

and thus $|e^{P(i\omega)}| = 1$. According to Definition 1, the Cauchy problem for (2.1.12) is well-posed. Also note that the change of variables $v = Uu$ reduces (2.1.12) to

$$v_t = UAU^* v_x = \begin{pmatrix} 1 & 0 \\ 0 & -1 \end{pmatrix} v_x;$$

we obtain two uncoupled scalar equations of the form discussed in Example 1.

Example 5. A so-called weakly hyperbolic system is given by

$$u_t = \begin{pmatrix} 1 & 1 \\ 0 & 1 \end{pmatrix} u_x - u.$$

*The notion of a hyperbolic system will be introduced below.

Here

$$e^{P(i\omega)t} = e^{-t}e^{i\omega t}\begin{pmatrix} 1 & i\omega t \\ 0 & 1 \end{pmatrix}, \qquad |e^{P(i\omega)t}| \sim e^{-t}(1+|\omega|t).$$

Thus the problem is ill-posed since the term $|\omega|t$ is not bounded independently of ω.

The growth of $|e^{P(i\omega)t}|$ with increasing $|\omega|$ is much less severe here than in the backward heat equation, where $|e^{P(i\omega)t}| = e^{\omega^2 t}$. In fact, the problem of Example 5 is *weakly well-posed*, which is defined as follows:

Definition 2. The Cauchy problem (2.1.11) is called weakly well-posed if there are constants α, K, q such that

$$|e^{P(i\omega)}| \leq K(1+|\omega|^q)e^{\alpha t}$$

for all $t \geq 0$ and all $\omega \in \mathbf{R}^s$.

In Section 2.2 we will discuss the difficulties involved with weak well-posedness: If a problem is only weakly well-posed but not well-posed, then a perturbation by a lower-order term can make the problem ill-posed in any sense; this makes it hard to go over to variable-coefficient problems.

Hyperbolic systems in 1D. Consider a first-order system

$$u_t = Au_x, \quad x \in \mathbf{R}, \quad t \geq 0, \tag{2.1.13}$$

where $A \in \mathbf{C}^{n,n}$. We show

Theorem 2.1.2. *The Cauchy problem for* (2.1.13) *is well-posed if and only if all eigenvalues of A are real and A has a complete set of eigenvectors.*

Proof. First assume that the eigenvalues of A are real and that there is a complete set of eigenvectors. There is a matrix S with

$$S^{-1}AS = \text{diag}\,(\lambda_1, \ldots, \lambda_n) =: \Lambda, \quad \lambda_j \in \mathbf{R}.$$

From $A = S\Lambda S^{-1}$ one obtains that

$$|e^{P(i\omega)t}| = |e^{i\omega At}| = |Se^{i\omega\Lambda t}S^{-1}| \leq |S||S^{-1}|,$$

since $|e^{i\omega\Lambda t}| = 1$. The estimate of Definition 1 holds with $\alpha = 0$, $K = |S||S^{-1}|$.

Conversely, assume the problem is well-posed. Let

$$\lambda = a + ib, \quad a, b \in \mathbf{R},$$

denote an eigenvalue of A. We must show $b = 0$. Since $e^{i\omega\lambda t}$ is an eigenvalue of $e^{P(i\omega)t}$, we obtain

$$|e^{P(i\omega)t}| \geq |e^{i\omega\lambda t}| = |e^{-\omega bt}|.$$

Clearly, if $b \in \mathbf{R}$, $b \neq 0$, we cannot have a bound by $Ke^{\alpha t}$ independent of ω. This shows that $b = 0$, i.e., that the eigenvalues of A are real.

Now suppose that A does not have a complete set of eigenvectors, and let

$$J_r = \begin{pmatrix} \lambda & 1 & & \\ & \ddots & \ddots & \\ & & \ddots & 1 \\ & & & \lambda \end{pmatrix} = \lambda I + B \ \in \mathbf{R}^{k,k}, \quad k \geq 2,$$

denote a (nontrivial) Jordan block of the Jordan matrix J of A. Then

$$|e^{i\omega J_r t}| = |e^{i\omega Bt}| \geq \frac{1}{(k-1)!} |\omega|^{k-1} t^{k-1}.$$

If $J = S^{-1}AS$, then

$$|e^{P(i\omega)t}| \geq \frac{1}{|S^{-1}||S|} |e^{i\omega Jt}| \geq c\,|\omega|^{k-1} t^{k-1},$$

in contradiction to the existence of a bound by $Ke^{\alpha t}$ independent of ω.

Definition 3. A first-order system $u_t = Au_x$ for which the Cauchy problem is well-posed is called *strongly hyperbolic*.

For first-order systems in any number of space dimensions, we will define strong hyperbolicity in the same way, namely by requesting well-posedness of the Cauchy problem. In Section 2.4 we will give an algebraic characterization generalizing the previous theorem.

Solution formula. Suppose $u_t = Au_x$ is strongly hyperbolic, and $S^{-1}AS = \Lambda = \operatorname{diag}(\lambda_j)$. If one introduces new variables v by $Sv = u$, then one obtains the diagonal system $v_t = \Lambda v_x$, which is decomposed into n scalar equations. An initial condition

$$u(x,0) = e^{i\omega x}\hat{f}(\omega)$$

transforms to

$$v(x,0) = e^{i\omega x}\hat{g}(\omega), \quad S\hat{g}(\omega) = \hat{f}(\omega).$$

Thus

$$v_j(x,t) = e^{i\omega(x+\lambda_j t)}\hat{g}_j(\omega), \quad j = 1, \dots, n,$$

and therefore

$$u(x,t) = \sum_{j=1}^{n} e^{i\omega(x+\lambda_j t)}\hat{g}_j(\omega) S_j,$$

where S_j denotes the j-th column of S. This shows that $u(x,t)$ is a sum of n waves which travel with the so-called characteristic speeds $-\lambda_1, \dots, -\lambda_n$.

Parabolic systems in 1D. We restrict ourselves here to second-order systems

$$u_t = Au_{xx} + Bu_x + Cu, \quad x \in \mathbf{R}, \quad t \geq 0,$$

where $A, B, C \in \mathbf{C}^{n,n}$.

Definition 4. The above system is called parabolic if the eigenvalues $\lambda_1, \dots, \lambda_n$ of A satisfy

$$\operatorname{Re} \lambda_j > 0, \quad j = 1, \dots, n.$$

Clearly, the heat equation $u_t = u_{xx}$ is a simple example. We want to apply Definition 1 and show

Theorem 2.1.3. *The Cauchy problem for a second-order parabolic system is well-posed.*

Proof. 1. By Schur's Theorem (Appendix 1) we can transform A by a unitary matrix U to upper-triangular form,

$$UAU^* = \begin{pmatrix} \lambda_1 & \alpha_{12} & \dots & \alpha_{1n} \\ & \ddots & & \alpha_{2n} \\ & & \ddots & \vdots \\ 0 & & & \lambda_n \end{pmatrix}, \quad \operatorname{Re} \lambda_j \geq \delta > 0.$$

If $D = \operatorname{diag}(1, d, \cdots, d^{n-1})$, $d > 0$, denotes a diagonal matrix, then

$$DUAU^*D^{-1} = \begin{pmatrix} \lambda_1 & \alpha_{12}/d & \dots \\ & \ddots & \vdots \\ 0 & & \lambda_n \end{pmatrix}.$$

Therefore, if we choose $d > 0$ sufficiently large and set $S = DU$, then

$$SAS^{-1} + (SAS^{-1})^* = \begin{pmatrix} \lambda_1 + \bar{\lambda}_1 & \alpha_{12}/d & \cdots & \\ \bar{\alpha}_{12}/d & \ddots & & \\ \vdots & & & \lambda_n + \bar{\lambda}_n \end{pmatrix} \geq \delta I.$$

The reason is that $\lambda_j + \bar{\lambda}_j \geq 2\delta > 0$, and if $d > 0$ is sufficiently large, then the influence of the outer-diagonal entries is as small as we please. We define the positive definite Hermitian matrix $H = S^*S$ and rewrite the above matrix inequality as

$$HA + A^*H = S^*SA + A^*S^*S = S^*\big(SAS^{-1} + S^{*-1}A^*S^*\big)S \geq \delta H.$$

2. Now consider the symbol

$$P(i\omega) = -\omega^2 A + i\omega B + C.$$

We obtain

$$HP(i\omega) + P^*(i\omega)H \leq -\omega^2\delta H + \text{const}\,(|\omega| + 1)H \leq 2\alpha H,$$

with α independent of ω .

To finish the proof of the theorem, we prove a lemma on matrix exponentials e^{Pt}. It can be applied to each symbol $P = P(i\omega)$ separately.

Lemma 2.1.4. *Let $P \in \mathbf{C}^{n,n}$, and let*

$$\frac{1}{K}I \leq H = H^* \leq KI, \quad K > 0.$$

If $HP + P^*H \leq 2\alpha H$, $\alpha \in \mathbf{R}$, then

$$|e^{Pt}| \leq Ke^{\alpha t}, \quad t \geq 0.$$

Proof. Suppose that $y(t)$ solves the initial value problem

$$y' = Py, \quad y(0) = y_0.$$

Then one obtains

$$\begin{aligned} \frac{d}{dt}\langle y(t), Hy(t)\rangle &= \langle y, Hy'\rangle + \langle y', Hy\rangle \\ &= \langle y, HPy\rangle + \langle Py, Hy\rangle \\ &= \langle y, (HP + P^*H)y\rangle \leq 2\alpha\langle y, Hy\rangle. \end{aligned}$$

Integration yields*

$$\langle y(t), Hy(t)\rangle \le e^{2\alpha t}\langle y_0, Hy_0\rangle.$$

Therefore,

$$\frac{1}{K}|y(t)|^2 \le \langle y(t), Hy(t)\rangle \le e^{2\alpha t}\langle y_0, Hy_0\rangle \le e^{2\alpha t}K|y_0|^2.$$

Hence we have shown that for all y_0,

$$|e^{Pt}y_0|^2 \le K^2 e^{2\alpha t}|y_0|^2,$$

and the lemma is proved.

2.2. Discussion of Concepts of Well-Posedness

In this section we introduce the space $\mathbf{M}_0$ of all functions $f = f(x)$ whose Fourier transforms are C^∞-smooth and have compact support. If an initial condition $u(x, 0) = f(x)$, $f \in \mathbf{M}_0$, is given, then the Cauchy problem for *any* constant-coefficient operator $P = P(\partial/\partial x)$ is solvable, and the solution is unique within a certain class. For example, we can solve the backward heat equation for initial data in $\mathbf{M}_0$. Well-posedness of the Cauchy problem means more, namely an estimate of the L_2-norm of the solution at all later times by the L_2-norm of the initial data. We will prove that such an estimate is possible if and only if the problem is well-posed according to the definition (using the symbol) given in the previous section.

One can weaken the concept of well-posedness and allow *derivative*-terms of the initial function f for estimates of the L_2-norm of the solution at later times. Again, this concept of *weak well-posedness* can be characterized in terms of estimates for the symbol $P(i\omega)$. However, if a problem is only weakly well-posed, then perturbations of the differential operator $P = P(\partial/\partial x)$ by suitable lower-order terms will lead to arbitrarily fast exponential explosion, and the perturbed problem is ill-posed in any sense. We will illustrate this result by an example.

2.2.1. Solution for Smooth Initial Data via Fourier Transform

The space $\mathbf{M}_0$. Let C_0^∞ denote the space of all functions $\phi : \mathbf{R}^s \to \mathbf{C}^n$ which have derivatives of all orders and have compact support; i.e., each $\phi \in C_0^\infty$ vanishes outside a bounded set in $\mathbf{R}^s$; the bounded set will depend on ϕ. By

*See Lemma 3.1.1 for a simple result on differential inequalities, which we are using here.

$\mathbf{M}_0$ we denote the space of all functions $f : \mathbf{R}^s \to \mathbf{C}^n$ which can be written in the form*

$$f(x) = \frac{1}{(2\pi)^{s/2}} \int_{\mathbf{R}^s} e^{i\langle\omega,x\rangle} \phi(\omega)\, d\omega, \quad \phi \in C_0^\infty. \tag{2.2.1}$$

One easily shows that $f \in C^\infty$, and one can differentiate under the integral sign since ϕ has compact support:

$$D^\nu f(x) = \frac{1}{(2\pi)^{s/2}} \int_{\mathbf{R}^s} (i\omega_1)^{\nu_1} \dots (i\omega_s)^{\nu_s} e^{i\langle\omega,x\rangle} \phi(\omega)\, d\omega.$$

Also, for $f \in \mathbf{M}_0$ the Fourier transform $\hat{f}(\omega)$ is defined, and

$$\hat{f}(\omega) = \frac{1}{(2\pi)^{s/2}} \int_{\mathbf{R}^s} e^{-i\langle\omega,x\rangle} f(x)\, dx = \phi(\omega), \quad \omega \in \mathbf{R}^s. \tag{2.2.2}$$

Therefore, the representation (2.2.1) is nothing but the Fourier representation of f.

For a function $f : \mathbf{R}^s \to \mathbf{C}^n$ we say that $f \in L_2$ if $x \to |f(x)|^2$ is (Lebesgue-) integrable over $\mathbf{R}^s$. On L_2 one defines an inner product and a norm by

$$(f, g) = \int_{\mathbf{R}^s} \langle f(x), g(x)\rangle dx, \quad \|f\| = (f, f)^{1/2}, \quad f, g \in L_2.$$

We remind the reader of **Parseval's relation**

$$(f, g) = (\hat{f}, \hat{g}), \quad \|f\| = \|\hat{f}\|.$$

Here we will need this relation only for functions $f, g \in \mathbf{M}_0$. (One can show that the space $\mathbf{M}_0$ is dense in L_2. Then $f \to \hat{f}$ is a densely defined bounded linear operator from $\mathbf{M}_0 \subset L_2 \to L_2$, and one can obtain the Fourier transform $\hat{f} \in L_2$ for any $f \in L_2$ by continuous extension; see Theorem 2.2.4 below. With this extension, Parseval's relation becomes valid for all $f, g \in L_2$.)

Solution formula. Consider the Cauchy problem

$$u_t = P(\partial/\partial x)u \equiv \sum_{|\nu| \le m} A_\nu \frac{\partial^{|\nu|}}{\partial x_1^{\nu_1} \cdots \partial x_s^{\nu_s}} u, \quad x \in \mathbf{R}^s, \quad t \ge 0, \tag{2.2.3}$$

$$u(x, 0) = f(x), \quad x \in \mathbf{R}^s, \tag{2.2.4}$$

where $f \in \mathbf{M}_0$; thus

$$f(x) = \frac{1}{(2\pi)^{s/2}} \int_{\mathbf{R}^s} e^{i\langle\omega,x\rangle} \hat{f}(\omega)\, d\omega, \quad \hat{f} \in C_0^\infty.$$

*The integral is defined componentwise.

Since a solution of the differential equation for initial data

$$e^{i\langle\omega,x\rangle}\,\hat{f}(\omega), \quad \omega\text{fixed},$$

is given by

$$e^{i\langle\omega,x\rangle}\,e^{P(i\omega)t}\,\hat{f}(\omega),$$

it follows that

$$u(x,t) = \frac{1}{(2\pi)^{s/2}}\int_{\mathbf{R}^s} e^{i\langle\omega,x\rangle}\,e^{P(i\omega)t}\,\hat{f}(\omega)\,d\omega \tag{2.2.5}$$

solves the given Cauchy problem. The reason is that the boundedness of the support of $\hat{f}$ ensures that we can differentiate under the integral sign. Before we can state a uniqueness result, we must make the *concept of a solution* more precise. A convenient concept* is the following:

Definition 1. A function $u = u(x,t)$ is called $\mathbf{M}_0$-solution of (2.2.3), (2.2.4) if

(i) $u(\cdot,t) \in \mathbf{M}_0$ for all $t \geq 0$;

(ii) the function $\hat{u}(\omega,t)$ is continuous, and $\hat{u}(\omega,t) = 0$ for $|\omega| > K$ with some K independent of t;

(iii) u is a classical solution; i.e., u_t exists and u satisfies (2.2.1), (2.2.2) at each point $x \in \mathbf{R}^s,\ t \geq 0$.

We prove:

Lemma 2.2.1. *For any* $f \in \mathbf{M}_0$ *the Cauchy problem* (2.2.3), (2.2.4) *has a unique* $\mathbf{M}_0$*-solution. It is given by* (2.2.5).

Proof. It is easy to show that (2.2.5) is an $\mathbf{M}_0$-solution; thus it remains to prove *uniqueness*. To this end, assume that u is an arbitrary $\mathbf{M}_0$-solution and note that

$$u_t(x,t) = P(\partial/\partial x)u(x,t) = \frac{1}{(2\pi)^{s/2}}\int_{\mathbf{R}^s} e^{i\langle\omega,x\rangle}P(i\omega)\hat{u}(\omega,t)\,d\omega.$$

Integration in t yields

$$u(x,t) - f(x) = \frac{1}{(2\pi)^{s/2}}\int_{\mathbf{R}^s} e^{i\langle\omega,x\rangle}P(i\omega)\int_0^t \hat{u}(\omega,\tau)\,d\tau\,d\omega.$$

*We do not aim for generality here, but merely want to illustrate that one can obtain existence and uniqueness results for ill-posed problems, too. The *estimates* are essential for well-posedness.

Since the Fourier representation of $u(x,t) - f(x)$ is unique, it follows that

$$\hat{u}(\omega,t) - \hat{f}(\omega) = P(i\omega)\int_0^t \hat{u}(\omega,\tau)\,d\tau.$$

Therefore,

$$\hat{u}_t(\omega,t) = P(i\omega)\hat{u}(\omega,t), \quad \hat{u}(\omega,0) = \hat{f}(\omega),$$

and we obtain

$$\hat{u}(\omega,t) = e^{P(i\omega)t}\hat{f}(\omega).$$

This shows that u has the representation (2.2.5), and uniqueness is proved.

Solution operator. The solution formula (2.2.5) shows that $u(\cdot,t)$ lies in $\mathbf{M}_0$ at each later time, and $u(\cdot,t)$ depends linearly on f. Thus we obtain linear operators $S_0(t) : \mathbf{M}_0 \to \mathbf{M}_0$ which assign to the initial data $u(\cdot,0) = f$ the solution

$$u(\cdot,t) = S_0(t)f$$

at time $t \geq 0$. Instead of determining $u(\cdot,t)$ directly from $u(\cdot,0)$ we can first determine

$$u(\cdot,t_1) = S_0(t_1)u(\cdot,0), \quad 0 \leq t_1 \leq t,$$

and use $u(\cdot,t_1)$ as new initial data to calculate $u(\cdot,t)$. We have

$$S_0(t)u(\cdot,0) = S_0(t-t_1)S_0(t_1)u(\cdot,0).$$

Thus the one-parameter family $\{S_0(t),\ t \geq 0\}$ of linear operators on $\mathbf{M}_0$ has the properties:

1. $S_0(0) = I$;
2. $S_0(t_1 + t_2) = S_0(t_2)S_0(t_1)$.

One says that the family $\{S_0(t),\ t \geq 0\}$ forms a *semigroup* on $\mathbf{M}_0$.

2.2.2. Estimates of the Solution by the Initial Data

If (2.2.3), (2.2.4) shall describe the evolution of a physical process, it is reasonable to require that the solutions are stable with respect to perturbations of the initial data, i.e., if $v(x,t)$ is the solution of $v_t = P(\partial/\partial x)v$ with perturbed initial data

$$v(x,0) = f(x) + g(x), \quad f,\, g \in \mathbf{M}_0,$$

then we must be able to estimate $v(x,t) - u(x,t)$ in terms of g. Linearity of the equation implies that such an estimate is possible if and only if one can estimate any solution $u(\cdot,t)$ in terms of its initial data $u(\cdot,0) = f$.

In this section we want to show how such estimates are related to bounds of the symbol $P(i\omega)$. Basically, the results follow from Parseval's relation. Throughout we denote by $u = u(x,t)$ the $\mathbf{M}_0$-solution of (2.2.3), (2.2.4) for initial data $f \in \mathbf{M}_0$.

If the problem is well-posed; i.e., if

$$|e^{P(i\omega)t}| \leq Ke^{\alpha t}, \quad \omega \in \mathbf{R}^s, \quad t \geq 0, \tag{2.2.6}$$

then (2.2.5) and Parseval's relation yield

$$\|u(\cdot,t)\| = \|e^{P(i\omega)t}\hat{f}(\omega)\| \leq Ke^{\alpha t}\|\hat{f}\| = Ke^{\alpha t}\|f\|.$$

In this way, we have an estimate of u at all later times by the initial data. The converse is also valid.

Theorem 2.2.2. *Given a constant-coefficient operator $P(\partial/\partial x)$. For any real K, α the following conditions are equivalent:*

1. *The symbols satisfy* (2.2.6).
2. *For all $u(\cdot,0) = f \in \mathbf{M}_0$ it holds that*

$$\|u(\cdot,t)\| \leq Ke^{\alpha t}\|f\|, \quad t \geq 0.$$

Proof. It remains to show that (2) implies (1). To this end, let $t \geq 0$ and $\omega_0 \in \mathbf{R}^s$ be fixed. There is a vector $v \in \mathbf{C}^n$, $|v| = 1$, with

$$|e^{P(i\omega_0)t}| = |e^{P(i\omega_0)t}v|.$$

For $\delta > 0$, define the function $\hat{f}(\omega)$ by

$$\hat{f}(\omega) = \begin{cases} v & \text{if } |\omega - \omega_0| \leq \delta, \\ 0 & \text{otherwise.} \end{cases}$$

(To be precise, we must approximate $\hat{f}$ by functions in C_0^∞.) For any $\epsilon > 0$ there is $\delta > 0$ with

$$\|e^{P(i\omega)t}\hat{f}(\omega)\| \geq (1-\epsilon)\,|e^{P(i\omega_0)t}|\|\hat{f}\|.$$

Using (2.2.5), we obtain

$$Ke^{\alpha t}\|f\| \geq \|u(\cdot,t)\| = \|e^{P(i\omega)t}\hat{f}(\omega)\| \geq (1-\epsilon)\,|e^{P(i\omega_0)t}|\|f\|.$$

Since $\epsilon > 0$ was arbitrary, the estimate of the symbol follows.

Now we allow derivative-terms of f to bound the solution at later times. For $q = 0, 1, 2, \ldots$ define

$$\|f\|_{H^q}^2 = \sum_{|\nu| \le q} \|D^\nu f\|^2, \quad f \in \mathbf{M}_0;$$

i.e., $\|\cdot\|_{H^q}$ measures all derivatives of order $\le q$. By Parseval's relation,

$$\|D^\nu f\| = \|\omega_1^{\nu_1} \ldots \omega_s^{\nu_s} \hat{f}(\omega)\|,$$

thus

$$\|f\|_{H^q}^2 = \frac{1}{(2\pi)^{s/2}} \int_{\mathbf{R}^s} \Big\{ \sum_{|\nu| \le q} \omega_1^{2\nu_1} \ldots \omega_s^{2\nu_s} \Big\} |\hat{f}(\omega)|^2 \, d\omega.$$

There is a constant $c = c_q$ independent of ω with

$$\frac{1}{c}(1 + |\omega|^q)^2 \le \sum_{|\nu| \le q} \omega_1^{2\nu_1} \ldots \omega_s^{2\nu_s} \le c(1 + |\omega|^q)^2, \qquad \omega \in \mathbf{R}^s.$$

Using this inequality, one obtains a characterization of weak well-posedness.

Theorem 2.2.3. *Given a constant-coefficient operator $P(\partial/\partial x)$; let $q \in \{1, 2, \ldots\}$ and $\alpha \in \mathbf{R}$. The following conditions are equivalent:*

1. *There is K_1 with*

$$(2.2.7) \qquad |e^{P(i\omega)t}| \le K_1 (1 + |\omega|^q)\, e^{\alpha t}, \quad \omega \in \mathbf{R}^s, \; t \ge 0.$$

2. *There is K_2 with*

$$\|u(\cdot, t)\| \le K_2\, e^{\alpha t}\, \|f\|_{H^q}, \quad t \ge 0,$$

for all $u(\cdot, 0) = f \in \mathbf{M}_0$.

To summarize, we can distinguish equations $u_t = P(\partial/\partial x)u$ of three different types. Accordingly, the Cauchy problem is

1. *well-posed*: the solutions satisfy $\|u(\cdot, t)\| \le K e^{\alpha t} \|u(\cdot, 0)\|$;
2. *weakly well-posed but not of type* 1: the solutions satisfy

$$(2.2.8) \qquad \|u(\cdot, t)\| \le K e^{\alpha t} \|u(\cdot, 0)\|_{H^q}$$

   for some positive integer q but not for $q = 0$;
3. *not weakly well-posed*: the symbols $|e^{P(i\omega)t}|$ grow faster than any polynomial in $|\omega|$. We say, there is *exponential explosion* and call the problem *ill-posed in any sense*.

Problems of type 1 or 2 can usually be treated with standard numerical methods. (Higher-order methods might be necessary for type 2 problems.) Since numerical calculations introduce round-off errors – which correspond to high wave-number oscillations – there is no hope to calculate solutions of type 3 problems with standard methods. (Admitting a data error and requesting additional restrictions for the solution, one can *regularize* ill-posed problems, however.) From this perspective, type 1 and type 2 problems are both adequate for applications. However, if one wants to go over to variable coefficients, then the concept of weak well-posedness leads to serious difficulties. These will be illustrated by the next two examples.

2.2.3. A Perturbation Leading to Exponential Explosion

Recall that Example 5 of Section 2.1.4 is weakly well-posed but not well-posed. If we add the zero-order term

$$\begin{pmatrix} 0 & 1 \\ -1 & 0 \end{pmatrix} u,$$

then the equation becomes

$$u_t = \begin{pmatrix} 1 & 1 \\ 0 & 1 \end{pmatrix} u_x + \begin{pmatrix} -1 & 1 \\ -1 & -1 \end{pmatrix} u. \tag{2.2.9}$$

Here the symbol

$$P(i\omega) = \begin{pmatrix} i\omega - 1 & i\omega + 1 \\ -1 & i\omega - 1 \end{pmatrix}$$

has the eigenvalues

$$\kappa_{1,2} = i\omega - 1 \pm \sqrt{-(1 + i\omega)}\,.$$

Hence, for large $|\omega|$, there is an eigenvalue with real part

$$\operatorname{Re} \kappa \sim |\omega|^{1/2}.$$

The Fourier transforms

$$\hat{u}(\omega, t) = e^{P(i\omega)t} \hat{f}(\omega)$$

of the solutions can grow like $e^{|\omega|^{1/2} t}$, and there is exponential explosion. This demonstrates that the estimate (2.2.7) with $q = 1$ is not invariant under perturbations by lower-order terms of the equation.

One can prove that this example is typical: If the solutions of a given differential equation satisfy an estimate (2.2.7) with some $q \geq 1$, but not with $q = 0$, then one can perturb the equation by a lower-order term such that the perturbed system shows exponential explosion.

2.2.4. Variability of Coefficients Corresponds to Perturbations

The system (2.2.9) can also serve as an example that *variability of coefficients* is related to adding lower-order terms. Consider

$$u_t = A(t)u_x - u, \quad x \in \mathbf{R}, \quad t \geq 0,$$

where

$$A(t) = U(t)\begin{pmatrix} 1 & 1 \\ 0 & 1 \end{pmatrix} U^{-1}(t), \quad U(t) = \begin{pmatrix} \sin t & \cos t \\ -\cos t & \sin t \end{pmatrix}.$$

If one freezes $A(t)$ at an arbitrary $t = t_0$, then one obtains a constant-coefficient equation, which behaves like Example 5, Section 2.1.4. Thus, for all frozen-coefficient problems one *cannot* estimate the L_2-norm of the solution by the L_2-norm of $u(\cdot, 0)$, but one *can* estimate the L_2-norm of the solution as in (2.2.8) with $q = 1$.

Now let us discuss the given *variable*-coefficient problem. The transformed function

$$v(x,t) = U^{-1}(t)\, u(x,t)$$

satisfies the constant-coefficient equation

$$v_t = \begin{pmatrix} 1 & 1 \\ 0 & 1 \end{pmatrix} v_x + \begin{pmatrix} -1 & 1 \\ -1 & -1 \end{pmatrix} v$$

discussed above. Thus the v-equation can show arbitrarily fast exponential growth, and – transforming back – the same holds for the given u-equation. Summarizing, to allow estimates of $\|u(\cdot, t)\|$ as in (2.2.8) with $q \geq 1$ is not useful if variable-coefficient problems are to be treated via localization.

2.2.5. *Extension of the Solution Operator $S_0(t)$

Up to this point we have only allowed initial data in $\mathbf{M}_0$. We shall now extend the admissible initial data to all functions $f \in L_2$ *provided the initial value problem is well-posed.* To this end, let $f \in L_2$ be given. There is a sequence $f_j \in \mathbf{M}_0$ with

$$\lim_{j \to \infty} \|f - f_j\| = 0.$$

*This section might be omitted on first reading. We will use – without proof – completeness of the space L_2 and density of $\mathbf{M}_0$ in L_2. Since C_0^∞ is dense in L_2 (mollification), the latter result can be shown by Fourier transformation.

From

$$\|S_0(t)f_j - S_0(t)f_k\| \le Ke^{\alpha t}\|f_j - f_k\| \to 0 \quad \text{for} \quad j,k \to 0$$

it follows that the sequence $u_j(\cdot,t) = S_0(t)f_j$ converges in L_2 for every fixed t. Also, the limit

$$u(\cdot,t) := \lim_{j\to\infty} S_0(t)f_j$$

does not depend on the specific choice of the approximating sequence f_j: If $\tilde{f}_j \in \mathbf{M}_0$ is another sequence with $\tilde{f}_j \to f$, then

$$\|S_0(t)f_j - S_0(t)\tilde{f}_j\| \le Ke^{\alpha t}\|f_j - \tilde{f}_j\| \to 0.$$

Hence, the construction defines a unique function $u(\cdot,t) \in L_2$ for given initial data $f \in L_2$:

$$u(\cdot,t) = \lim_{j\to\infty} S_0(t)f_j.$$

The function u is called the *generalized solution* of the initial value problem (2.2.3), (2.2.4). We write

$$u(\cdot,t) = S(t)f, \quad S(t): L_2 \to L_2,$$

and call the one parameter family $\{S(t),\ t \ge 0\}$ the semigroup of generalized solution operators.

For fixed t, the above construction is nothing but the usual extension of a densely defined bounded linear operator to the whole space. With arguments as given above one can prove the following theorem of functional analysis:

Theorem 2.2.4. *Let B_1, B_2 denote normed spaces, let M denote a dense subspace of B_1, and let B_2 be complete. If $S_0 : M \to B_2$ is a bounded linear operator, then there is a unique bounded linear operator $S : B_1 \to B_2$ with $Sf = S_0 f$ for all $f \in M$. The operator S is called the extension of S_0.*

By our construction, the generalized solution $u(x,t)$ is just an L_2-function with respect to x for each fixed t. It is often possible, however, to obtain smoothness properties of $u(x,t)$ with respect to x and t by further investigations. Let us consider two simple examples.

Example 1. Consider the differential equation

$$u_t + u_x = 0$$

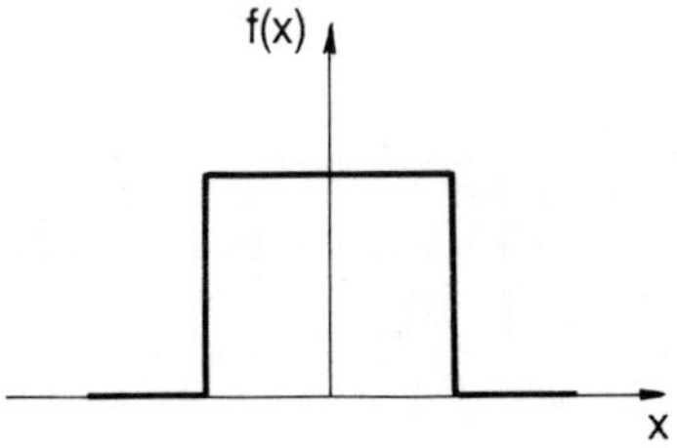

FIGURE 2.2.1. Discontinuous initial function.

with initial data

$$u(x,0) = f(x) = \begin{cases} 1 & \text{for } |x| \le 1, \\ 0 & \text{for } |x| > 1. \end{cases}$$

We write $f(x)$ in the form

$$f(x) = \frac{1}{(2\pi)^{1/2}} \int_{-\infty}^{+\infty} e^{i\omega x} \hat{f}(\omega)\, d\omega, \quad \hat{f}(\omega) = \frac{1}{(2\pi)^{1/2}} \int_{-1}^{+1} e^{-i\omega x}\, dx$$

and approximate f by

$$f_j(x) = \frac{1}{(2\pi)^{1/2}} \int_{-j}^{+j} e^{i\omega x} \hat{f}(\omega)\, d\omega.$$

(The functions f_j have Fourier transforms $\hat{f}_j$ with compact support. The piecewise smooth function $\hat{f}_j$ can be approximated with arbitrary accuracy by functions in C_0^∞.) For the solution with initial data f_j it holds that

$$u_j(x,t) = \frac{1}{(2\pi)^{1/2}} \int_{-j}^{+j} e^{i\omega(x-t)} \hat{f}(\omega)\, d\omega = f_j(x-t),$$

and therefore

$$u(x,t) = \lim_{j\to\infty} u_j(x,t) = \lim_{j\to\infty} f_j(x-t) = f(x-t).$$

Hence, the generalized solution consists of the box traveling with speed 1 to the right. Here u is not smooth, but at least $t \to u(\cdot,t) \in L_2$ is continuous.

Example 2. Consider the heat equation

$$u_t = u_{xx}$$

subject to the initial condition given in Example 1. Here we obtain

$$u_j(x,t) = \frac{1}{(2\pi)^{1/2}} \int_{-j}^{+j} e^{-\omega^2 t} e^{i\omega x} \hat{f}(\omega)\, d\omega.$$

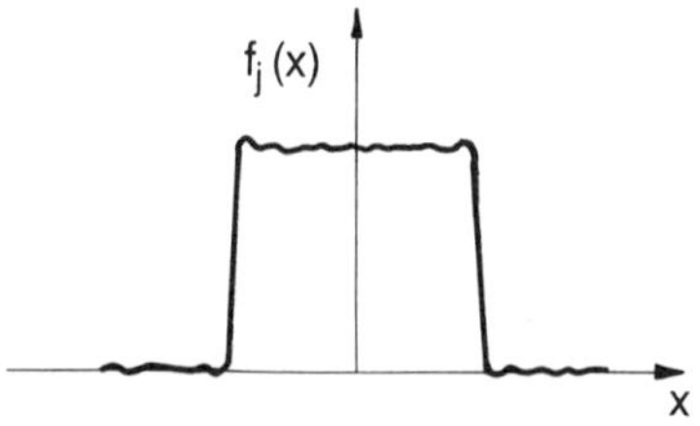

FIGURE 2.2.2. Smooth approximation.

Hence, in any interval $0 < \delta \leq t < \infty$ the sequence $u_j(x,t)$ converges *uniformly* to

$$u(x,t) = \lim_{j\to\infty} u_j(x,t) = \frac{1}{(2\pi)^{1/2}} \int_{-\infty}^{+\infty} e^{-\omega^2 t} e^{i\omega x} \hat{f}(\omega)\, d\omega.$$

In this case, the generalized solution is a C^∞-function for $x \in \mathbf{R}$, $t > 0$, which satisfies the differential equation in the classical (i.e., pointwise) sense for $t > 0$. The discontinuities of the initial data disappear for $t > 0$. This behavior is typical for parabolic equations.

2.3. Algebraic Characterization of Well-Posedness

According to Definition 1, Section 2.1, the Cauchy problem for a constant-coefficient system $u_t = P(\partial/\partial x)$ is well-posed if and only if there is $\alpha \in \mathbf{R}$ such that

$$|e^{(P(i\omega)-\alpha I)t}| \leq K$$

with K independent of $\omega \in \mathbf{R}^s$ and $t \geq 0$. Thus, for fixed α, one has to consider the set of matrices

$$\mathbf{F} = \{A = A(\omega) = P(i\omega) - \alpha I \,:\, \omega \in \mathbf{R}^s\}$$

and establish a uniform bound of the matrix exponentials:

$$|e^{At}| \leq K, \quad K \text{ independent of } A \in \mathbf{F} \text{ and } t \geq 0.$$

In the main result of this section, Theorem 2.3.2 below, we will characterize the validity of such an estimate by other algebraic conditions.

The case of a single matrix. A simple result on matrix exponentials is shown first.

Lemma 2.3.1. *For any $A \in \mathbf{C}^{n,n}$ the following conditions are equivalent:*

1. *There is a constant K such that $|e^{At}| \le K$ for all $t \ge 0$.*
2. *All eigenvalues κ of the matrix A have a real part $\operatorname{Re}\kappa \le 0$. Furthermore, if J_r is a Jordan block of the Jordan matrix $J = SAS^{-1}$ which corresponds to an eigenvalue κ with $\operatorname{Re}\kappa = 0$, then J_r has dimension 1×1. (In other words, if κ is an eigenvalue with $\operatorname{Re}\kappa = 0$, then the dimension of its eigenspace equals the multiplicity of κ as a root of the characteristic polynomial of A.)*

Proof. First note that

$$e^{At} = e^{S^{-1}JSt} = S^{-1}e^{Jt}S.$$

Also, if

$$J_r = \begin{pmatrix} \kappa & 1 & & \\ & \ddots & \ddots & \\ & & \ddots & 1 \\ & & & \kappa \end{pmatrix}$$

is an arbitrary Jordan block, then the exponential $e^{J_r t}$ stays bounded for $t \ge 0$ if and only if either $\operatorname{Re}\kappa < 0$ or ($\operatorname{Re}\kappa = 0$ and J_r has dimension 1×1). Thus the result follows.

The Matrix Theorem and its proof. Let $\mathbf{F}$ denote an infinite set of matrices $A \in \mathbf{C}^{n,n}$. The *uniform* boundedness

$$|e^{At}| \le K, \quad K \text{ independent of } A \in \mathbf{F} \text{ and } t \ge 0,$$

is not as easily discussed. It is not sufficient to request condition 2 of the previous lemma for each $A \in \mathbf{F}$ separately because $|S^{-1}||S|$ can depend on A and can become arbitrarily large. The characterization given next is useful if one wants to derive necessary and sufficient conditions for well-posedness.

Theorem 2.3.2. *Let $\mathbf{F}$ denote a set of matrices $A \in \mathbf{C}^{n,n}$. The following four conditions are equivalent:*

1. *There is a constant K_1 with $|e^{At}| \le K_1$ for all $A \in \mathbf{F}$ and all $t \ge 0$.*
2. *For all $A \in \mathbf{F}$ and all $s \in \mathbf{C}$ with $\operatorname{Re} s > 0$ the matrix $A - sI$ is nonsingular, and there is a constant K_2 such that*

$$|(A - sI)^{-1}| \le \frac{K_2}{\operatorname{Re} s}, \qquad A \in \mathbf{F}, \qquad \operatorname{Re} s > 0. \tag{2.3.1}$$

(This condition is known as the **Resolvent Condition**.*)*

3. *There are constants* K_{31}, K_{32} *with the following property: For each* $A \in \mathbf{F}$ *there is a transformation* $S = S(A)$ *with*

$$|S^{-1}| + |S| \le K_{31} \tag{2.3.2}$$

such that SAS^{-1} *is upper triangular,*

$$SAS^{-1} = \begin{pmatrix} \kappa_1 & b_{12} & \dots & b_{1n} \\ & \kappa_2 & & b_{2n} \\ & & \ddots & \vdots \\ 0 & & & \kappa_n \end{pmatrix}, \tag{2.3.3}$$

the diagonal is ordered,

$$0 \ge \operatorname{Re}\kappa_1 \ge \operatorname{Re}\kappa_2 \ge \cdots \ge \operatorname{Re}\kappa_n, \tag{2.3.4}$$

and the upper diagonal elements satisfy the estimate

$$|b_{ij}| \le K_{32}|\operatorname{Re}\kappa_i|, \quad 1 \le i < j \le n. \tag{2.3.5}$$

4. *There is a positive constant* K_4 *with the following property: For each* $A \in \mathbf{F}$ *there exists a Hermitian matrix* $H = H(A)$ *with*

$$K_4^{-1} I \le H \le K_4 I \quad \text{and} \quad HA + A^*H \le 0.$$

For our applications, the most interesting implication is (4) $\Rightarrow$ (1). It allows us to show well-posedness by constructing a suitable Hermitian matrix $H = H(\omega)$ for each symbol $P(i\omega)$. This part of the Matrix Theorem follows immediately from Lemma 2.1.4. The only difficult part in the proof of the theorem is to show (2) $\Rightarrow$ (3). This part will only be used to show *necessary* conditions for well-posedness, and its proof might be omitted on a first reading.

Proof of Theorem 2.3.2. We show the implications

$$(1) \Rightarrow (2) \Rightarrow (3) \Rightarrow (4) \Rightarrow (1)$$

"(1) $\Rightarrow$ (2)": If $A \in \mathbf{F}$ then $|e^{At}| \le K_1$ for all $t \ge 0$, thus by Lemma 2.3.1 all eigenvalues κ of A have a real part $\operatorname{Re}\kappa \le 0$. Therefore, if $\operatorname{Re}s > 0$, the matrix $A - sI$ is nonsingular and

$$(A - sI)^{-1} = -\int_0^\infty e^{(A-sI)t}\,dt.$$

Furthermore,

$$|(A - sI)^{-1}| \le \int_0^\infty |e^{At}||e^{-st}|\,dt \le K_1 \int_0^\infty e^{-(\operatorname{Re}s)t}\,dt = \frac{K_1}{\operatorname{Re}s}.$$

This shows that (2) holds with $K_2 = K_1$.

"(2) ⇒ (3)": For any matrix $A \in \mathbf{C}^{n,n}$ there is a unitary transformation U such that UAU^{-1} is upper triangular; the ordering of the diagonal entries can also be prescribed. This is Schur's Theorem, see Appendix 1. Since unitary transformations do not change the Resolvent Condition (2.3.1), it is no restriction to assume that the matrices $A \in \mathbf{F}$ already have the form

$$A = \begin{pmatrix} \kappa_1 & a_{12} & \dots & a_{1n} \\ & \kappa_2 & & a_{2n} \\ & & \ddots & \vdots \\ 0 & & & \kappa_n \end{pmatrix}, \qquad 0 \geq \operatorname{Re}\kappa_1 \geq \dots \geq \operatorname{Re}\kappa_n. \tag{2.3.6}$$

It suffices to show the following lemma by induction on n.

Lemma 2.3.3. *Given a set* $\mathbf{F}$ *of matrices* $A \in \mathbf{C}^{n,n}$ *of the form* (2.3.6) *which satisfy the Resolvent Condition* (2.3.1), *there are constants* K_{31}, K_{32} *depending only on* n *and* K_2 *which have the following property: For each* $A \in \mathbf{F}$ *there is a transformation* $S = S(A)$ *of the form*

$$S = \begin{pmatrix} 1 & s_{12} & \dots & s_{1n} \\ & 1 & & s_{2n} \\ & & \ddots & \vdots \\ 0 & & & 1 \end{pmatrix}$$

with $|S| + |S^{-1}| \leq K_{31}$ *such that the transformed matrix* SAS^{-1} *(see* (2.3.3)*) fullfills the estimates* (2.3.5).

Proof. The statement of the lemma is obvious for $n = 1$. To simplify the induction step below, we first consider the case $n = 2$. Thus let

$$A = \begin{pmatrix} \kappa_1 & a_{12} \\ 0 & \kappa_2 \end{pmatrix}, \qquad 0 \geq \operatorname{Re}\kappa_1 \geq \operatorname{Re}\kappa_2.$$

We want to transform A to

$$SAS^{-1} = \begin{pmatrix} \kappa_1 & b_{12} \\ 0 & \kappa_2 \end{pmatrix} \quad \text{with} \quad |b_{12}| \leq K_{32}|\operatorname{Re}\kappa_1|.$$

Note that $|m_{ij}| \leq |M|$ for any matrix M, and that

$$(A - sI)^{-1} = \begin{pmatrix} (\kappa_1 - s)^{-1} & x_{12} \\ 0 & (\kappa_2 - s)^{-1} \end{pmatrix}, \quad x_{12} = \frac{-a_{12}}{(\kappa_1 - s)(\kappa_2 - s)}.$$

Hence assumption (2.3.1) yields the bound

$$\left|\frac{-a_{12}}{(\kappa_1 - s)(\kappa_2 - s)}\right| \leq \frac{K_2}{\operatorname{Re} s}, \qquad \operatorname{Re} s > 0,$$

which we can rewrite as

$$|a_{12}| \le K_2 \frac{|\kappa_1 - s|}{\text{Re}\, s} |\kappa_2 - s|, \qquad \text{Re}\, s > 0.$$

For $s \to -\overline{\kappa}_1$, one obtains that

$$|a_{12}| \le 2K_2|\kappa_2 + \overline{\kappa}_1|. \tag{2.3.7}$$

If $\kappa_2+\overline{\kappa}_1 = 0$, we do not have to transform A. Thus we define the transformation matrix

$$S = \begin{pmatrix} 1 & -\gamma \\ 0 & 1 \end{pmatrix},$$

where

$$\gamma = \frac{a_{12}}{\kappa_2 + \overline{\kappa}_1} \quad \text{for} \quad \kappa_2 + \overline{\kappa}_1 \neq 0, \qquad \gamma = 0 \quad \text{for} \quad \kappa_2 + \overline{\kappa}_1 = 0$$

and obtain

$$SAS^{-1} = \begin{pmatrix} \kappa_1 & 2\gamma \text{Re}\, \kappa_1 \\ 0 & \kappa_2 \end{pmatrix} =: \begin{pmatrix} \kappa_1 & b_{12} \\ 0 & \kappa_2 \end{pmatrix}.$$

Here

$$|b_{12}| = 2|\gamma||\text{Re}\, \kappa_1| \le 4K_2|\text{Re}\, \kappa_1|$$

by (2.3.7). A bound $|S| + |S^{-1}| \le K_{31}$ also follows from $|\gamma| \le 2K_2$, and the result of the lemma is shown for $n = 2$.

Now let $n \ge 3$, and assume the statement of the lemma is true for all sets of matrices of order $n-1$. Let $\mathbf{F}$ denote the given set of n by n matices A of the form described above. Any $A \in \mathbf{F}$ can be partitioned as

$$A = \begin{pmatrix} A_1 & a_1 \\ 0 & \kappa_n \end{pmatrix}, \qquad A_1 \in \mathbf{C}^{n-1,n-1}.$$

Since $(A - sI)^{-1}$ is of the form

$$(A - sI)^{-1} = \begin{pmatrix} (A_1 - sI)^{-1} & * \\ 0 & (\kappa_n - s)^{-1} \end{pmatrix},$$

the Resolvent Condition (2.3.1) also holds for the set of $n-1$ by $n-1$ matrices $\{A_1, A \in \mathbf{F}\}$. By the induction hypothesis, there are constants K'_{31}, K'_{32} – depending only on n and K_2 – and transformations

$$S = S(A_1) = \begin{pmatrix} 1 & s_{12} & \dots & s_{1n-1} \\ & 1 & & s_{2n-1} \\ & & \ddots & \vdots \\ 0 & & & 1 \end{pmatrix}$$

with $|S| + |S^{-1}| \le K'_{31}$ such that the transformed matrix

$$SA_1S^{-1} = \begin{pmatrix} \kappa_1 & a'_{12} & \dots & a'_{1n-1} \\ & \kappa_2 & & a'_{2n-1} \\ & & \ddots & \vdots \\ 0 & & & \kappa_{n-1} \end{pmatrix}$$

satisfies the estimates

$$|a'_{ij}| \le K'_{32}|\text{Re}\,\kappa_i|, \qquad 1 \le i < j \le n-1. \tag{2.3.8}$$

With

$$S_1 = \begin{pmatrix} S & 0 \\ 0 & 1 \end{pmatrix}$$

we transform the matrix A itself and find

$$S_1AS_1^{-1} = \begin{pmatrix} S & 0 \\ 0 & 1 \end{pmatrix}\begin{pmatrix} A_1 & a_1 \\ 0 & \kappa_n \end{pmatrix}\begin{pmatrix} S^{-1} & 0 \\ 0 & 1 \end{pmatrix} = \begin{pmatrix} SA_1S^{-1} & Sa_1 \\ 0 & \kappa_n \end{pmatrix}$$
$$= \begin{pmatrix} \kappa_1 & a'_{12} & \dots & a'_{1n} \\ & \kappa_2 & & a'_{2n} \\ & & \ddots & \vdots \\ 0 & & & \kappa_n \end{pmatrix} =: A'.$$

Except for the last column, the elements of A' are known to satisfy the desired estimates. The set of matrices $\{A',\, A \in \mathbf{F}\}$ again fulfills the Resolvent Condition:

$$|(A' - sI)^{-1}| = |(S_1AS_1^{-1} - sI)^{-1}| \le |S_1||S_1^{-1}||(A - sI)^{-1}| \le \frac{K'_2}{\text{Re}\, s}.$$

We partition A' in the form

$$A' = \begin{pmatrix} \kappa_1 & a'_2 \\ 0 & A'_2 \end{pmatrix}, \qquad A'_2 \in \mathbf{C}^{n-1,n-1}.$$

As before, we can apply the induction hypothesis to the matrices A'_2 and find transformations $S' = S'(A'_2)$ with

$$\begin{pmatrix} 1 & 0 \\ 0 & S' \end{pmatrix} \begin{pmatrix} \kappa_1 & a_2' \\ 0 & A_2' \end{pmatrix} \begin{pmatrix} 1 & 0 \\ 0 & S'^{-1} \end{pmatrix} = \begin{pmatrix} \kappa_1 & a_2' S'^{-1} \\ 0 & S' A_2' S'^{-1} \end{pmatrix}$$

$$= \begin{pmatrix} \kappa_1 & a_{12}'' & \dots & a_{1n}'' \\ & \kappa_2 & & a_{2n}'' \\ & & \ddots & \vdots \\ 0 & & & \kappa_n \end{pmatrix} =: A''.$$

Here

$$|a_{ij}''| \le K_{32}'' |\mathrm{Re}\, \kappa_i|, \qquad 2 \le i < j \le n.$$

Furthermore,

$$(a_{12}'', \cdots, a_{1n}'') = a_2' S'^{-1}.$$

Using the upper-triangular form of S'^{-1} and the fact that the elements of

$$a_2' = (a_{12}', \cdots, a_{1n}')$$

except a_{1n}' have already been estimated (see (2.3.8)), we also have that

$$|a_{1j}''| \le K_{32}'' |\mathrm{Re}\, \kappa_1|, \qquad 1 \le j \le n-1.$$

Except for a_{1n}'', all outer-diagonal entries of A'' are known to satisfy the desired estimates. We now apply an additional transformation of the form $A''' = S_3 A'' S_3^{-1}$ with

$$S_3 = I - \gamma \begin{pmatrix} 0 & \dots & 0 & 1 \\ 0 & \dots & 0 & 0 \\ \vdots & & \vdots & \vdots \\ 0 & \dots & 0 & 0 \end{pmatrix},$$

(2.3.9) $$\gamma = \frac{a_{1n}''}{\kappa_n + \overline{\kappa}_1} \quad \text{for} \quad \kappa_n + \overline{\kappa}_1 \ne 0, \qquad \gamma = 0 \quad \text{for} \quad \kappa_n + \overline{\kappa}_1 = 0.$$

This is motivated by the 2 by 2 case discussed above. Note that

$$S_3^{-1} = I + \gamma \begin{pmatrix} 0 & \dots & 0 & 1 \\ 0 & \dots & 0 & 0 \\ \vdots & & \vdots & \vdots \\ 0 & \dots & 0 & 0 \end{pmatrix}$$

and that

$$S_3 A'' S_3^{-1} = \begin{pmatrix} \kappa_1 & a''_{12} & \dots & a''_{1n-1} & 2\gamma \operatorname{Re} \kappa_1 \\ & \kappa_2 & \dots & \dots & a''_{2n} \\ & & \ddots & & \vdots \\ & & & \ddots & \vdots \\ 0 & & & & \kappa_n \end{pmatrix} \quad \text{if} \quad \overline{\kappa}_1 + \kappa_n \neq 0.$$

It remains to prove an inequality of the form

$$|a''_{1n}| \leq \text{const}\, |\overline{\kappa}_1 + \kappa_n|$$

in order to bound γ. To this end, note that A'' satisfies again the resolvent condition,

$$|(A'' - sI)^{-1}| \leq \frac{K''_2}{\operatorname{Re} s} \quad \text{for} \quad \operatorname{Re} s > 0.$$

Let $(A'' - sI)^{-1} = (c_{ij})$, and compute c_{1n} by Cramer's rule:

$$|c_{1n}| = \left\{ \prod_{i=1}^{n} |\kappa_i - s|^{-1} \right\} \left| \det \begin{pmatrix} a''_{12} & \dots & \dots & a''_{1n} \\ \kappa_2 - s & \ddots & & \vdots \\ & \ddots & \ddots & \vdots \\ 0 & & \kappa_{n-1} - s & a''_{n-1,n} \end{pmatrix} \right|$$

$$= \left\{ \prod_{i=1}^{n} |\kappa_i - s|^{-1} \right\} |D_1 + D_2|$$

with

$$D_1 = \det \begin{pmatrix} 0 & \dots & 0 & a''_{1n} \\ \kappa_2 - s & * & \dots & * \\ & \ddots & \ddots & \vdots \\ 0 & & \kappa_{n-1} - s & * \end{pmatrix} = a''_{1n} \left\{ \prod_{i=2}^{n-1} |\kappa_i - s| \right\}$$

and

$$D_2 = \det \begin{pmatrix} a''_{12} & \dots & a''_{1n-1} & 0 \\ \kappa_2 - s & * & \dots & * \\ & \ddots & \ddots & \vdots \\ 0 & & \kappa_{n-1} - s & * \end{pmatrix}.$$

To treat D_2, we can use the proven estimates

$$|a''_{ij}| \leq K''_{32} |\, \kappa_i|, \qquad (i, j) \neq (1, n),$$

and find

$$\left\{\prod_{i=1}^{n}|\kappa_i - s|^{-1}\right\}|D_2| = \frac{1}{|\kappa_n - s|}\left|\det\begin{pmatrix} \frac{a''_{12}}{|\kappa_1-s|} & \cdots & \frac{a''_{1n-1}}{|\kappa_1-s|} & 0 \\ 1 & * & \cdots & \frac{a''_{2n}}{|\kappa_2-s|} \\ & \ddots & \ddots & \vdots \\ 0 & & 1 & \frac{a''_{n-1,n}}{|\kappa_{n-1}-s|}\end{pmatrix}\right|$$

$$\leq \frac{K_2'''}{|\kappa_n - s|}\,.$$

Therefore,

$$\frac{|a''_{1n}|}{|\kappa_1 - s||\kappa_n - s|} = \frac{|D_1|}{\prod_{i=1}^{n}|\kappa_i - s|} \leq |c_{1n}| + \frac{|D_2|}{\prod_{i=1}^{n}|\kappa_i - s|}$$

$$\leq \frac{K_2''}{\operatorname{Re} s} + \frac{K_2'''}{\operatorname{Re} s} = \frac{K_2''''}{\operatorname{Re} s} \quad \text{for} \quad \operatorname{Re} s > 0.$$

As in the 2 by 2 case we let $s \to -\overline{\kappa}_1$ and obtain the bound

$$|a''_{1n}| \leq 2K_2''''|\kappa_n + \overline{\kappa}_1|.$$

This gives a uniform estimate for the quantity γ defined in (2.3.9) and ends the proof of Lemma 2.3.3. Hence the implication "2. $\Rightarrow$ 3." of Theorem 2.3.2 is proved. We proceed with the proof of the theorem.

"3. $\Rightarrow$ 4.": Let $A \in \mathbf{F}$ be arbitrary and let $S = S(A)$ be determined as in 3.; i.e.,

$$B = SAS^{-1} = \begin{pmatrix} \kappa_1 & b_{12} & \dots & b_{1n} \\ & \kappa_2 & & b_{2n} \\ & & \ddots & \vdots \\ 0 & & & \kappa_n \end{pmatrix}, \qquad 0 \geq \operatorname{Re}\kappa_1 \geq \cdots \geq \operatorname{Re}\kappa_n,$$

where $|S| + |S^{-1}| \leq K_{31}$ and

$$|b_{ij}| \leq K_{32}|\operatorname{Re}\kappa_i|, \qquad 1 \leq i < j \leq n.$$

Define a diagonal matrix

$$D = \operatorname{diag}(1, d, d^2, \cdots, d^{n-1}), \qquad d \geq 1, \tag{2.3.10}$$

and set

$$C = DBD^{-1} = (DS)A(DS)^{-1};$$

thus

$$C = \begin{pmatrix} \kappa_1 & b_{12}/d & b_{13}/d^2 & \cdots \\ & \kappa_2 & b_{23}/d & \cdots \\ & & \ddots & \vdots \\ 0 & & & \kappa_n \end{pmatrix} =: \begin{pmatrix} \kappa_1 & c_{12} & c_{13} & \cdots \\ & \kappa_2 & c_{23} & \cdots \\ & & \ddots & \vdots \\ 0 & & & \kappa_n \end{pmatrix}.$$

We note that

$$|c_{ij}| \le d^{-1}|b_{ij}| \le d^{-1}K_{32}|\mathrm{Re}\,\kappa_i|.$$

As a consequence, let us show that the Hermitian matrix $C + C^*$ is negative definite, if $d = d(n, K_{32})$ is sufficiently large. The i-th row of $C + C^*$ reads

$$(\overline{c}_{1i}, \overline{c}_{2i}, \dots, \overline{c}_{i-1,i}, 2\,\mathrm{Re}\,\kappa_i, c_{i,i+1}, \dots, c_{in}).$$

Using the ordering $0 \ge \mathrm{Re}\,\kappa_1 \ge \cdots \ge \mathrm{Re}\,\kappa_n$, we find that the sum of the absolute values of the outer-diagonal elements in the i-th row of $C + C^*$ is bounded by

$$(n-1)d^{-1}K_{32}|\mathrm{Re}\,\kappa_i|.$$

Thus, by Gerschgorin's Circle Theorem, all eigenvalues of $C + C^*$ are nonpositive if

$$d > \frac{1}{2}nK_{32}.$$

Now set $S_1 = DS$, where D is defined in (2.3.10) and d fulfills the estimate above. The Hermitian matrix

$$H = H(A) = S_1^* S_1$$

satisfies

$$|H| \le |S||S^*||D|^2 \le K_4, \qquad |H^{-1}| \le |S^{-1}||S^{*-1}| \le K_4.$$

Furthermore,

$$\begin{aligned} HA + A^*H &= S_1^*S_1A + A^*S_1^*S_1 = S_1^*(S_1AS_1^{-1} + S_1^{*-1}A^*S_1^*)S_1 \\ &= S_1^*(C + C^*)S_1 \le 0, \end{aligned}$$

and condition 4 is proved.

"(4) ⇒ (1)": This implication follows immediately from Lemma 2.1.4 with $\alpha = 0$. Thus we have proved Theorem 2.3.2.

Applications to the question of well-posedness. Let us note two simple implications of the theorem. As before, we consider the Cauchy problem for a constant-coefficient system $u_t = P(\partial/\partial x)u$. The problem is well-posed if and only if there are constants α, $K \in \mathbf{R}$ with

$$|e^{(P(i\omega)-\alpha I)t}| \leq K \text{ for all } \ \omega \in \mathbf{R}^s \text{ and all } \ t \geq 0.$$

This holds if and only if for each $\omega \in \mathbf{R}^s$ there is a Hermitian matrix $H(\omega) \in \mathbf{C}^{n,n}$ with

$$K_4^{-1} I \leq H(\omega) \leq K_4 I,$$

$$H(\omega)\big(P(i\omega) - \alpha I\big) + \big(P^*(i\omega) - \alpha I\big)H(\omega) \leq 0,$$

where K_4 does not depend on ω. For later reference we summarize this result:

Corollary 2.3.4. *The Cauchy problem for $u_t = P(\partial/\partial x)u$ is well-posed if and only if for each $\omega \in \mathbf{R}^s$ there is a Hermitian matrix $H(\omega) \in \mathbf{C}^{n,n}$ with*

$$K_4^{-1} I \leq H(\omega) \leq K_4 I \ \text{ and } \ H(\omega)P(i\omega) + P^*(i\omega)H(\omega) \leq 2\alpha H(\omega),$$

where K_4 and α are independent of ω.

Let us note again that the "difficult" part of the Matrix Theorem is only needed to ensure the existence of $H(\omega)$ for a well-posed problem. The converse result, namely that the existence of $H(\omega)$ implies well-posedness, is elementary and follows immediately from Lemma 2.1.4.

In another application of the Matrix Theorem we show that well-posedness does not depend on the zero-order term. (This result should be compared with the example in Section 2.2.3, which demonstrated that *weak* well-posedness does depend on the zero-order term.)

Lemma 2.3.5. *Let $P(\partial/\partial x)$ denote a constant-coefficient operator, let $B \in \mathbf{C}^{n,n}$, and let $P_0(\partial/\partial x) = P(\partial/\partial x) + B$. The Cauchy problem is well-posed for $u_t = Pu$ if and only if it is well-posed for $u_t = P_0 u$.*

Proof. Assume that the Cauchy problem is well-posed for $u_t = Pu$; thus

$$H(\omega)P(i\omega) + P^*(i\omega)H(\omega) \leq 2\alpha H(\omega).$$

Here we use the notations of Corollary 2.3.4. Consequently,

$$\begin{aligned} H(\omega)\big(P(i\omega) + B\big) + \big(P^*(i\omega) + B^*\big)H(\omega) &\leq 2\alpha H(\omega) + H(\omega)B + B^*H(\omega) \\ &\leq 2\alpha H(\omega) + \big(|B| + |B^*|\big)\, K_4 I \\ &\leq \Big\{2\alpha + \big(|B| + |B^*|\big)\, K_4^2\Big\}\, H(\omega). \end{aligned}$$

Another application of Corollary 2.3.4 shows the well-posedness of the Cauchy problem for $P_0 = P + B$.

Remark. Suppose that the matrices $H(\omega)$ are constructed for a given operator $P(\partial/\partial x)$. Then the above corollary shows the well-posedness of $u_t = Pu + Bu$ for any matrix B. To obtain this result, only the elementary part 4. $\Rightarrow$ 1. of the Matrix Theorem is needed.

2.4. Hyperbolic and Parabolic Systems

In this section we define strong hyperbolicity and parabolicity for constant-coefficient equations in any number of space dimensions. The Cauchy problem for these equations is well-posed. As an application, we consider the compressible Euler equations linearized at a constant flow. One obtains a system which is strongly hyperbolic if $dr/d\rho > 0$, where $p = r(\rho)$ is the equation of state. If one adds viscosity, i.e., goes over to the Navier-Stokes equations, then the linearized system is neither parabolic nor hyperbolic, but "almost" parabolic. This motivates us to treat certain *mixed* systems in Section 2.5; these can be considered as coupled hyperbolic-parabolic equations.

2.4.1. Hyperbolic Systems

Consider a first-order equation in s space dimensions,

$$(2.4.1) \qquad u_t = P(\partial/\partial x)u \equiv \sum_{j=1}^{s} A_j \frac{\partial u}{\partial x_j}, \qquad A_j \in \mathbf{C}^{n,n}.$$

We want to characterize all equations of the above form for which the Cauchy problem is well-posed. Note that the symbol

$$P(i\omega) = i \sum_{j=1}^{s} \omega_j A_j, \qquad \omega \in \mathbf{R}^s,$$

depends in a linear way on the length $|\omega|$ of ω: for $\omega \neq 0$, we set $\omega' = \omega/|\omega|$ and obtain $P(i\omega) = |\omega| P(i\omega')$. This simple observation and Theorem 2.3.2 lead to

Theorem 2.4.1. *The Cauchy problem for the first-order equation* (2.4.1) *is well-posed if and only if the following two conditions hold:*

1. *For all $\omega' \in \mathbf{R}^s$, $|\omega'| = 1$, all eigenvalues of $P(i\omega')$ are purely imaginary.*
2. *There is a constant K_{31}, and for each $\omega' \in \mathbf{R}^s$, $|\omega'| = 1$, there is a transformation $S(\omega')$ with*

$$|S(\omega')| + |S^{-1}(\omega')| \leq K_{31}$$

such that the transformed matrix

$$S(\omega')P(i\omega')S^{-1}(\omega')$$

has diagonal form.

Proof. First assume conditions (1) and (2) to hold. For $\omega \neq 0$, the diagonal matrix

$$S(\omega')P(i\omega)S^{-1}(\omega') = |\omega|S(\omega')P(i\omega')S^{-1}(\omega') =: |\omega|\Lambda(\omega')$$

has purely imaginary entries. Therefore,

$$\left|e^{P(i\omega)t}\right| \leq K_{31}^2 \left|e^{|\omega|\Lambda(\omega')t}\right| = K_{31}^2.$$

Thus the problem is well-posed. (The wave-vector $\omega = 0$ has the symbol $P(0) = 0$ which causes no problem for well-posedness.)

Now assume conversely that the problem is well-posed, and let

$$a + ib, \quad a, b \in \mathbf{R},$$

denote an eigenvalue of $P(i\omega')$. We first show that $a = 0$. The matrix $P(i\omega)$, $\omega = |\omega|\omega'$, has the eigenvalue $|\omega|(a + ib)$ with real part $a|\omega|$. If $a > 0$ then

$$|e^{P(i\omega)t}| \geq e^{a|\omega|t}$$

cannot be bounded by $Ke^{\alpha t}$ with α, K independent of ω. If $a < 0$ then we consider $P(-i|\omega|\omega')$, and also arrive at a contradiction. This shows that all eigenvalues of $P(i\omega')$ are purely imaginary.

To prove the second condition, note that well-posedness yields

$$|e^{|\omega|P(i\omega')t}| \leq Ke^{\alpha t}$$

with some K, α independent of ω and $t \geq 0$. Therefore,

$$\left|e^{\left(P(i\omega')-(\alpha/|\omega|)I\right)|\omega|t}\right| \leq K.$$

We fix $t' = |\omega|t$ and obtain for $|\omega| \to \infty$,

$$|e^{P(i\omega')t'}| \leq K, \quad t' \geq 0.$$

Now we apply Theorem 2.3.2 to the set of matrices

$$P(i\omega'), \quad \omega' \in \mathbf{R}^s, \quad |\omega'| = 1,$$

and use condition 3. of that theorem. Since we have already shown that Re $\kappa = 0$ for the eigenvalues κ of $P(i\omega')$, the estimate (2.3.5) implies that $SP(i\omega')S^{-1}$ is diagonal. This finishes the proof of the theorem.

Concepts of hyperbolicity. Theorem 2.4.1 characterizes those first-order (constant-coefficient) equations for which the Cauchy problem is well-posed. These equations $u_t = Pu$ are called *strongly hyperbolic*. The definition of other concepts of hyperbolicity, which play a role in applications, is summarized in

Definition 1. A first-order equation $u_t = P(\partial/\partial x)u$ with

$$P(\partial/\partial x) = \sum_{j=1}^{s} A_j \frac{\partial}{\partial x_j}, \quad A_j \in \mathbf{C}^{n,n},$$

is called:
weakly hyperbolic if for all $\omega \in \mathbf{R}^s$ all eigenvalues of $P(i\omega)$ are purely imaginary;
strongly hyperbolic if the conditions of Theorem 2.4.1 are met;
symmetric hyperbolic if $A_j = A_j^*$, $j = 1, \cdots, s$;
strictly hyperbolic if for all $\omega \in \mathbf{R}^s$, $\omega \neq 0$, all eigenvalues of $P(i\omega)$ are purely imaginary and *distinct*.

As we have seen, the Cauchy problem for a weakly hyperbolic equation is generally not well-posed. In the case of a symmetric hyperbolic equation, the symbol has the form $P(i\omega) = i\sum \omega_j A_j$ with a Hermitian matrix $\sum \omega_j A_j$. Therefore, $P(i\omega)$ can be diagonalized by a *unitary* transformation, and the Cauchy problem is well-posed. If the equation is strictly hyperbolic, then there is a complete set of eigenvectors for $P(i\omega')$, $|\omega'| = 1$, and the eigenvectors can be chosen as analytic functions of ω'. If $S^{-1}(\omega')$ contains the eigenvectors as columns, then $S(\omega')P(i\omega')S^{-1}(\omega')$ is diagonal, and one obtains a bound

$$|S(\omega')| + |S^{-1}(\omega')| \leq K_{31}.$$

Consequently, strictly hyperbolic equations are strongly hyperbolic, and the Cauchy problem is well-posed.

No exponential growth. The solutions of a strongly hyperbolic system (2.4.1) (without zero-order term) do not show any exponential growth: For each ω there is a transformation $S = S(\omega)$ such that $SP(i\omega)S^{-1} = \Lambda$ is diagonal with purely imaginary entries. Thus, for the Hermitian matrix $H = S^*S$ it holds that

$$\begin{aligned} HP(i\omega) + P^*(i\omega)H &= S^*SP(i\omega) + P^*(i\omega)S^*S \\ &= S^*\{SP(i\omega)S^{-1} + S^{*-1}P^*(i\omega)S^*\}S \\ &= S^*\{\Lambda + \Lambda^*\}S = 0. \end{aligned}$$

Lemma 2.1.4 applies with $\alpha = 0$, and therefore $|e^{P(i\omega)t}| \leq K$. Here K is independent of ω and of $t \geq 0$.

For later reference we formulate an algebraic result, which summarizes a part of the above considerations.

Lemma 2.4.2. *For each $A \in \mathbf{C}^{n,n}$ the following two conditions are equivalent:*

1. *The eigenvalues of A are real, and A has a complete set of eigenvectors.*
2. *There exists a positive definite Hermitian matrix H with $HA = A^*H$.*

Proof. First assume that (1) holds and set $P = iA$. There is S such that SPS^{-1} is diagonal with purely imaginary entries. We define $H = S^*S$ and obtain, as above,

$$0 = HP + P^*H = i(HA - A^*H).$$

Conversely, let $HA = A^*H,\ H = H^* > 0$. We can write $H = H^{1/2}H^{1/2}$, $H^{1/2} = (H^{1/2})^*$, and obtain

$$H^{1/2}AH^{-1/2} = H^{-1/2}A^*H^{1/2} = \left(H^{1/2}AH^{-1/2}\right)^*.$$

For the Hermitian matrix $H^{1/2}AH^{-1/2}$ there is a unitary matrix U such that

$$UH^{1/2}AH^{-1/2}U^* =: D$$

is real diagonal. Consequently, if we set $S = UH^{1/2}$, then $SAS^{-1} = D$, and (1) is shown.

Smoothness of $H(\omega)$. For general strongly hyperbolic systems, the matrices $H = H(\omega)$ with $HP(i\omega) + P^*(i\omega)H = 0$ cannot be chosen as globally smooth functions of $\omega \neq 0$. This leads to technical difficulties if one wants to go over to variable coefficients. We will comment on this further in Section 3.3.1 below. Of course, if the system is *symmetric* hyperbolic, then $P(i\omega) + P^*(i\omega) = 0$, and one can choose $H(\omega) = I$. Also in the *strictly* hyperbolic case the matrix $H(\omega)$ can be chosen as a smooth function of $\omega \neq 0$, because the eigenvectors of $P(i\omega)$ depend smoothly on the matrix elements. However, strictly hyperbolic systems hardly ever appear in applications. One can prove, for example, that a strictly hyperbolic system in three space variables must have at least dimension $n = 7$.

2.4.2. The Compressible Euler Equations Linearized at Constant Flow

The full equations without forcing read

$$\rho \frac{D}{Dt}\mathbf{u} + \operatorname{grad} p = 0, \quad \frac{D}{Dt}\rho + \rho \operatorname{div} \mathbf{u} = 0, \quad p = r(\rho),$$

where

$$\frac{D}{Dt} = \frac{\partial}{\partial t} + u\frac{\partial}{\partial x} + v\frac{\partial}{\partial y} + w\frac{\partial}{\partial z}.$$

Let $\mathbf{U} = (U, V, W)$, R, P denote a *constant state* of $\mathbf{u}$, ρ, p with $P = r(R)$; i.e., $\mathbf{U}$, R, P are independent of $\mathbf{x}$ and t.

With – supposedly small – corrections $\mathbf{u}'$, ρ', p' we substitute the ansatz

$$\mathbf{u} = \mathbf{U} + \mathbf{u}', \quad \rho = R + \rho', \quad p = P + p'$$

into the above equations; after neglecting all terms which are quadratic in the corrections we obtain the *linear* equations

$$R\frac{D}{Dt}\mathbf{u}' + \operatorname{grad} p' = 0, \quad \frac{D}{Dt}\rho' + R \operatorname{div} \mathbf{u}' = 0, \quad p' = \frac{dr}{d\rho}(R)\rho',$$

where D/Dt denotes now the constant-coefficient operator

$$\frac{D}{Dt} = \frac{\partial}{\partial t} + U\frac{\partial}{\partial x} + V\frac{\partial}{\partial y} + W\frac{\partial}{\partial z}.$$

To simplify the notation, we drop $'$ in the expressions for the perturbation terms $\mathbf{u}'$, etc. If we set

$$\frac{dr}{d\rho}(R) =: \kappa$$

and use the equation $p = \kappa\rho$ to eliminate the pressure from the momentum equations, we find that

$$\frac{D}{Dt}\mathbf{u} + \frac{\kappa}{R}\operatorname{grad} \rho = 0, \quad \frac{D}{Dt}\rho + R \operatorname{div} \mathbf{u} = 0.$$

In matrix form, this first-order system reads

$$
\begin{aligned}
(2.4.2)\qquad & \frac{\partial}{\partial t}\begin{pmatrix} u \\ v \\ w \\ \rho \end{pmatrix} + \begin{pmatrix} U & 0 & 0 & \kappa/R \\ 0 & U & 0 & 0 \\ 0 & 0 & U & 0 \\ R & 0 & 0 & U \end{pmatrix} \frac{\partial}{\partial x}\begin{pmatrix} u \\ v \\ w \\ \rho \end{pmatrix} \\
& + \begin{pmatrix} V & 0 & 0 & 0 \\ 0 & V & 0 & \kappa/R \\ 0 & 0 & V & 0 \\ 0 & R & 0 & V \end{pmatrix} \frac{\partial}{\partial y}\begin{pmatrix} u \\ v \\ w \\ \rho \end{pmatrix} \\
& + \begin{pmatrix} W & 0 & 0 & 0 \\ 0 & W & 0 & 0 \\ 0 & 0 & W & \kappa/R \\ 0 & 0 & R & W \end{pmatrix} \frac{\partial}{\partial z}\begin{pmatrix} u \\ v \\ w \\ \rho \end{pmatrix} = 0.
\end{aligned}
$$

Let us assume $\kappa = c^2 > 0$. Then the system can be symmetrized by a simple scaling. Using the new variable $\tilde{\rho} = c\rho/R$, we obtain the *symmetric hyperbolic* system

$$
\begin{aligned}
& \frac{\partial}{\partial t}\begin{pmatrix} u \\ v \\ w \\ \tilde{\rho} \end{pmatrix} + \begin{pmatrix} U & 0 & 0 & c \\ 0 & U & 0 & 0 \\ 0 & 0 & U & 0 \\ c & 0 & 0 & U \end{pmatrix} \frac{\partial}{\partial x}\begin{pmatrix} u \\ v \\ w \\ \tilde{\rho} \end{pmatrix} + \begin{pmatrix} V & 0 & 0 & 0 \\ 0 & V & 0 & c \\ 0 & 0 & V & 0 \\ 0 & c & 0 & V \end{pmatrix} \frac{\partial}{\partial y}\begin{pmatrix} u \\ v \\ w \\ \tilde{\rho} \end{pmatrix} \\
& + \begin{pmatrix} W & 0 & 0 & 0 \\ 0 & W & 0 & 0 \\ 0 & 0 & W & c \\ 0 & 0 & c & W \end{pmatrix} \frac{\partial}{\partial z}\begin{pmatrix} u \\ v \\ w \\ \tilde{\rho} \end{pmatrix} = 0.
\end{aligned}
$$

Since this system leads to a well-posed Cauchy problem, the same is true for the equations (2.4.2). The assumption

$$\kappa = \frac{dr}{d\rho}(R) = c^2 > 0$$

is crucial. If $\kappa < 0$ then, for example, the matrix

$$\begin{pmatrix} W & 0 & 0 & 0 \\ 0 & W & 0 & 0 \\ 0 & 0 & W & \kappa/R \\ 0 & 0 & R & W \end{pmatrix}$$

has the nonreal eigenvalues

$$W \pm i\sqrt{-\kappa}.$$

Hence the system (2.4.2) is not hyperbolic, and the Cauchy problem is ill-posed. For $\kappa = 0$ the Cauchy problem is ill-posed, too.

2.4.3. Parabolic Systems

Consider an even-order operator of the form

$$P(\partial/\partial x) = P_{2m}(\partial/\partial x) + Q(\partial/\partial x),$$

where

$$P_{2m}(\partial/\partial x) = \sum_{|\nu|=2m} A_\nu \frac{\partial^{|\nu|}}{\partial x_1^{\nu_1} \cdots \partial x_s^{\nu_s}}$$

and

$$Q(\partial/\partial x) = \sum_{|\nu|\leq 2m-1} A_\nu \frac{\partial^{|\nu|}}{\partial x_1^{\nu_1} \cdots \partial x_s^{\nu_s}}.$$

$P_{2m}(\partial/\partial x)$ is called the *principle part* of $P(\partial/\partial x)$. Parabolicity of $u_t = P(\partial/\partial x)u$ is defined in terms of the principle part of P as follows:

Definition 2. The equation $u_t = P(\partial/\partial x)u$ is called *parabolic* if for all $\omega \in \mathbf{R}^s$ the eigenvalues $\kappa_j(\omega)$, $j = 1, \ldots, n$, of $P_{2m}(i\omega)$ satisfy

$$\text{Re}\,\kappa_j(\omega) \leq -\delta|\omega|^{2m}, \quad j = 1, \cdots, n, \tag{2.4.3}$$

with some $\delta > 0$ independent of ω.

A simple but important example is the equation $u_t = \Delta u$ with the **Laplacian**

$$\Delta = \frac{\partial^2}{\partial x_1^2} + \cdots + \frac{\partial^2}{\partial x_s^2}.$$

Here $P_2(i\omega) = -(\omega_1^2 + \cdots + \omega_s^2)$, and one can take $\delta = 1$. We show

Theorem 2.4.3. *The Cauchy problem for a parabolic system $u_t = P(\partial/\partial x)u$ is well-posed.*

Proof. The proof proceeds along the same lines as the proof of Theorem 2.1.3; the Matrix Theorem is not needed. First consider the operator $P_{2m}(\partial/\partial x)$ with the symbol

$$P_{2m}(i\omega) = |\omega|^{2m} P_{2m}(i\omega'), \quad \omega' = \omega/|\omega|.$$

Transform $P_{2m}(i\omega')$ to upper-triangular form by a unitary matrix $U = U(\omega')$:

$$UP_{2m}(i\omega')U^* = \begin{pmatrix} \kappa_1 & b_{12} & \dots & b_{1n} \\ & \ddots & & b_{2n} \\ & & \ddots & \vdots \\ 0 & & & \kappa_n \end{pmatrix}.$$

The elements of $P_{2m}(i\omega')$, $|\omega'| = 1$, are uniformly bounded; therefore, for some constant K it holds that

$$|b_{jk}| = |b_{jk}(\omega')| \leq K, \qquad 1 \leq j < k \leq n.$$

The parabolicity assumption yields

$$\operatorname{Re} \kappa_j(\omega') \leq -\delta, \quad j = 1, \dots, n.$$

We set $D = \operatorname{diag}(1, d, \dots, d^{n-1})$, $d > 0$, $S = DU$, and obtain, for sufficiently large d,

$$SP_{2m}(i\omega')S^{-1} + \big(SP_{2m}(i\omega')S^{-1}\big)^* \leq -\delta I.$$

Here d can be chosen independently of ω', $|\omega'| = 1$. As in the proof of Theorem 2.1.3, we set $H = H(\omega') = S^*S$. Then

$$HP(i\omega) + P^*(i\omega)H \leq -\delta|\omega|^{2m}H + \text{const}\big(|\omega|^{2m-1} + 1\big)H \leq 2\alpha H,$$

and the result follows from Lemma 2.1.4.

Roughly speaking, in the present context of well-posedness, the main feature of parabolic equations is the following: The dissipativity of the principle part – which is expressed by the estimate (2.4.3) – will force all high wave-number components $e^{i\langle\omega,x\rangle}$, $|\omega|$ large, to decay in time, no matter what the lower-order terms of $P(\partial/\partial x)$ look like. The formal analog in the above proof is that we can dominate the term $\text{const}\,(|\omega|^{2m-1} + 1)$ by $\delta|\omega|^{2m}$ if $|\omega|$ is sufficiently large.

2.5. Mixed Systems and the Compressible N-S Equations Linearized at Constant Flow

The compressible N-S equations without forcing read

$$\rho\frac{D}{Dt}\mathbf{u} + \operatorname{grad} p = \mathbf{P}, \qquad \frac{D}{Dt}\rho + \rho \operatorname{div}\mathbf{u} = 0, \quad p = r(\rho),$$

where

$$\mathbf{P} = \mu\Delta\mathbf{u} + (\mu + \mu')\operatorname{grad}(\operatorname{div}\mathbf{u}), \qquad \mu > 0, \quad \mu' \geq 0.$$

As in Section 2.4.2, we let $\mathbf{U} = (U, V, W)$, R, P denote a constant state with $P = r(R)$ and use an ansatz as described in 2.4.2. Neglecting all terms which are quadratic in the corrections, we find *linear* equations for $\mathbf{u}'$, ρ', p'. Again, we drop $'$ in the notation and eliminate the pressure using

$$p = \kappa\rho, \quad \kappa = \frac{dr}{d\rho}(R).$$

As a result, we obtain the linear constant-coefficient equations

$$\frac{D}{Dt}\mathbf{u} + \frac{\kappa}{R}\operatorname{grad}\rho = \nu\Delta\mathbf{u} + (\nu+\nu')\operatorname{grad}(\operatorname{div}\mathbf{u}), \quad \nu = \mu/R, \quad \nu' = \mu'/R,$$

$$\frac{D}{Dt}\rho + R\operatorname{div}\mathbf{u} = 0.$$

These read in matrix form

$$\frac{\partial}{\partial t}\begin{pmatrix} u \\ v \\ w \\ \rho \end{pmatrix} + \left\{A_1\frac{\partial}{\partial x} + A_2\frac{\partial}{\partial y} + A_3\frac{\partial}{\partial z}\right\}\begin{pmatrix} u \\ v \\ w \\ \rho \end{pmatrix}$$

$$= \nu\begin{pmatrix} \Delta & 0 & 0 & 0 \\ 0 & \Delta & 0 & 0 \\ 0 & 0 & \Delta & 0 \\ 0 & 0 & 0 & 0 \end{pmatrix}\begin{pmatrix} u \\ v \\ w \\ \rho \end{pmatrix}$$

$$+ (\nu+\nu')\begin{pmatrix} D_1^2 & D_1D_2 & D_1D_3 & 0 \\ D_1D_2 & D_2^2 & D_2D_3 & 0 \\ D_1D_3 & D_2D_3 & D_3^2 & 0 \\ 0 & 0 & 0 & 0 \end{pmatrix}\begin{pmatrix} u \\ v \\ w \\ \rho \end{pmatrix}.$$

Here $D_1 = \partial/\partial x$, etc. The matrices A_1, A_2, A_3 can be read off from (2.4.2). Let us denote the second-order operator on the right side of the above equation by $\tilde{\mathbf{P}}_2 = \tilde{\mathbf{P}}_2(\partial/\partial\mathbf{x})$; thus $\tilde{\mathbf{P}}_2$ acts on $(u, v, w, \rho)^T$. Obviously, the symbol

$$\tilde{\mathbf{P}}_2(i\omega), \quad \omega = (\omega_1, \omega_2, \omega_3),$$

has 0 as an eigenvalue. The reason is that the continuity equation does not contain a second-order term. Consequently, the above system is *not* parabolic.

We now rewrite the above system in block form and separate the momentum equations from the continuity equation:

$$\begin{aligned}\frac{\partial}{\partial t}\begin{pmatrix} \mathbf{u} \\ \rho \end{pmatrix} &= -\begin{pmatrix} UI & * \\ * & U \end{pmatrix}\frac{\partial}{\partial x}\begin{pmatrix} \mathbf{u} \\ \rho \end{pmatrix} - \begin{pmatrix} VI & * \\ * & V \end{pmatrix}\frac{\partial}{\partial y}\begin{pmatrix} \mathbf{u} \\ \rho \end{pmatrix} \\ &\quad - \begin{pmatrix} WI & * \\ * & W \end{pmatrix}\frac{\partial}{\partial z}\begin{pmatrix} \mathbf{u} \\ \rho \end{pmatrix} + \begin{pmatrix} P_2 & 0 \\ 0 & 0 \end{pmatrix}\begin{pmatrix} \mathbf{u} \\ \rho \end{pmatrix}.\end{aligned} \tag{2.5.1}$$

The second-order operator P_2 acts only on $\mathbf{u}$, not on $(\mathbf{u}, \rho)$.

Let us show that $\mathbf{u}_t = P_2\mathbf{u}$ is parabolic. The symbol is real and symmetric; it reads

$$P_2(i\omega) = -\nu(\omega_1^2 + \omega_2^2 + \omega_3^2)\begin{pmatrix}1 & 0 & 0\\ 0 & 1 & 0\\ 0 & 0 & 1\end{pmatrix} - (\nu + \nu')\begin{pmatrix}\omega_1^2 & \omega_1\omega_2 & \omega_1\omega_3\\ \omega_1\omega_2 & \omega_2^2 & \omega_2\omega_3\\ \omega_1\omega_3 & \omega_2\omega_3 & \omega_3^2\end{pmatrix}$$

$$= -\nu|\omega|^2 I - (\nu + \nu')\begin{pmatrix}\omega_1 & 0 & 0\\ 0 & \omega_2 & 0\\ 0 & 0 & \omega_3\end{pmatrix}\begin{pmatrix}1 & 1 & 1\\ 1 & 1 & 1\\ 1 & 1 & 1\end{pmatrix}\begin{pmatrix}\omega_1 & 0 & 0\\ 0 & \omega_2 & 0\\ 0 & 0 & \omega_3\end{pmatrix}.$$

For any $y \in \mathbf{R}^3$ it holds that

$$y^T P_2(i\omega)y = -\nu|\omega|^2|y|^2 - (\nu + \nu')(\omega_1 y_1, \omega_2 y_2, \omega_3 y_3)\begin{pmatrix}1 & 1 & 1\\ 1 & 1 & 1\\ 1 & 1 & 1\end{pmatrix}\begin{pmatrix}\omega_1 y_1\\ \omega_2 y_2\\ \omega_3 y_3\end{pmatrix}$$

$$= -\nu|\omega|^2|y|^2 - (\nu + \nu')(\omega_1 y_1 + \omega_2 y_2 + \omega_3 y_3)^2 \leq -\nu|\omega|^2|y|^2.$$

It follows that all eigenvalues of the matrix $P_2(i\omega)$ are $\leq -\nu|\omega|^2$. Hence, according to Definition 2, Section 2.4.3, the equation $\mathbf{u}_t = P_2\mathbf{u}$ is parabolic.

To study the *full* linear system for $(\mathbf{u}, \rho)$ given above, we consider the system in its block form,

$$\frac{\partial}{\partial t}\begin{pmatrix}\mathbf{u}\\ \rho\end{pmatrix} = \begin{pmatrix}P_2 & 0\\ 0 & P_1\end{pmatrix}\begin{pmatrix}\mathbf{u}\\ \rho\end{pmatrix} + \begin{pmatrix}R_{11} & R_{12}\\ R_{21} & 0\end{pmatrix}\begin{pmatrix}\mathbf{u}\\ \rho\end{pmatrix}.$$

Here P_1 is the first-order scalar operator

$$P_1(\partial/\partial\mathbf{x})\rho = -(U\frac{\partial}{\partial x}\rho + V\frac{\partial}{\partial y}\rho + W\frac{\partial}{\partial z}\rho)$$

as can be read off from (2.5.1). Clearly, the scalar equation $\rho_t = P_1\rho$ (with real coefficients) is strongly hyperbolic. The operators R_{ij} are of order one. Their precise form will not be important for the discussion.

Since both P_2 and P_1 lead to well-posed Cauchy problems, it is clear that the Cauchy problem for the completely uncoupled system

$$\frac{\partial}{\partial t}\begin{pmatrix}\mathbf{u}\\ \rho\end{pmatrix} = \begin{pmatrix}P_2 & 0\\ 0 & P_1\end{pmatrix}\begin{pmatrix}\mathbf{u}\\ \rho\end{pmatrix}$$

is also well-posed. We want to show that the coupling term

$$R_1 = \begin{pmatrix}R_{11} & R_{12}\\ R_{21} & 0\end{pmatrix}$$

does not destroy well-posedness. This follows from the result proved in the next theorem, where we consider a slightly more general situation. It might be worth while to note again that – on the Fourier side – the discussion is purely algebraic; the relevant estimate for the coupling term becomes quite transparent.

Theorem 2.5.1. *Consider vector functions*

$$w(x,t) = \begin{pmatrix} u(x,t) \\ v(x,t) \end{pmatrix}, \qquad x \in \mathbf{R}^s, \quad t \geq 0,$$

where $u(x,t)$ has m components and $v(x,t)$ has n components. Let $P = P(\partial/\partial x)$ denote a constant-coefficient differential operator whose corresponding block form is

$$P = \begin{pmatrix} P_2 & 0 \\ 0 & P_1 \end{pmatrix} + \begin{pmatrix} R_{11} & R_{12} \\ R_{21} & 0 \end{pmatrix} + R_0.$$

Assume that $u_t = P_2 u$ is second-order parabolic, that $v_t = P_1 v$ is first-order strongly hyperbolic, that the R_{jk} are of order one, and that R_0 is of order zero. Then the Cauchy problem for $w_t = P(\partial/\partial x)w$ is well-posed.

Proof. We can assume that P_2 coincides with its principle part. With suitable constants $\delta > 0,\ K_4 > 0$ and suitable Hermitian matrices $H_1(\omega)$, $H_2(\omega)$ we have, for all $\omega \in \mathbf{R}^s$,

$$\begin{aligned} H_2(\omega)P_2(i\omega) + P_2^*(i\omega)H_2(\omega) &\leq -\delta|\omega|^2 I, \\ H_1(\omega)P_1(i\omega) + P_1^*(i\omega)H_1(\omega) &= 0, \\ K_4^{-1} I \leq H_1(\omega),\ H_2(\omega) &\leq K_4 I. \end{aligned}$$

We define the Hermitian matrix

$$H(\omega) = \begin{pmatrix} H_2(\omega) & 0 \\ 0 & H_1(\omega) \end{pmatrix}$$

and consider the quadratic form belonging to

$$Q(\omega) := H(\omega)P(i\omega) + P^*(i\omega)H(\omega).$$

For all $\hat{u} \in \mathbf{C}^m$, $\hat{v} \in \mathbf{C}^n$, all $\omega \in \mathbf{R}^s$, and some constant K it holds that

$$\begin{aligned} \left\langle \begin{pmatrix} \hat{u} \\ \hat{v} \end{pmatrix}, Q(\omega) \begin{pmatrix} \hat{u} \\ \hat{v} \end{pmatrix} \right\rangle &\leq \left\langle \hat{u}, \left\{ H_2(\omega)P_2(i\omega) + P_2^*(i\omega)H_2(\omega) \right\} \hat{u} \right\rangle \\ &\quad + K|\omega| \left\{ |\hat{u}||\hat{v}| + |\hat{u}|^2 \right\} + K\left\{ |\hat{u}|^2 + |\hat{v}|^2 \right\} \\ &\leq -\delta|\omega|^2|\hat{u}|^2 + K|\omega|\left\{ |\hat{u}||\hat{v}| + |\hat{u}|^2 \right\} + K\left\{ |\hat{u}|^2 + |\hat{v}|^2 \right\} \\ &=: \text{r.h.s.} \end{aligned}$$

For any two real numbers a, b we know that $2ab \le a^2 + b^2$, and therefore,

$$2|\omega||\hat{u}||\hat{v}| = 2\epsilon|\omega||\hat{u}|\frac{1}{\epsilon}|\hat{v}| \le \epsilon^2|\omega|^2|\hat{u}|^2 + \frac{1}{\epsilon^2}|\hat{v}|^2, \quad \epsilon > 0.$$

Using this inequality, we arrive at the estimate

$$\text{r.h.s.} \le -\delta|\omega|^2|\hat{u}|^2 + \frac{K}{2}\epsilon^2|\omega|^2|\hat{u}|^2 + \frac{K}{2}\frac{1}{\epsilon^2}|\hat{v}|^2 + K|\omega||\hat{u}|^2 + K\{|\hat{u}|^2 + |\hat{v}|^2\}.$$

If we choose $\epsilon^2 = \delta/K$ then

$$\text{r.h.s.} \le \{-\frac{\delta}{2}|\omega|^2 + K|\omega|\}|\hat{u}|^2 + \frac{K}{2}\frac{1}{\epsilon^2}|\hat{v}|^2 + K\{|\hat{u}|^2 + |\hat{v}|^2\}.$$

Hence, with some constant K_1 independent of ω, $\hat{u}$, $\hat{v}$ it holds that

$$\left\langle \begin{pmatrix} \hat{u} \\ \hat{v} \end{pmatrix}, Q(\omega) \begin{pmatrix} \hat{u} \\ \hat{v} \end{pmatrix} \right\rangle \le K_1\{|\hat{u}|^2 + |\hat{v}|^2\} \le K_1 \left| \begin{pmatrix} \hat{u} \\ \hat{v} \end{pmatrix} \right|^2,$$

and therefore,

$$Q(\omega) \le K_1 I \le K_1 K_4 H(\omega).$$

Now well-posedness follows from Corollary 2.3.4.

2.6. Properties of Constant-Coefficient Equations

As mentioned previously, the switching between a variable-coefficient problem and the constant-coefficient equations obtained by freezing coefficients introduces lower-order terms. Thus, it is natural to ask which constant-coefficient equations can be perturbed by arbitrary lower-order terms without destroying well-posedness. We shall show that *only* strongly hyperbolic and parabolic systems have this property. (Of course, for other equations, a restricted class of perturbing lower-order terms might not destroy well-posedness.)

Another technically important result of this section can be described as follows: If the Cauchy problem for $u_t = P(\partial/\partial x)u$ is well-posed, then one can construct an inner product and a norm

$$(u, v)_H = (u, \mathbf{H}v), \quad \|u\|_H = (u, \mathbf{H}u)^{1/2},$$

such that

$$\|u(\cdot, t)\|_H \le e^{\alpha t}\, \|u(\cdot, 0)\|_H. \tag{2.6.1}$$

In other words, the estimate of Theorem 2.2.2, which expresses well-posedness, holds with $K = 1$ if we replace the L_2-norm by $\|\ \ \|_H$. If one uses the L_2-norm only, then – in general – a constant $K > 1$ is required; however, if this is the case, the solution-estimate becomes locally (in time) useless: it does not even express continuity in time. Therefore, the construction of norms $\| \quad \|_H$ with property (2.6.1) is important if one wants to treat variable-coefficient problems by localization.

Furthermore, we will establish **Duhamel's Principle** for inhomogeneous equations

$$u_t = P(\partial/\partial x)u + F(x,t).$$

2.6.1. Perturbation by Lower-Order Terms and Well-Posedness

Consider the Cauchy problem for a system

$$u_t = P(\partial/\partial x)u, \quad P = P_m + Q, \tag{2.6.2}$$

where

$$P_m = \sum_{|\nu|=m} A_\nu D^\nu, \quad Q = \sum_{|\nu|\le m-1} A_\nu D^\nu, \quad D^\nu = \frac{\partial^{|\nu|}}{\partial x_1^{\nu_1}\dots\partial x_s^{\nu_s}}.$$

If the principle part P_m is of first order and strongly hyperbolic or of even order and parabolic, then the Cauchy problem for $P = P_m + Q$ is well-posed, no matter what the lower-order terms Q look like. For the strongly hyperbolic case the operator Q is of order zero, and the result follows from Lemma 2.3.5; the parabolic case has been treated in Theorem 2.4.3. We show the following converse:

Theorem 2.6.1. *Consider the Cauchy problem for* (2.6.2) *with some fixed principle part P_m, and assume it is well-posed for any choice of the lower-order terms Q. Then either $m = 1$ and* (2.6.2) *is strongly hyperbolic or m is even and* (2.6.2) *is parabolic.*

Proof. For $m = 1$, strong hyperbolicity and well-posedness are the same; see Theorem 2.4.1 and the definition following the theorem. Thus we can assume $m \ge 2$ and must show that m is even and (2.6.2) is parabolic.

First assume $m \ge 3$ to be odd. We show that the real parts of the eigenvalues of $P_m(i\omega)$ are zero. Since the Cauchy problem is well-posed for $Q \equiv 0$, we have

$$|e^{P_m(i\omega)t}| \le Ke^{\alpha t}. \tag{2.6.3}$$

As before, let $\omega = |\omega|\,\omega'$, $|\omega'| = 1$, and let κ denote an eigenvalue of $P_m(i\omega')$. Then $P_m(i\omega)$ has $|\omega|^m\kappa$ as an eigenvalue and (since m is odd) $P_m(-i\omega)$ has $-|\omega|^m\kappa$ as an eigenvalue. The estimate (2.6.3) implies that Re $\kappa = 0$.

Now let

$$Q(\partial/\partial x) = -\frac{\partial^2}{\partial x_1^2};$$

thus

$$P(i\omega) = P_m(i\omega) + \omega_1^2 I.$$

The real part of each eigenvalue of $P(i\omega)$ equals ω_1^2. This function of ω is unbounded, and therefore the Cauchy problem for P cannot be well-posed.

Second, assume that m is even, but $u_t = P_m u$ is not parabolic. Let $\kappa_j(\omega')$ denote the eigenvalues of $P_m(i\omega')$, $|\omega'| = 1$. The function

$$\max_j \ \text{Re}\ \kappa_j(\omega')$$

depends continuously on ω', and if $u_t = P_m u$ is *not* parabolic, then there exists a vector η, $|\eta| = 1$, and an eigenvalue $\kappa(\eta)$ of $P_m(i\eta)$ with

$$\text{Re}\ \kappa(\eta) \geq 0. \tag{2.6.4}$$

Let

$$Q(\partial/\partial x) = -i \sum_{j=1}^{s} \eta_j \frac{\partial}{\partial x_j};$$

the symbol of $P = P_m + Q$ reads

$$P(i\omega) = P_m(i\omega) + \left(\sum_{j=1}^{s} \eta_j \omega_j \right) I.$$

For the wave-vectors

$$\omega = |\omega|\ \eta, \quad |\omega| \ \text{large},$$

where η is chosen with (2.6.4), the real parts of the eigenvalues of $P(i\omega)$ cannot be bounded. This contradicts well-posedness.

2.6.2. Symmetrization and Energy Norms

Suppose that the Cauchy problem for $u_t = P(\partial/\partial x)u$ is well-posed where P is a general constant-coefficient operator; see (2.2.3). There are constants α and K with

$$\left| e^{\left(P(i\omega)-\alpha I\right)t} \right| \leq K \quad \text{for all } \omega \in \mathbf{R}^s,\ t \geq 0.$$

According to the Matrix Theorem 2.3.2, one can construct Hermitian matrices $H(\omega)$ with

$$K_4^{-1} I \leq H(\omega) \leq K_4 I, \quad H(\omega)\ P(i\omega) + P(i\omega)^* H(\omega) \leq 2\alpha H(\omega).$$

Here $K_4 > 0$ is independent of $\omega \in \mathbf{R}^s$. Using these matrices, we define the linear (pseudo-differential) operator $\mathbf{H}$ mapping L_2 into L_2 by

$$(\mathbf{H}u)(x) = \frac{1}{(2\pi)^{s/2}} \int_{\mathbf{R}^s} e^{i\langle \omega, x\rangle} H(\omega)\hat{u}(\omega)\, d\omega, \quad x \in \mathbf{R}^s.$$

(The symbol $P(i\omega)$ depends analytically on ω, and an examination of the constructive proof of the Matrix Theorem shows that $H(\omega)$ can be chosen as piecewise smooth. Therefore, the above integral exists for $u \in \mathbf{M}_0$. For general $u \in L_2$, the image $\mathbf{H}u$ can be obtained by the usual extension process; see Theorem 2.2.4.) The operator $\mathbf{H}$ is used to define an inner product and a norm by

$$(u, v)_H = (u, \mathbf{H}v), \quad \|u\|_H = (u, u)_H^{1/2}.$$

Theorem 2.6.2. *The constructed norm is equivalent to the L_2-norm; i.e.,*

$$K_4^{-1}\|u\|^2 \leq \|u\|_H^2 \leq K_4\|u\|^2.$$

Furthermore, if $u(x,t)$ denotes a ($\mathbf{M}_0$ or generalized) solution of the Cauchy problem for $u_t = Pu$, then

$$\|u(\cdot, t)\|_H \leq e^{\alpha t}\, \|u(\cdot, 0)\|_H, \quad t \geq 0.$$

Proof. With Parseval's relation one finds that

$$\begin{aligned}(u, \mathbf{H}u) = (\hat{u}, \widehat{\mathbf{H}u}) &= \int_{\mathbf{R}^s} \langle \hat{u}(\omega), H(\omega)\hat{u}(\omega)\rangle\, d\omega \\ &\leq K_4 \int_{\mathbf{R}^s} \langle \hat{u}(\omega), \hat{u}(\omega)\rangle\, d\omega = K_4\|u\|^2,\end{aligned}$$

and similarly $(u, \mathbf{H}u) \geq K_4^{-1}\|u\|^2$.

The following equations hold for any $\mathbf{M}_0$-solution $u = u(x,t)$:

$$\begin{aligned}
\frac{d}{dt}\big(u(x,t),\mathbf{H}u(x,t)\big) &= \big(u_t(x,t),\mathbf{H}u(x,t)\big) + \big(u(x,t),\mathbf{H}u_t(x,t)\big)\\
&= \big(Pu(x,t),\mathbf{H}u(x,t)\big) + \big(u(x,t),\mathbf{H}Pu(x,t)\big)\\
&= \big(P(i\omega)\hat{u}(\omega,t),H(\omega)\hat{u}(\omega,t)\big)\\
&\quad + \big(\hat{u}(\omega,t),H(\omega)P(i\omega)\hat{u}(\omega,t)\big)\\
&= \Big(\hat{u}\big(\omega,t\big),\big(H(\omega)P(i\omega)+P^*(i\omega)H(\omega)\big)\hat{u}(\omega,t)\Big)\\
&\le 2\alpha\big(\hat{u}(\omega,t),H(\omega)\hat{u}(\omega,t)\big) = 2\alpha\big(u(x,t),\mathbf{H}u(x,t)\big).
\end{aligned}$$

Thus the solution-estimate follows. For general initial data $u(\cdot,0) \in L_2$, the result follows by approximation.

The inner product $(u,v)_H$ is called an *energy inner product* for $P = P(\partial/\partial x)$ and $\|\ \|_H$ is called an *energy norm*. The transformation by $H(\omega)$ in Fourier space is called *symmetrization*. The above proof shows that the operator P satisfies the following estimate for all sufficiently smooth functions $w = w(x),\ \ w \in L_2$:

$$(w,Pw)_H + (Pw,w)_H \le 2\alpha(w,w)_H.$$

For this reason the operator P is called semibounded w.r.t. the inner product $(\cdot,\cdot)_H$. With this definition we can summarize our result in the following way.

Theorem 2.6.3. *The Cauchy problem for $u_t = P(\partial/\partial x)u$ is well-posed if and only if the operator P is semibounded w.r.t. an inner product $(\cdot,\cdot)_H$ which corresponds to a norm equivalent to the L_2-norm.*

The above construction of the operator $\mathbf{H}$ is important in the general theory of partial differential equations. Many proofs concerning the well-posedness of linear problems with *variable* coefficients proceed technically by constructing suitable inner products such that the differential operators with frozen coefficients become semibounded. For the Navier-Stokes equations one can generally work with the usual L_2-inner product.

2.6.3. Inhomogeneous Systems: Duhamel's Principle

We start with an *ordinary* initial value problem

(2.6.5) $$u'(t) = A(t)u(t) + F(t),\quad t \ge 0,\quad u(0) = u_0.$$

Here $A(t) \in \mathbf{C}^{n,n}$ and $F(t) \in \mathbf{C}^n$ are assumed to be continuous in t. For any $\tau \geq 0$ consider the homogeneous system

$$v'(t) = A(t)v(t), \quad t \geq \tau, \quad v(\tau) = v_0.$$

The solution depends linearly on v_0; thus we can write

$$v(t) = S(t,\tau)v_0, \quad t \geq \tau.$$

This defines the solution operator $S(t,\tau) \in \mathbf{C}^{n,n}$. Duhamel's principle for (2.6.5) states

Lemma 2.6.4. *The solution of the inhomogeneous system* (2.6.5) *is given by*

$$u(t) = S(t,0)u_0 + \int_0^t S(t,\tau)F(\tau)\,d\tau. \tag{2.6.6}$$

Proof. By definition of the solution operator $S(t,\tau)$ we have that

$$A(t)S(t,\tau)v_0 = A(t)v(t) = v'(t) = \frac{\partial}{\partial t}\, S(t,\tau)v_0,$$

and therefore

$$\frac{\partial}{\partial t}\, S(t,\tau) = A(t)S(t,\tau).$$

Since $S(t,t) = I$, we find, by differentiation of (2.6.6),

$$\begin{aligned} u'(t) &= A(t)S(t,0)u_0 + \int_0^t A(t)S(t,\tau)F(\tau)\,d\tau + F(t) \\ &= A(t)u(t) + F(t). \end{aligned}$$

This proves the lemma.

Consider a Cauchy problem

$$\begin{aligned} u_t &= P(\partial/\partial x)u + F(x,t), \quad && x \in \mathbf{R}^s, \quad t \geq 0, \\ u(x,0) &= f(x), && x \in \mathbf{R}^s, \quad f \in L_2, \end{aligned} \tag{2.6.7}$$

and assume the problem is well-posed for $F \equiv 0$. The homogeneous system

$$\begin{aligned} v_t &= P(\partial/\partial x)v, \quad && x \in \mathbf{R}^s, \quad t \geq \tau, \\ v(x,\tau) &= g(x), && x \in \mathbf{R}^s, \quad g \in L_2, \end{aligned}$$

is solved by

$$v(\cdot,t) = S(t-\tau)g,$$

where $S(\xi)$ is the generalized solution operator defined in Section 2.2.5. Then Duhamel's principle states that (2.6.7) is solved by

$$u(\cdot,t) = S(t)f + \int_0^t S(t-\tau)F(\cdot,\tau)\,d\tau. \tag{2.6.8}$$

On the right-hand side, one has to integrate the L_2-valued function $\tau \to S(t-\tau)F(\cdot,\tau)$ from $\tau = 0$ to $\tau = t$. With a proper concept of a generalized solution and a proper definition of the integral, one can indeed prove formula (2.6.8) under mild assumptions on F. We will restrict ourselves here to $\mathbf{M}_0$-solutions (see Section 2.2.1) and assume that

i) $f \in \mathbf{M}_0$;
ii) $F(\cdot,t) \in \mathbf{M}_0$ for all $t \geq 0$;
iii) the function $\hat{F}(\omega,t)$ is continuous, and $\hat{F}(\omega,t) = 0$ for $|\omega| > K$ with some K independent of t.

Then Fourier transformation yields

$$\hat{u}_t(\omega,t) = P(i\omega)\hat{u}(\omega,t) + \hat{F}(\omega,t), \quad \hat{u}(\omega,0) = \hat{f}(\omega).$$

By Duhamel's principle for ordinary differential equations, Lemma 2.6.4,

$$\hat{u}(\omega,t) = e^{P(i\omega)t}\hat{f}(\omega) + \int_0^t e^{P(i\omega)(t-\tau)}\hat{F}(\omega,\tau)\,d\tau.$$

Transforming back, we obtain

$$u(x,t) = \{S_0(t)f\}(x) + \int_0^t \{S_0(t-\tau)F(\cdot,\tau)\}(x)\,d\tau, \quad x \in \mathbf{R}^s, \; t \geq 0. \tag{2.6.9}$$

Here we have changed the order of integration in x and t; this is justified under our assumptions on F. If we now *define*

$$\Big\{\int_0^t S_0(t-\tau)F(\cdot,\tau)\,d\tau\Big\}(x)$$

as denoting the integral on the right-hand side of (2.6.9), we can drop the variable x in (2.6.9) and write the equality of functions

$$u(\cdot,t) = S_0(t)f + \int_0^t S_0(t-\tau)F(\cdot,\tau)\,d\tau.$$

Thus we have shown Duhamel's principle (2.6.8) under our restrictive assumptions.

We can also derive solution-estimates for inhomogeneous equations. Let $(\cdot,\cdot)_H$ denote an inner product which makes P semibounded, and let $\|\ \|_H$ denote the corresponding norm. Under the same assumptions on f and F as above,

$$\|u(\cdot,t)\|_H \le e^{\alpha t}\,\|f\|_H + \int_0^t e^{\alpha(t-\tau)}\|F(\cdot,\tau)\|_H\,d\tau$$

$$\le e^{\alpha t}\|f\|_H + \max_{0\le\tau\le t} e^{\alpha(t-\tau)} \int_0^t \|F(\cdot,\tau)\|_H\,d\tau.$$

More general data f and F can be treated by approximation.

It is not difficult to deduce that

$$\|u(\cdot,t)\|^2 \le K(T)\Big\{\|f\|^2 + \int_0^t \|F(\cdot,\tau)\|^2\,d\tau\Big\}$$

for $0 \le t \le T$. An inequality of the above type will be used below as the *definition* of well-posedness for problems with variable coefficients.

2.7. The Spatially Periodic Cauchy Problem: A Summary for Variable Coefficients

Thus far we have assumed a constant-coefficient operator and initial data on the whole space. The purpose of this section is to state briefly the corresponding results for the spatially periodic case. For constant coefficients, the proofs could be given as above simply by replacing Fourier transforms with Fourier expansions.

For later reference we state definitions and results for *variable*-coefficient equations. The proofs are carried out in Chapter 3 for one space dimension and in Sections 6.1 and 6.2 for more than one space dimension. In the last section of this chapter we present two counterexamples to a naive localization principle.

2.7.1. Solution Concept and Well-Posedness

A function $v = v(x)$, $x \in \mathbf{R}^s$, is 1-periodic in the j-th coordinate if

$$v(x) = v(x + e_j) \quad \text{for all } x \in \mathbf{R}^s,$$

$$e_j = (0,\ldots,0,1,0,\ldots,0), \qquad 1 \text{ in coordinate } j.$$

We call $v = v(x)$, $x \in \mathbf{R}^s$, 1-periodic in x if v is 1-periodic in each coordinate x_j, $j = 1,\ldots,s$. Also, a function $v(x,t)$ is called 1-periodic in x if $x \to v(x,t)$ is 1-periodic in x for each fixed t. These functions are also called 1-periodic, for short; we will not deal with periodicity in time.

If $u, v : \mathbf{R}^s \to \mathbf{C}^n$ are C^∞-functions which are 1-periodic in x, then their L_2-inner product and norm are defined by

$$(u, v) = \int_0^1 \cdots \int_0^1 \langle u(x), v(x) \rangle \, dx_1 \ldots dx_s, \quad \|u\| = (u, u)^{1/2}.$$

We refer to the following problem as the **spatially periodic Cauchy problem**. Suppose that

$$A_\nu = A_\nu(x, t), \quad F = F(x, t), \quad f(x)$$

are C^∞-functions of their arguments, which are 1-periodic in x. We try to find a C^∞-solution $u = u(x, t)$ of

$$\begin{aligned} u_t &= P(x, t, \partial/\partial x)u + F(x, t) \\ &\equiv \sum_{|\nu| \le m} A_\nu(x, t) D^\nu u + F(x, t), \ x \in \mathbf{R}^s, \quad t \ge 0, \\ u(x, 0) &= f(x), \ x \in \mathbf{R}^s, \end{aligned}$$

which is 1-periodic in x.

Definition 1. The above spatially periodic Cauchy problem is called well-posed if:

i) for each $f = f(x)$ and each $F = F(x, t)$ (satisfying the above conditions) there exists a unique solution $u = u(x, t)$ (satisfying the above conditions);
ii) for each $T > 0$ there is a constant $K(T)$ independent of f, F with

$$\|u(\cdot, t)\|^2 \le K(T)\{\|f\|^2 + \int_0^t \|F(\cdot, \tau)\|^2 \, d\tau\}, \quad 0 \le t \le T.$$

If the coefficients $A_\nu = A_\nu(x, t)$ are constant, then one can use the symbol $P(i\omega)$ to decide the question of well-posedness. Similarly to Theorem 2.2.2, one obtains:

The spatially periodic Cauchy problem with a constant-coefficient operator $P(\partial/\partial x)$ is well-posed if and only if there are constants K and α such that

$$\sup\left\{|e^{P(i\omega)t}| : \omega \in \mathbf{R}^s\right\} \le K e^{\alpha t} \text{ for all } t \ge 0.$$

(One could restrict the components of ω to be integer-multiples of 2π. This would not change the possibility of an estimate of the above form, however.)

2.7.2. Strongly Hyperbolic and Second-Order Parabolic Equations

Hyperbolic systems. Consider a first-order operator with variable coefficients

$$P(x,t,\partial/\partial x) = \sum_{j=1}^{s} A_j(x,t)\,\frac{\partial}{\partial x_j}.$$

We define its symbol by

$$P(x,t,i\omega) = i\sum_{j=1}^{s} \omega_j A_j(x,t), \quad \omega \in \mathbf{R}^s, \quad |\omega| = 1.$$

Suppose all frozen-coefficient problems

$$u_t = P(x_0,t_0,\partial/\partial x)u$$

are strongly hyperbolic as defined in Section 2.4.1. It follows from Lemma 2.4.2 that there is a positive definite Hermitian matrix

$$H(x_0,t_0,\omega) = H^*(x_0,t_0,\omega) > 0$$

with

$$H(x_0,t_0,\omega)\;P(x_0,t_0,i\omega) + P^*(x_0,t_0,i\omega)\;H(x_0,t_0,i\omega) = 0.$$

We define strong hyperbolicity of the variable-coefficient equation

(2.7.1) $$u_t = P(x,t,\partial/\partial x)u \equiv \sum_{j=1}^{s} A_j(x,t)\,\frac{\partial u}{\partial x_j}$$

as follows:

Definition 2. The equation (2.7.1) is called *strongly hyperbolic* if there exists a Hermitian matrix function

$$H(x,t,\omega) > 0, \quad x \in \mathbf{R}^s, \; t \geq 0, \quad \omega \in \mathbf{R}^s, \quad |\omega| = 1,$$

which is C^∞-smooth in all arguments, is 1-periodic in x, and satisfies

$$H(x,t,\omega)\;P(x,t,i\omega) + P^*(x,t,i\omega)\;H(x,t,\omega) = 0.$$

(If a zero-order term and a forcing function are added in (2.7.1), strong hyperbolicity is defined in the same way.)

The matrix function $H(x,t,\omega)$ is called a *symmetrizer*. Thus, except for the smoothness and periodicity of the symmetrizer, we have defined strong hyperbolicity by adopting the constant-coefficient concept.

The most important cases are *symmetric* hyperbolic systems, where

$$A_j(x,t) = A_j^*(x,t), \quad j = 1,\dots,s.$$

For such systems the symbol is always antisymmetric, and one can take $H \equiv I$ as the trivial symmetrizer. The hyperbolic systems which occur in the context of the Euler equations are symmetric hyperbolic after a simple transformation of the density; a first example was given already in Section 2.4.2. One can show

Theorem 2.7.1. *The spatially periodic Cauchy problem for a strongly hyperbolic equation*

$$u_t = \sum_{j=1}^{s} A_j(x,t)\,\frac{\partial u}{\partial x_j} + B(x,t)u + F(x,t)$$

is well-posed.

For a proof in one space dimension, see Section 3.3.1; for the general case, see Section 6.2.

Parabolic systems. Now consider a second-order equation

$$u_t = P(x,t,\partial/\partial x)u + F(x,t), \tag{2.7.2}$$

where

$$P(x,t,\partial/\partial x)u = \sum_{|\nu|\le 2} A_\nu(x,t)D^\nu u.$$

The principle part of P is

$$P_2(x,t,\partial/\partial x) = \sum_{j,k=1}^{s} A_{jk}(x,t)\,\frac{\partial^2}{\partial x_j \partial x_k},$$

and we define its symbol

$$P_2(x,t,i\omega) = -\sum_{j,k=1}^{s} \omega_j\omega_k\, A_{jk}(x,t), \quad \omega \in \mathbf{R}^s, \quad |\omega| = 1.$$

Let

$$\kappa_\ell = \kappa_\ell(x,t,\omega), \quad \ell = 1,\dots,n,$$

denote the eigenvalues of $P_2(x,t,i\omega)$. Suppose all frozen-coefficient equations of (2.7.2) (with $F \equiv 0$) are parabolic, as defined in Section 2.4.3. Then, for each fixed (x,t), there is $\delta(x,t) > 0$ with

$$\operatorname{Re}\, \kappa_\ell(x,t,\omega) \le -\delta(x,t) \text{ for all } \omega \in \mathbf{R}^s, \quad |\omega| = 1, \quad \ell = 1,\ldots,n.$$

Smoothness and periodicity in x imply that we can choose a uniform $\delta = \delta(T) > 0$ for $0 \le t \le T$. Therefore, the following definition of parabolicity requires nothing but parabolicity of all frozen-coefficient problems.

Definition 3. The equation (2.7.2) is called parabolic if for each $T > 0$ there is $\delta(T) > 0$ with

$$\operatorname{Re}\, \kappa_\ell(x,t,\omega) \le -\delta(T) \quad \text{for all } \omega \in \mathbf{R}^s, \quad |\omega| = 1, \quad 0 \le t \le T, \quad \ell = 1,\ldots,n.$$

One can show

Theorem 2.7.2. *The spatially periodic Cauchy problem for a second-order parabolic system* (2.7.2) *is well-posed.*

This result will be proved in Sections 3.1 and 3.2 for one space dimension and in Section 6.1 for more than one space dimension. If we ignore the requirement of the *smoothness* of the symmetrizer in the strongly hyperbolic case, we can summarize Theorems 2.7.1, 2.7.2 by saying: *for strongly hyperbolic and second-order parabolic systems the localization principle is valid.* These variable-coefficient systems inherit their well-posedness from the frozen-coefficient equations. It is not known whether the requirement of smoothness of the symmetrizer in the strongly hyperbolic case is really necessary.

In the next section we present two counterexamples to localization for equations which are neither hyperbolic nor parabolic.

2.7.3. Counterexamples to Localization

Our first example is a 2π-periodic system for which all frozen-coefficient equations are well-posed; the variable-coefficient system is ill-posed nevertheless.

Example 1. Consider the second-order system

$$u_t = iU^*(x)\begin{pmatrix} \alpha & \beta \\ 0 & \gamma \end{pmatrix} U(x)u_{xx},$$

where

$$U(x) = \begin{pmatrix} \cos x & \sin x \\ -\sin x & \cos x \end{pmatrix} \quad \text{and} \quad \alpha,\ \beta, \gamma \in \mathbf{R}, \quad \alpha \ne \gamma, \quad \beta \ne 0.$$

For any frozen-coefficient problem

$$v_t = iU^*(x_0)\begin{pmatrix} \alpha & \beta \\ 0 & \gamma \end{pmatrix} U(x_0)v_{xx}$$

we can introduce new variables

$$w := S\,U(x_0)v, \quad S\begin{pmatrix} \alpha & \beta \\ 0 & \gamma \end{pmatrix} S^{-1} = \begin{pmatrix} \alpha & 0 \\ 0 & \gamma \end{pmatrix},$$

and obtain

$$w_t = i\begin{pmatrix} \alpha & 0 \\ 0 & \gamma \end{pmatrix} w_{xx}.$$

The Cauchy problems for w and for v are well-posed. Now we introduce

$$v(x,t) = U(x)u(x,t)$$

as a new variable into the given variable-coefficient problem; then we obtain

$$\begin{aligned} v_t = Uu_t &= i\begin{pmatrix} \alpha & \beta \\ 0 & \gamma \end{pmatrix} U(U^*v)_{xx} \\ &= i\begin{pmatrix} \alpha & \beta \\ 0 & \gamma \end{pmatrix} \left(v_{xx} + 2UU_x^*v_x + UU_{xx}^*v\right) \\ &= i\begin{pmatrix} \alpha & \beta \\ 0 & \gamma \end{pmatrix} v_{xx} + 2\begin{pmatrix} i\beta & -i\alpha \\ i\gamma & 0 \end{pmatrix} v_x - i\begin{pmatrix} \alpha & \beta \\ 0 & \gamma \end{pmatrix} v. \end{aligned}$$

The system for v has constant coefficients, and therefore we can decide the question of well-posedness. The eigenvalues κ_1, κ_2 of the symbol $P(i\omega)$ are the solutions of

$$0 = \det\bigl(P(i\omega) - \kappa\bigr) = \det\begin{pmatrix} -i\alpha\omega^2 - 2\beta\omega - i\alpha - \kappa & -i\omega^2\beta + 2\alpha\omega - i\beta \\ -2\gamma\omega & -i\gamma\omega^2 - i\gamma - \kappa \end{pmatrix}.$$

Thus,

$$\text{Re}\ (\kappa_1 + \kappa_2) = -2\beta\omega,$$

and consequently the problem is ill-posed for $\beta \neq 0$.

In the following **Schrödinger equation** the converse happens: The variable-coefficient problem is well-posed although the problems with frozen coefficients are ill-posed.

Example 2. Consider the scalar equation

$$u_t = ip(x)u_{xx} + ip_x(x)u_x$$

where $p(x)$ is a real, smooth 1-periodic function with $p(x) \geq p_0 > 0$. The Cauchy problem for a frozen-coefficient equation

$$v_t = ip(x_0)v_{xx} + ip_x(x_0)v_x$$

is ill-posed if $p_x(x_0) \neq 0$, because

$$\text{Re}\,\big(-ip(x_0)\omega^2 - p_x(x_0)\omega\big) = -p_x(x_0)\omega$$

is not bounded from above. We only sketch the proof that the given variable-coefficient equation leads to a well-posed spatially periodic Cauchy problem. The equation can be written as

$$u_t = i(pu_x)_x.$$

If we assume that $u = u(x,t)$ is a smooth 1-periodic solution, then

$$\frac{d}{dt}\big(u(\cdot,t), u(\cdot,t)\big) = (u_t, u) + (u, u_t) = i\big\{(pu_x, u_x) - (u_x, pu_x)\big\} = 0.$$

Thus we obtain *a priori*

$$\|u(\cdot,t)\| = \|u(\cdot,0)\|, \quad t \geq 0.$$

To obtain the existence of a solution, we can consider the parabolic equations

$$u_t = \epsilon u_{xx} + i(pu_x)_x, \quad \epsilon > 0,$$

and send $\epsilon \to 0$. This technique to prove existence is illustrated in Chapter 3.

Notes on Chapter 2

The first paper dealing with the Cauchy problem for general systems of partial differential equations with constant coefficients is due to Petrovskii (1937). He uses Hadamard's (1921) definition of well-posedness, which is equivalent to our definition of weak well-posedness. He proved that the Cauchy problem for $u_t = P(\partial/\partial x)u$ is well-posed if and only if the eigenvalues κ of the symbol $P(i\omega)$ satisfy an inequality

$$\text{Re}\,\kappa(\omega) \leq C_1 \log(1 + |\omega|) + C_2, \quad C_1, C_2 \text{ constants.}$$

Later Gårding (1951) proved that one can choose $C_1 = 0$.

The Matrix Theorem is proved in Kreiss (1959); its application to the Cauchy problem is discussed in Kreiss (1963). The latter paper also shows that the Cauchy problem for hyperbolic first-order systems is well-posed if the eigenvalues of $P(i\omega)$ are purely imaginary and their algebraic multiplicity is constant. Our perturbation example in Section 2.2.3 is typical. Yamaguti and Kasahara (1959) have proved the following theorem: If the Cauchy problem for a first-order system is weakly well-posed for all lower-order perturbations, then it is well-posed. In connection with our counterexamples to localization, W.G.

Strang (1966) gave a necessary condition for the well-posedness of Cauchy problems: If the Cauchy problem is well-posed for a system $u_t = P(x, \partial/\partial x)u$ with x-variable coefficients then the problem is also well-posed for all problems $u_t = P_m(x_0, \partial/\partial x)u$ with frozen coefficients. Here P_m denotes the principle part of P.

3 Linear Variable-Coefficient Cauchy Problems in 1D

In this chapter we treat second-order parabolic and first-order strongly hyperbolic systems in one space dimension. Instead of considering the Cauchy problem with initial data in L_2, we deal with problems which are 1-periodic in x. The periodic problem has the technical advantage that the behavior at $x = \pm\infty$ need not be specified, but the arguments for initial data in L_2 would be essentially the same.

Assuming the existence of a smooth solution for the parabolic equation, we first prove estimates of the solution and its derivatives. In Section 3.2 we write down a simple difference scheme and prove analogous estimates for it. By sending the step-size to zero, we obtain a solution for parabolic systems in a rather elementary and constructive way. Strongly hyperbolic equations are treated by adding a small second-order term, whose coefficient is sent to zero. In a similar fashion as in the constant-coefficient case, we also treat certain mixed hyperbolic-parabolic systems and give an application to the linearized N-S equations.

In the parabolic case we use as a guiding principle: first, *assume* the existence of a solution and show estimates for it and its derivatives; second, write down a difference scheme and show analogous estimates which *imply* the existence of a solution. This principle is very useful for equations of different type, also. We demonstrate this in Section 3.6 with an application to the linearized **Korteweg-**

de Vries equation. The linear **Schrödinger equation** will be treated as a limit of parabolic equations.

The estimates derived for the parabolic systems in Section 3.1 are very elementary. They do not express the smoothing property of the parabolic operator. In some applications this smoothing is important, however, and it will be shown in Section 3.2.6. Also, we demonstrate some important properties of strongly hyperbolic systems in Section 3.3: there is a finite speed of propagation, and discontinuities travel along the characteristics.

3.1. A Priori Estimates for Strongly Parabolic Problems

We consider a second-order system

(3.1.1)
$$u_t = A_2 u_{xx} + A_1 u_x + A_0 u + F =: P(x,t,\partial/\partial x)u + F, \qquad x \in \mathbf{R}, \quad t \geq 0,$$

together with an initial condition

(3.1.2)
$$u(x,0) = f(x), \quad x \in \mathbf{R}.$$

The matrices

$$A_j = A_j(x,t) \in \mathbf{C}^{n,n}, \quad j = 0,1,2,$$

and the vector functions

$$F = F(x,t),\ f = f(x) \in \mathbf{C}^n$$

are assumed to be of class C^∞ for simplicity. It is an essential assumption, however, that all functions are taken as 1-periodic in x for each fixed t. Furthermore, to begin with, we assume that

$$A_2(x,t) + A_2^*(x,t) \geq 2\delta I \quad \text{for all } (x,t), \qquad \text{with } \delta > 0 \text{ fixed.}$$

This is slightly more restrictive than the assumption of parabolicity for all frozen-coefficient problems. If $A_2 + A_2^*$ satisfies a lower bound of the above form, the equation (3.1.1) is called *strongly parabolic.*

In this section we will *assume* that $u(x,t)$ is a C^∞-solution of the above problem which is 1-periodic in x for each fixed t. The existence of such a solution will be proven in Section 3.2. Here we will derive estimates for u and its derivatives.

Notations. For vectors $u, v \in \mathbf{C}^n$ and matrices $A \in \mathbf{C}^{n,n}$ we remind the reader of the notations

$$\langle u, v\rangle = \sum_{i=1}^{n} \overline{u}_i v_i, \quad |u| = \langle u, u\rangle^{1/2}, \quad |A| = \max\{|Au|, \, |u| = 1\}.$$

For (smooth) 1-periodic vector functions $u = u(x)$, $v = v(x)$ our basic inner-product and norm are

$$(u, v) = \int_0^1 \langle u(x), v(x)\rangle \, dx, \qquad \|u\| = (u, u)^{1/2}.$$

For nonnegative integers p we also use the **Sobolev** inner-product and norm given by

$$(u, v)_{H^p} = \sum_{j=0}^{p} \left(\frac{d^j u}{dx^j}, \frac{d^j v}{dx^j}\right), \qquad \|u\|_{H^p} = (u, u)_{H^p}^{1/2}.$$

For (smooth) 1-periodic matrix functions $A = A(x)$ let

$$|A|_\infty = \max\{|A(x)| : 0 \le x \le 1\}.$$

It will be frequently used that

$$(u, Av) \le \|u\| \|Av\| \;\le\; |A|_\infty \|u\| \|v\|,$$

and similarly that

$$(Au, v) \;\le\; |A|_\infty \|u\| \|v\|.$$

If $u = u(x, t)$ and $v = v(x, t)$ are functions of (x, t) we often abbreviate

$$(u, v) = (u(\cdot, t), v(\cdot, t)) = \int_0^1 \langle u(x, t), v(x, t)\rangle \, dx,$$

and thus leave off the arguments to simplify the notation.
The main tool we will use is "integration by parts": If $u, v \in C^1$ are 1-periodic, then

$$(u_x, v) = \int_0^1 \langle u_x, v\rangle \, dx = -\int_0^1 \langle u, v_x\rangle \, dx = -(u, v_x).$$

3.1.1. The Basic Energy Estimate

Let $T > 0$ denote an arbitrary but fixed time. For $0 \le t \le T$ we have

$$
\begin{aligned}
\frac{d}{dt}(u,u) &= (u_t,u)+(u,u_t)\\
&= (A_2u_{xx},u)+(A_1u_x,u)+(A_0u,u)+(F,u)\\
&\quad + (u,A_2u_{xx})+(u,A_1u_x)+(u,A_0u)+(u,F)\\
&\le (A_2u_{xx},u)+(u,A_2u_{xx})+c_1\{\|u\|\|u_x\|+\|u\|^2+\|F\|^2\}.
\end{aligned}
$$

Since $A_2u_{xx}=(A_2u_x)_x - A_{2x}u_x$, integration by parts yields

$$(A_2u_{xx},u) = -(A_2u_x,u_x)-(A_{2x}u_x,u) = -(u_x,A_2^*u_x)-(A_{2x}u_x,u),$$

and similarly

$$(u,A_2u_{xx}) = -(u_x,A_2u_x)-(u,A_{2x}u_x).$$

Therefore,

$$
\begin{aligned}
\text{(3.1.3)}\qquad \frac{d}{dt}(u,u) &\le -\big(u_x,(A_2+A_2^*)u_x\big)+c_2\{\|u\|\|u_x\|+\|u\|^2+\|F\|^2\}\\
&\le -2\delta\|u_x\|^2+c_2\{\|u\|\|u_x\|+\|u\|^2+\|F\|^2\}.
\end{aligned}
$$

For any two real numbers a,b we know that $ab\le \frac{1}{2}a^2+\frac{1}{2}b^2$; thus for any $\sigma\ne 0$,

$$ab=(\sigma a)(\frac{b}{\sigma})\le \frac{\sigma^2}{2}a^2+\frac{1}{2\sigma^2}b^2,$$

and therefore,

$$\|u\|\|u_x\| \le \frac{\sigma^2}{2}\|u_x\|^2+\frac{1}{2\sigma^2}\|u\|^2.$$

Choosing $\sigma=\sigma(\delta,c_2)>0$ appropriately, we find from (3.1.3) that

$$\text{(3.1.4)}\qquad \frac{d}{dt}(u,u)\le c_3\{\|u\|^2+\|F\|^2\}\qquad \text{for}\quad 0\le t\le T.$$

The derivation shows that c_1, c_2, c_3 only depend on δ and on bounds for

$$\text{(3.1.5)}\qquad |A_0|_\infty,\ |A_1|_\infty,\ |A_{2x}|_\infty$$

in the interval $0\le t\le T$. Using the abbreviations

$$y(t)=\|u(\cdot,t)\|^2,\quad \phi(t)=\|F(\cdot,t)\|^2,$$

we have shown the differential inequality

$$y'(t)\le c_3y(t)+c_3\phi(t),\quad 0\le t\le T.$$

Now (a simple version of) **Gronwall's Lemma**, formulated next, allows us to estimate y(t).

Lemma 3.1.1. *Suppose $y \in C^1[0,T]$, $\psi \in C[0,T]$ satisfy*

$$y'(t) \le cy(t) + \psi(t), \quad 0 \le t \le T,$$

for some $c \ge 0$. Then

$$y(t) \le e^{ct}\{y(0) + \int_0^t |\psi(\tau)|\,d\tau\}, \quad 0 \le t \le T.$$

Proof. For the function $z(t) = e^{-ct}y(t)$ it holds that

$$z'(t) = -ce^{-ct}y(t) + e^{-ct}y'(t) \le e^{-ct}\psi(t).$$

Thus integration yields

$$z(t) - z(0) \le \int_0^t |\psi(\tau)|\,d\tau,$$

and the desired estimate for $y(t) = e^{ct}z(t)$ follows.

An application of this result gives us

Lemma 3.1.2. *Let C_1 denote a constant which bounds the norms* (3.1.5) *in $0 \le t \le T$. There is another constant C_2 which depends only on C_1, δ, and T such that*

$$\|u(\cdot,t)\|^2 \le C_2\{\|f\|^2 + \int_0^t \|F(\cdot,\tau)\|^2\,d\tau\} \quad \text{in } 0 \le t \le T.$$

Proof. If $c = c_3$ denotes the constant in (3.1.4), the result follows from the previous lemma with $C_2 = e^{cT}\max\{1, c\}$.

3.1.2. Estimates for Space-Derivatives of u

If we differentiate the given differential equation (3.1.1) with respect to x we find that, for $v = u_x$,

$$\begin{aligned} v_t = u_{tx} &= A_2u_{xxx} + A_{2x}u_{xx} + A_1u_{xx} + A_{1x}u_x + A_0u_x + A_{0x}u + F_x \\ &= A_2v_{xx} + (A_{2x} + A_1)v_x + (A_{1x} + A_0)v + G \end{aligned}$$

with

$$G = A_{0x}u + F_x.$$

The function v satisfies the initial condition $v(x,0) = u_x(x,0) = f_x(x)$. Since we have an estimate for $\|u\|$ and thus for G, we can apply the previous lemma and obtain an estimate for $v = u_x$:

$$\|u_x(\cdot,t)\|^2 \le C_3 \left\{ \|f\|^2 + \|f_x\|^2 + \int_0^t \left[\|F(\cdot,\tau)\|^2 + \|F_x(\cdot,\tau)\|^2\right] d\tau \right\}$$
$$\text{in } 0 \le t \le T.$$

Here C_3 depends only on δ, on T, and on a bound for

$$|A_0|_\infty, \ |A_1|_\infty, \ |A_{jx}|_\infty, \quad j = 0,1,2,$$

in $0 \le t \le T$. It is obvious that we can proceed in the same way and further obtain estimates for u_{xx}, u_{xxx}, etc. Just differentiate the above equation for $v = u_x$ with respect to x, etc.

Lemma 3.1.3. *Given a positive integer p and a time $T > 0$, then the solution of* (3.1.1), (3.1.2) *satisfies*

$$\|u(\cdot,t)\|_{H^p}^2 \le C \left\{ \|f\|_{H^p}^2 + \int_0^t \|F(\cdot,\tau)\|_{H^p}^2 \, d\tau \right\} \quad \text{in } 0 \le t \le T.$$

Here C depends only on p, on T, on δ, and on a bound for the coefficients $A_j(x,t)$ and their derivatives of order $\le p$ in $0 \le t \le T$.

3.1.3. Estimates for Time-Derivatives and Mixed Derivatives

Using the given differential equation (3.1.1), we can always express time derivatives of u by space derivatives. For example, if the differential equation reads

$$u_t = Au_{xx} + F,$$

then differentiation gives us

$$\begin{aligned} u_{tt} &= Au_{txx} + A_t u_{xx} + F_t \\ &= AAu_{xxxx} + 2AA_x u_{xxx} + AA_{xx}u_{xx} + AF_{xx} + A_t u_{xx} + F_t. \end{aligned}$$

To express u_{ttt}, we need six space derivatives of u, four space derivatives of F, two space derivatives of F_t, and the function F_{tt}. In general, we can express q time derivatives of u by $2q$ space derivatives of u and $2(q-1-k)$ space derivatives of

$$\frac{\partial^k}{\partial t^k} F, \quad k = 0, \ldots, q-1.$$

Further differentiation with respect to x allows us to express mixed derivatives of u also. Since the space derivatives of u are already estimated in Lemma 3.1.3, we have

Theorem 3.1.4. *Given any nonnegative integers p and q and a time $T > 0$, there is a constant C with*

$$\left\| \frac{\partial^q}{\partial t^q} u(\cdot, t) \right\|_{H^p}^2$$

$$\leq C\|f\|_{H^{2q+p}}^2 + C \int_0^t \left\{ \|F(\cdot, \tau)\|_{H^{2q+p}}^2 + \sum_{k=1}^{q-1} \left\| \frac{\partial^k}{\partial t^k} F(\cdot, \tau) \right\|_{H^{2(q-1-k)+p}}^2 \right\} d\tau$$

in $0 \leq t \leq T$. The constant C is independent of f and F, and only depends on δ, on T, and on a bound for the derivatives of the coefficients A_0, A_1, A_2 of order $\leq 2q + p$ in $0 \leq t \leq T$. For $p = q = 0$ one also needs a bound for A_{2x}.

3.2. Existence for Parabolic Problems via Difference Approximations

In this section we prove existence of a C^∞-solution $u(x, t)$ of the space-periodic initial value problem (3.1.1), (3.1.2). As in Section 3.1, the main assumption is

$$A_2(x, t) + A_2^*(x, t) \geq 2\delta I \quad \text{for all } (x, t), \qquad \text{with } \delta > 0 \text{ fixed.}$$

We relax this condition in Section 3.2.5, however, where we treat general parabolic problems. All functions are assumed to be 1-periodic in x, and the coefficients of the differential equation are assumed to be C^∞-smooth. First we also assume that the initial function is in C^∞; general L_2-initial data and the smoothing property of parabolic equations are treated in Section 3.2.6.

It follows immediately from the basic energy estimate stated in Lemma 3.1.1 that the Cauchy problem (3.1.1), (3.1.2) has at most one classical solution, i.e., at most one solution in $C^1(t) \cap C^2(x)$. If there were two solutions u, v then $w = u - v$ would be a solution of (3.1.1), (3.1.2) with $F \equiv f \equiv 0$. Then our basic energy estimate implies $w \equiv 0$.

3.2.1. The Main Result and Outline of the Proof

We want to prove the existence of solutions using a difference approximation. For that reason we introduce a gridlength $h = 1/N$, N a natural number, gridpoints $x_\nu = \nu h$, $\nu \in \mathbf{Z}$, and gridfunctions with function values $v_\nu := v(x_\nu) \in \mathbf{C}^n$.

We shall always assume that the gridfunctions are 1-periodic; i.e.,

$$v_\nu = v_{N+\nu} \quad \text{for all } \nu \in \mathbf{Z} \quad (\text{note } Nh = 1).$$

As usual, the translation operator E is defined by

$$(Ev)_\nu := v_{\nu+1}.$$

Clearly, $w := Ev$ is also 1-periodic. The j-th power of E (j an integer) is

$$(E^j v)_\nu = v_{\nu+j}, \quad \nu \in \mathbf{Z},$$

and we shall use the notation

$$E^0 =: I.$$

With the help of the translation operator E, we define the forward and backward divided difference operators D_+, D_-:

(3.2.1) $$hD_+ := E - I, \quad hD_- := I - E^{-1} = hE^{-1}D_+;$$

i.e.,

$$h(D_+v)_\nu = v_{\nu+1} - v_\nu, \quad h(D_-v)_\nu = v_\nu - v_{\nu-1}.$$

The usual second-order accurate approximations to $\partial/\partial x$ and $\partial^2/\partial x^2$ are

(3.2.2) $$D_0 = \frac{1}{2}(D_+ + D_-) \text{ and } D_+D_- = h^{-2}(E - 2I + E^{-1}),$$

respectively. If $v = v(x)$ is a smooth function of $x \in \mathbf{R}$, then its restriction to the grid is $(v_\nu)_{\nu\in\mathbf{Z}} = (v(x_\nu))_{\nu\in\mathbf{Z}}$. This gridfunction is also denoted by v, for simplicity. A similar notation applies to matrix functions $A = A(x)$. Often $A = A(x)$ will be a 1-periodic matrix function and Av a gridfunction with function values

$$(Av)_\nu = A(x_\nu)v(x_\nu).$$

We need some simple error formulae for the above difference operators. E.g., if we apply the forward difference operator D_+ to the restriction of a smooth function v, then Taylor expansion gives us

(3.2.3) $$(D_+v)_\nu = \frac{v(x_{\nu+1}) - v(x_\nu)}{h} = v'(x_\nu) + \frac{h}{2}v''(x_\nu) + \frac{h^2}{6}v'''(x_\nu) + \frac{h^3}{24}\psi_\nu,$$

where

$$|\psi_\nu| \le \max_{x_\nu \le \xi \le x_{\nu+1}} |v^{(4)}(\xi)|.$$

Correspondingly,

$$(D_-v)_\nu = v'(x_\nu) - \frac{h}{2}v''(x_\nu) + \frac{h^2}{6}v'''(x_\nu) + \frac{h^3}{24}\bar{\psi}_\nu,$$

(3.2.4)
$$(D_0v)_\nu = v'(x_\nu) + \frac{h^2}{6}v'''(x_\nu) + \frac{h^4}{24}\phi_\nu,$$

$$(D_+D_-v)_\nu = v''(x_\nu) + \frac{h^2}{12}\bar{\phi}_\nu$$

where

$$|\bar{\psi}_\nu| \le \max_{x_{\nu-1}\le\xi\le x_\nu} |v^{(4)}(\xi)|, \quad |\phi_\nu| \le \max_{x_{\nu-1}\le\xi\le x_{\nu+1}} |v^{(5)}(\xi)|,$$

$$|\bar{\phi}_\nu| \le \max_{x_{\nu-1}\le\xi\le x_{\nu+1}} |v^{(4)}(\xi)|.$$

These error formulae will be applied below.

We now describe a simple difference analog of (3.1.1), (3.1.2), which we use for our existence argument. Only space is discretized, time is left continuous. The difference analog reads

(3.2.5)
$$dv_\nu/dt = A_{2\nu}D_+D_-v_\nu + A_{1\nu}D_0v_\nu + A_{0\nu}v_\nu + F_\nu, \quad \nu \in \mathbf{Z},$$
$$v_\nu(0) = f_\nu, \quad \nu \in \mathbf{Z}.$$

The equations (3.2.5) represent an *infinite* system of ordinary differential equations for $v(t) = (v_\nu(t))_{\nu\in\mathbf{Z}}$ subject to initial conditions. However, we are only interested in 1-periodic solutions and therefore we need to consider (3.2.5) only for $\nu = 0, 1, 2, ..., N-1$. Using the periodicity conditions

$$v_{-1}(t) = v_{N-1}(t), \quad v_0(t) = v_N(t),$$

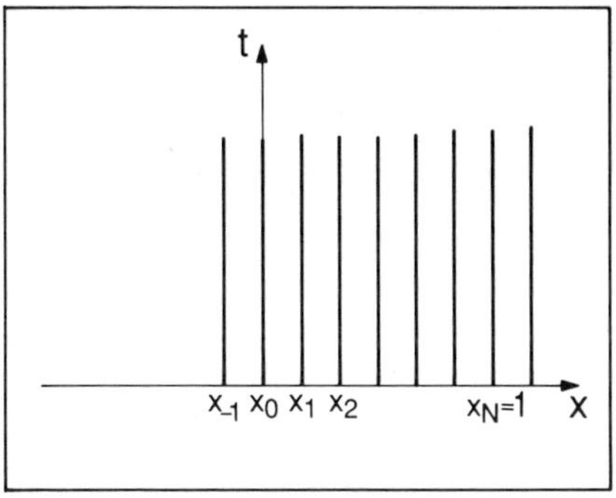

FIGURE 3.2.1. Gridlines for difference scheme.

we can eliminate $v_N(t)$ and $v_{-1}(t)$ and obtain a system of N differential equations (with initial conditions) for the N unknowns $v_0(t), ..., v_{N-1}(t)$. Any solution of this initial value problem generates the corresponding solution of (3.2.5) by periodic extension. Therefore, the existence of a unique periodic solution of (3.2.5) follows from standard theorems for ordinary differential equations. It is clear that the solution $v(t) = v^h(t)$ depends on the step-size h, but we often suppress this in our notation.

We are going to show in Section 3.2.3 that we can estimate all difference-differential quotients $D_+^p d^q v^h(t)/dt^q$ independently of h. Then we shall use a theorem from approximation theory which states that we can interpolate the gridfunction $v = v^h$ with respect to x in such a way that the interpolant $w^h(x,t)$ is as smooth as v^h; i.e., we can estimate all derivatives $\partial^{p+q} w^h(x,t)/\partial x^p \partial t^q$ in terms of $D_+^p d^q v^h(t)/dt^q$. The next step is to show that w^h satisfies (3.1.1), (3.1.2) with F, f replaced by $F + hF_1^h, f + hf_1^h$. For $h \to 0$, the functions w^h converge to the solution of (3.1.1), (3.1.2).

3.2.2. The Basic Energy Estimate for the Solutions of the Difference Approximation

Before we derive the basic energy estimate for the system (3.2.5), let us first introduce a discrete scalar product and norm and let us establish some simple rules. If v, w are 1-periodic gridfunctions, then we define their discrete scalar product and norm by

$$(v, w)_h := \sum_{\nu=0}^{N-1} \langle v_\nu, w_\nu \rangle h, \quad \|v\|_h^2 := (v, v)_h.$$

The scalar product for gridfunctions has the same general properties as the L_2-scalar product of functions of a continuous variable; i.e., it is a sesquilinear form, and we have the rules

$$(v, w)_h = \overline{(w, v)_h},$$

$$|(v, w)_h| \le \|v\|_h \, \|w\|_h, \quad \|v + w\|_h \le \|v\|_h + \|w\|_h.$$

Also,

$$\|E^j u\|_h = \|u\|_h, \quad \|D_+ u\|_h = \|D_- u\|_h, \quad \|D_0 u\|_h \le \|D_+ u\|_h,$$

$$(A^* v, w)_h = (v, Aw)_h \le |A|_{\infty,h} \|u\|_h \, \|v\|_h,$$

where

$$|A|_{\infty,h} := \max_\nu |A_\nu|.$$

Note that

$$|A|_{\infty,h} \le \max_x |A(x)| = |A|_\infty$$

is bounded independently of h if $A = A(x)$ is a smooth 1-periodic matrix function.

Another simple rule is: If $A_\nu + A_\nu^* \ge 2\delta I$ for all ν, then

$$(u, (A + A^*)u)_h \ge 2\delta \|u\|_h^2. \tag{3.2.6}$$

Corresponding to integration by parts, there are rules of summation by parts in the discrete case. For periodic gridfunctions v, w,

$$(v, D_- w)_h = -(D_+ v, w)_h, \quad (v, D_0 w)_h = -(D_0 v, w)_h.$$

This is easily checked:

$$\begin{aligned}
(v, D_- w)_h &= \sum_{\nu=0}^{N-1} \langle v_\nu, w_\nu \rangle - \sum_{\nu=0}^{N-1} \langle v_\nu, w_{\nu-1} \rangle \\
&= \sum_{\nu=0}^{N-1} \langle v_\nu, w_\nu \rangle - \sum_{\nu=-1}^{N-2} \langle v_{\nu+1}, w_\nu \rangle \\
&= \sum_{\nu=0}^{N-1} \langle v_\nu, w_\nu \rangle - \sum_{\nu=0}^{N-1} \langle v_{\nu+1}, w_\nu \rangle - \langle v_0, w_{-1} \rangle + \langle v_N, w_{N-1} \rangle \\
&= -\sum_{\nu=0}^{N-1} \langle v_{\nu+1} - v_\nu, w_\nu \rangle = -(D_+ v, w)_h.
\end{aligned}$$

The second relation follows from the first. The identity

$$h(D_+ Av)_\nu = A_{\nu+1} v_{\nu+1} - A_\nu v_\nu = A_\nu (v_{\nu+1} - v_\nu) + (A_{\nu+1} - A_\nu) v_{\nu+1}$$

gives us a discrete analog of Leibniz' product rule:

$$D_+(Av) = A(D_+ v) + (D_+ A)(Ev).$$

Now we can derive the basic energy estimate. The solution $v = v^h$ of the system (3.2.5) satisfies

$$\begin{aligned}
\frac{d}{dt}(v, v)_h &= (v, \frac{dv}{dt})_h + (\frac{dv}{dt}, v)_h \\
&\le (v, A_2 D_+ D_- v)_h + (A_2 D_+ D_- v, v)_h \\
&\quad + 2|A_1|_\infty \|v\|_h \|D_+ v\|_h + 2|A_0|_\infty \|v\|_h^2 + 2\|v\|_h \|F\|_h.
\end{aligned} \tag{3.2.7}$$

Here

$$(v, A_2 D_+ D_- v)_h = (v, D_+(A_2 D_- v))_h - R_0 = -(D_- v, A_2 D_- v)_h - R_0,$$

$$R_0 = (v, (D_+ A_2) D_+ v)_h;$$

thus

$$|R_0| \le |A_{2x}|_\infty \|v\|_h \, \|D_+ v\|_h.$$

(Observe that $(D_+ A_2)_\nu = h^{-1} \int_{x_\nu}^{x_{\nu+1}} A_{2x}\, dx$.) Correspondingly,

$$(A_2 D_+ D_- v, v)_h = -(D_- v, A_2^* D_- v)_h + \bar{R}_0,$$

where $\bar{R}_0$ satisfies a similar estimate as R_0.

Introducing these expressions into (3.2.7) and observing (3.2.6), we obtain an inequality of the same type as for the continuous case, and therefore the following basic energy estimate holds:

Lemma 3.2.1. *Let C_1 denote a constant which bounds the norms* (3.1.5) *in* $0 \le t \le T$. *There is another constant C_2 which depends only on C_1, δ, and T such that*

$$\|v(t)\|_h^2 \le C_2 \left\{ \|f\|_h^2 + \int_0^t \|F(\cdot, \tau)\|_h^2 d\tau \right\} \quad \text{for } 0 \le t \le T\,.$$

3.2.3. Estimates for Higher Order Divided Differences

As an auxiliary result we first show

Lemma 3.2.2. *For any smooth* 1*-periodic function g it holds that*

$$|D_+^p g|_{\infty,h} \le |d^p g/dx^p|_\infty.$$

Proof. For $p = 0$ the estimate is true. Assume it is valid up to the order $p-1$. Let

$$f(x) = \frac{1}{h} \int_0^h g'(x + \xi)\, d\xi;$$

thus $f(x_\nu) = (D_+ g)_\nu$. Therefore,

$$|(D_+^p g)_\nu| = |(D_+^{p-1} f)(x_\nu)| \;\le\; |d^{p-1} f/dx^{p-1}|_\infty \;\le\; |d^p g/dx^p|_\infty.$$

This proves the lemma.

We begin with an estimate for $w := D_+ v$ where v is the solution of (3.2.5), as before. Applying D_+ to (3.2.5) and using the discrete Leibniz' rule, we find the following system for w:

(3.2.8)
$$
\begin{aligned}
dw_\nu/dt = {} & A_{2\nu}D_+D_-w_\nu + (D_+A_{2\nu} + A_{1\nu})D_0w_\nu \\
& + (A_{0\nu} + (D_+A_{1\nu})E)w_\nu + F_\nu^{(1)},
\end{aligned}
$$

where

$$F_\nu^{(1)} := (D_+A_{0\nu})Ev_\nu + D_+F_\nu.$$

The system (3.2.8) is of the same form as (3.2.5) and – using Lemma 3.2.2 to estimate coefficients – we obtain an estimate for $w = D_+v$, which is of the same type as for the continuous case. Repeating this process, we obtain estimates for all forward difference quotients $D_+^p v$. As in the continuous case, we can also differentiate with respect to time and obtain bounds for $D_+^p d^q v(t)/dt^q$. If one introduces the discrete analog of the continuous H^p-norm by

$$(u,v)_{h,H^p} = \sum_{j=0}^{p} (D_+^j u, D_+^j v)_h, \quad \|u\|_{h,H^p}^2 = (u,u)_{h,H^p},$$

one can summarize the result as follows:

Theorem 3.2.3. *The estimate of Theorem* 3.1.4 *holds for* $v = v^h$ *if one replaces the* H^p*-norm by its discrete version. The constant is independent of the step-size* h.

3.2.4. Convergence as $h \to 0$

The following result about Fourier interpolation is proved in Appendix 2.

Theorem 3.2.4. *Let* $h = 1/(2m+1)$ *where* m *is a positive integer. If* $v^h(x)$ *is a 1-periodic gridfunction, then there is a unique Fourier polynomial*

$$w^h(x) = \sum_{k=-m}^{m} a_k^h e^{2\pi i k x}, \quad a_k^h \in \mathbf{C}^n,$$

which interpolates v^h *at all gridpoints:*

$$w^h(x_\nu) = v^h(x_\nu), \ \nu \in \mathbf{Z}.$$

For any $p = 0, 1, 2, \ldots$ *it holds that*

$$\left\|\frac{d^p w^h}{dx^p}\right\|^2 \le \left(\frac{\pi}{2}\right)^{2p} \|D_+^p v^h\|_h^2.$$

It is remarkable that the constant in the estimate does not depend on h. In our application, the gridfunction $v^h(x,t)$ depends smoothly on t. If $w^h(x,t)$ denotes the Fourier interpolant w.r.t. x, then w^h is a C^∞-function of (x,t) since the coefficients a_k^h are C^∞-functions of t. (This follows from the construction of the interpolant in Appendix 2.) Also, differentiation of w^h w.r.t. t gives us the Fourier interpolant of dv^h/dt. Therefore, Theorems 3.2.3 and 3.2.4 imply bounds

$$\max\left\{\|\frac{\partial^{p+q}}{\partial x^p \partial t^q} w^h(\cdot,t)\| \,:\, 0 \le t \le T\right\} \;\le\; C(p,q,T), \tag{3.2.9}$$

with constants independent of h. To obtain bounds in maximum norm, the following result – an example of a **Sobolev inequality** – will be applied. (More general Sobolev inequalities, which we need below, are proved in Appendix 3.)

Lemma 3.2.5. *Suppose* $u \in C^1[0,1]$. *Then*

$$|u|_\infty^2 \le \|u\|^2 + 2\|u\|\,\|u_x\|.$$

Proof. There are points x_0, x_1 with

$$\min\{|u(x)| :\; 0 \le x \le 1\} = |u(x_0)|,$$

$$\max\{|u(x)| :\; 0 \le x \le 1\} = |u(x_1)| = |u|_\infty.$$

Let $x_0 < x_1$ for definiteness. Then

$$|u(x_1)|^2 - |u(x_0)|^2 = \int_{x_0}^{x_1} \{\frac{d}{dx}|u(x)|^2\}\, dx$$

$$= \int_{x_0}^{x_1} \{\langle u, u_x\rangle + \langle u_x, u\rangle\}\, dx \le 2\|u\|\,\|u_x\|.$$

Since $|u(x_0)| \le \|u\|$, the result follows.

Using the estimates (3.2.9) and the previous lemma, we obtain bounds for all derivatives of $w^h(x,t)$ in maximum norm:

$$\max\{|\frac{\partial^{p+q}}{\partial x^p \partial t^q} w^h(x,t)| :\;\; x \in \mathbf{R},\; 0 \le t \le T\} \;\le\; C_0(p,q,T)$$

with constants independent of h. For short, we say that the family of functions w^h is *uniformly smooth* in each finite time interval.

Let us show next that $w^h(x,t)$ solves the Cauchy problem (3.1.1), (3.1.2) up to terms of order h.

Lemma 3.2.6. *The functions* $F_1^h(x,t)$, $f_1^h(x)$ *defined by*

$$w_t^h(x,t) = P(x,t,\partial/\partial x)w^h(x,t) + F(x,t) + hF_1^h(x,t),$$

$$w^h(x,0) = f(x) + hf_1^h(x)$$

satisfy estimates

$$|F_1^h(x,t)| \le C, \quad x \in \mathbf{R}, \quad 0 \le t \le T,$$
$$|f_1^h(x)| \le C, \quad x \in \mathbf{R},$$

where C does not depend on h.

Proof. Let $x_\nu \le x < x_{\nu+1}$. Then

$$\begin{aligned} w_t^h(x,t) &= w_t^h(x_\nu,t) + \int_{x_\nu}^x w_{xt}^h(\xi,t)\,d\xi \\ &=: w_t^h(x_\nu,t) + h\varphi_1^h(x,t) \\ &= dv_\nu^h(t)/dt + h\varphi_1^h(x,t), \\ w_x^h(x,t) &= D_0 w^h(x_\nu,t) + h\varphi_2^h(x,t) \\ &= D_0 v_\nu^h(t) + h\varphi_2^h(x,t), \\ w_{xx}^h(x,t) &= D_+D_- v_\nu^h(t) + h\varphi_3^h(x,t), \end{aligned}$$

where $|\varphi_j^h(x,t)| \le \text{const},\ j = 1,2,3$. If we use these expressions to replace w^h in the above definition of F_1^h and observe the equation (3.2.5) for v^h, then the estimate for F_1^h follows. Also,

$$w^h(x,0) - f(x) = \int_{x_\nu}^x \{w_x^h(\xi,0) - f_x(\xi)\}\,d\xi = hf_1^h(x), \quad |f_1^h(x)| \le C,$$

and the lemma is proved.

Since the family of functions $w^h = w^h(x,t)$ is uniformly smooth in $0 \le t \le T$, we can apply the result of Appendix 4 (which is based on the Arzela-Ascoli Theorem) and obtain the following: There is a 1-periodic C^∞-function $u = u(x,t),\ x \in \mathbf{R},\ 0 \le t \le T$, and a sequence $h = h_j \to 0$ such that

$$\max\{|\frac{\partial^{p+q}}{\partial x^p \partial t^q} w^h(x,t) - \frac{\partial^{p+q}}{\partial x^p \partial t^q} u(x,t)| : x \in \mathbf{R},\ 0 \le t \le T\} \to 0$$
$$\text{as} \quad h = h_j \to 0$$

for all derivatives $\partial^{p+q}/\partial x^p \partial t^q$. By the previous lemma it follows that u solves the given initial value problem in $0 \le t \le T$. Here $T > 0$ is arbitrary, and thus we have shown

Theorem 3.2.7. *The strongly parabolic initial value problem* (3.1.1), (3.1.2) *with* 1*-periodic (in x) C^∞-data has a unique C^∞-solution $u = u(x,t)$ which is* 1*-periodic in x.*

Remark. To obtain *existence* of a solution u, it was sufficient to establish convergence for some *subsequence* $w^h, h = h_j \to 0$, via the Arzela-Ascoli Theorem. In view of numerical applications one might ask, however, whether in general $w^h \to u$ as $h \to 0$. This convergence does indeed hold as follows from the basic energy estimate of Section 3.1.1 applied to the difference $w^h - u$. Actually, the (nonconstructive) Arzela-Ascoli argument can be avoided altogether in an existence proof: From the basic energy estimate it follows that w^h is a Cauchy sequence in L_2 as $h = h_j \to 0$, and smoothness of the limit function can be deduced from the uniform (in h) decay rate of the Fourier coefficients of w^h. Such a uniform rate follows from the uniform smoothness.

3.2.5. General Parabolic Systems

Thus far we have considered parabolic systems (3.1.1) under the restriction

$$A_2(x,t) + A_2^*(x,t) \geq 2\delta I, \quad \delta > 0 \text{ fixed.}$$

We want to replace this by the weaker assumption that all frozen-coefficient problems are parabolic in the sense of the definition given in Section 2.4.3. To this end, we fix $T > 0$ and assume there exists $\delta > 0$ independent of x, t with

$$\text{(3.2.10)} \qquad \operatorname{Re}\kappa \geq \delta > 0 \text{ for all eigenvalues } \kappa \text{ of } A(x,t), \quad 0 \leq t \leq T.$$

(The assumption $A_2 + A_2^* \geq 2\delta I$ implies $\operatorname{Re}\kappa \geq \delta$ for all eigenvalues κ of A_2.) A variable-coefficient problem (3.1.1) satisfying the eigenvalue condition (3.2.10) is called *parabolic*.

The discussion given below of a *variable*-coefficient problem – via properties of equations with constant coefficients – is very important in a general theory of partial differential equations. Basically the technique is the following: Well-posedness of each constant-coefficient problem – obtained by freezing coefficients at an arbitrary point $P = (x_0, t_0)$ – is expressed via the Matrix Theorem in Fourier space. One obtains existence of a certain Hermitian matrix H_P for each point $P = (x_0, t_0)$. These matrices H_P are used to define time dependent norms $\| \ \|_{H(t)}$, which are all equivalent to the L_2-norm. Applying $\| \ \|_{H(t)}$ instead of the L_2-norm, one can again prove the basic energy estimate, where the energy is now measured in the new norm. In the present situation we do not have to rely on the Matrix Theorem to construct H_P. Instead, using the first part of the proof of Theorem 2.1.3, we have

Lemma 3.2.8. *Let A denote a constant matrix and let $\operatorname{Re}\kappa \geq \delta > 0$ for all eigenvalues κ of A. There exists a Hermitian matrix $H = H^* \geq I$ such that*

$$HA + A^*H \geq \delta H.$$

Note that the bound $H \geq I$ can be enforced by multiplying H with a scalar, if necessary. We proceed with the construction of the norms $\| \ \|_{H(t)}$. The idea of the construction is due to Gårding (1953).

Lemma 3.2.9. *Under assumption* (3.2.10) *there exists a smooth* 1*-periodic matrix function* $H(x,t) = H^*(x,t) \geq I$ *such that*

$$H(x,t)A_2(x,t) + A_2^*(x,t)H(x,t) \geq \frac{\delta}{2}H(x,t). \tag{3.2.11}$$

The norms $\| \ \|_{H(t)}$ *defined by*

$$\|u\|_{H(t)}^2 = (u,u)_{H(t)}, \quad (u,v)_{H(t)} = \int_0^1 \langle u(x), H(x,t)v(x)\rangle \, dx$$

are equivalent to the L_2*-norm.*

Proof. For each point $P = (x_0, t_0)$ we construct a matrix H_P with

$$H_P A_2(x_0,t_0) + A_2^*(x_0,t_0)H_P \geq \delta H_P, \quad H_P = H_P^* \geq I,$$

using the previous lemma. By continuity, the point P has a neighborhood $N(P)$ with

$$H_P A_2(x,t) + A_2^*(x,t)H_P \geq \frac{\delta}{2}H_P$$

for all $(x,t) \in N(P)$. By the Heine-Borel Theorem there are finitely many points $P_1, \cdots, P_J$ whose neighborhoods $N(P_j)$ cover the set $[0,1] \times [0,T]$, $T > 0$ being fixed. Choose a partition of unity $\phi_j = \phi_j(x,t)$ of 1-periodic functions subordinate to these neighborhoods:

$$\phi_j \in C^\infty, \ \phi_j \geq 0, \ \sum_j \phi_j = 1, \ \text{support}\, \phi_j \text{ contained in } N(P_j).$$

(To be precise, the Heine-Borel Theorem and the Partition of unity argument are being applied to the compact space $S^1 \times [0,T]$ where S^1 denotes the circle. Functions on this space are identified with 1-periodic functions.) With the help of the locally supported functions ϕ_j, define the matrix function

$$H(x,t) = \sum_j \phi_j(x,t) H_{P_j},$$

and note that

1. $I \leq H(x,t) \leq \text{const}\, I$;
2. $\phi_j(x,t)H_{P_j}A_2(x,t) + A_2^*(x,t)\phi_j(x,t)H_{P_j} \geq \frac{\delta}{2}\phi_j(x,t)H_{P_j}$.

The first property ensures that $\| \ \|_{H(t)}$ is equivalent to the L_2-norm; the second property implies the estimate (3.2.11) if we sum over j.

To illustrate how the basic energy estimate can be obtained under the condition (3.2.10), we consider an equation

$$u_t = Au_{xx}.$$

We have

$$\begin{aligned}
\frac{d}{dt}(u,u)_{H(t)} &= \frac{d}{dt}\int_0^1 \langle u(x,t), H(x,t)u(x,t)\rangle\, dx \\
&= (u_t, Hu) + (u, Hu_t) + (u, H_t u) \\
&= (Au_{xx}, Hu) + (u, HAu_{xx}) + (u, H_t u) \\
&\leq -(u_x, (HA + A^*H)u_x) + \text{const}\{\|u\|_{H(t)}\|u_x\|_{H(t)} + \|u\|^2_{H(t)}\} \\
&\leq -\frac{\delta}{2}\|u_x\|^2_{H(t)} + \text{const}\{\|u\|_{H(t)}\|u_x\|_{H(t)} + \|u\|^2_{H(t)}\}.
\end{aligned}$$

Thus the proof of the basic energy estimate can be finished in the same way as above. The other estimates follow similarly. We summarize the result in

Theorem 3.2.10. *The spatially periodic Cauchy problem* (3.1.1), (3.1.2) *is well-posed if the system is pointwise parabolic.*

3.2.6. Smoothing Properties of Parabolic Systems

For simplicity we neglect lower-order terms and consider a strongly parabolic system

(3.2.12) $$u_t = Au_{xx}, \quad A + A^* \geq 2\delta I, \quad A \in C^\infty.$$

If the initial data $u(\cdot, 0) = f$ are given in L_2, there is a unique *generalized* solution u since the problem is well-posed. We want to show in this section that $u \in C^\infty$ for $t > 0$ even if the initial data are just in L_2 and have no smoothness properties. To show this, we first derive sharper *a priori* estimates for the solutions to initial data in C^∞. Recall that

$$\frac{d}{dt}\|u\|^2 \leq -\delta\|u_x\|^2 + \text{const}\|u\|^2.$$

This gives us the usual bound for $\|u(\cdot, t)\|$, which depends only on $\|u(\cdot, 0)\|$. The function $v = u_x$ satisfies

$$v_t = (Av_x)_x,$$

and therefore

$$\frac{d}{dt}(tv, v) = (v + tv_t, v) + (tv, v_t) = (v, v) + t(v_t, v) + t(v, v_t)$$

$$= (v, v) - t(v_x, (A + A^*)v_x) \leq (v, v) - 2\delta t\|v_x\|^2$$

$$= \|u_x\|^2 - 2\delta t\|v_x\|^2.$$

Thus

$$\delta^{-1}\frac{d}{dt}\|u\|^2 + \frac{d}{dt}(tv, v) \leq -2\delta t\|v_x\|^2 + \text{const}\|u\|^2 \leq \text{const}\|u\|^2.$$

Integration from 0 to t yields

$$t\|v(\cdot, t)\|^2 \leq \text{const}\,\{\int_0^t \|u(\cdot, \tau)\|^2 d\tau + \|u(\cdot, 0)\|^2\}.$$

Hence we have a bound for $\|u_x(\cdot, t)\|$ which depends only on $\|u(\cdot, 0)\|$ and not on $\|u_x(\cdot, 0)\|$. This process can be continued. We obtain bounds for all expressions $t^p\|\partial^p u/\partial x^p\|^2$ which do not depend on derivatives of the initial data, but only on the L_2-norm of $u(\cdot, 0) = f$.

Now consider the case where the initial data $u(\cdot, 0) \in L_2$ are not necessarily smooth. We construct the corresponding generalized solution as the limit of regular solutions $u^{(j)}(x, t)$ with initial data

$$u^{(j)}(\cdot, 0) \in C^\infty, \quad \lim_{j\to\infty} \|u^{(j)}(\cdot, 0) - u(\cdot, 0)\| = 0.$$

For $0 < \tau \leq t \leq T$ we find that

$$\|u^{(j)}(\cdot, t)\|_{H^P} \leq C,$$

with a constant C independent of j. Similar estimates also hold for all time derivatives of $u^{(j)}$. Using the Sobolev inequality formulated in Lemma 3.2.5, one finds that

$$\left|\frac{\partial^{p+q}}{\partial x^p \partial t^q} u^{(j)}(x, t)\right| \leq C(p, q, \tau, T) \quad \text{for} \quad 0 < \tau \leq t \leq T.$$

This implies smoothness of the limit function u, and we have proven

Theorem 3.2.11. *Consider* (3.2.12) *with initial data in L_2. The corresponding generalized solution is a function belonging to C^∞ for $t > 0$.*

The same result holds for general parabolic systems.

3.3. Hyperbolic Systems: Existence and Properties of Solutions

In this section we consider linear strongly hyperbolic systems in one space dimension and first show the existence of a unique solution to the spatially periodic Cauchy problem. The principle for obtaining existence is as follows: The addition of a small second-order term to the equations leads to a parabolic system for which the existence of a solution has been shown. We estimate the solutions and their derivatives independently of the coefficient of the second-order term. Thus we can send the coefficient to zero and obtain a solution of the hyperbolic problem. Furthermore, we show some important properties of hyperbolic systems: There is no smoothing, and information travels with a finite speed.

3.3.1. Symmetric Hyperbolic and Strongly Hyperbolic Systems; General Existence Theorems

We consider the Cauchy problem for a hyperbolic system

$$u_t = Au_x + Bu + F, \quad u(x,0) = f(x), \tag{3.3.1}$$

where the coefficients $A = A(x,t)$, $B = B(x,t)$, the forcing $F = F(x,t)$, and the initial function $f = f(x)$ are C^∞-smooth and 1-periodic in x. Also, we assume first that $A = A^*$ is Hermitian and call the system *symmetric hyperbolic*. The assumption of symmetry will be relaxed later. We want to prove that (3.3.1) has a unique smooth 1-periodic solution. To this end, let $\epsilon > 0$ be a small constant destined to vanish, and consider instead of (3.3.1) the parabolic system

$$v_t = \epsilon v_{xx} + Av_x + Bv + F, \quad v(x,0) = f(x), \quad \epsilon > 0. \tag{3.3.2}$$

According to Theorem 3.2.7, its solution $v = v^\epsilon$ exists, and we will prove estimates of v^ϵ and its derivatives independently of ϵ. More precisely,

Lemma 3.3.1. *For every finite time interval* $0 \le t \le T$ *there are constants* K_p, $p = 0, 1, 2, ...,$ *with*

$$\|v(\cdot,t)\|^2_{H^p} \le K_p\Big\{\|f\|_{H^p} + \int_0^t \|F(\cdot,\tau)\|^2_{H^p}\,d\tau\Big\}.$$

Here the constant K_0 *depends only on* T *and on bounds for* A, A_x, B*; for* $p \ge 1$, *the constant* K_p *depends only on* T *and on bounds for* A, B *and their derivatives of order* $\le p$.

Proof. We have

$$\frac{d}{dt}\|v\|^2 \leq -2\epsilon\|v_x\|^2 + (v, Av_x) + (Av_x, v) + 2|B|_\infty\|v\|^2 + 2\|v\|\,\|F\|.$$

Integration by parts gives us

$$(v, Av_x) = (Av, v_x) = -((Av)_x, v) = -(Av_x, v) - (A_x v, v),$$

and therefore

$$\frac{d}{dt}\|v\|^2 \leq (2|B|_\infty + |A_x|_\infty)\|v\|^2 + 2\|v\|\,\|F\|.$$

This shows the desired estimate for $\|v\|^2$.

The function $w = v_x$ satisfies

$$w_t = \epsilon w_{xx} + Aw_x + (A_x + B)w + B_x v + F_x. \tag{3.3.3}$$

We have already a bound for v, and therefore we can consider $B_x v + F_x = \tilde{F}$ as a new forcing function. Thus (3.3.3) is of the same form as (3.3.2), and we obtain the corresponding estimate. Repeating this process, we obtain the estimates for the higher derivatives, and the lemma is proved.

Bounds for time derivatives and mixed derivatives of $v = v^\epsilon$ can be shown as previously by using the differential equation.

We now prove that (3.3.1) has a unique solution and start with uniqueness. If u and v are two smooth solutions then their difference $w = u - v$ satisfies the homogeneous system (3.3.1). By the estimate of the above lemma (with $\epsilon = 0$) we have

$$\frac{d}{dt}\|w\|^2 \leq \text{ const }\|w\|^2, \quad w(x,0) = 0;$$

hence $w \equiv 0$, and uniqueness is shown. To prove existence, we consider a sequence of solutions v^ϵ for $\epsilon \to 0$. The functions v^ϵ are uniformly smooth, and by the general argument given in Appendix 4 we can select a subsequence which converges along with all its derivatives. The limit is clearly a solution of (3.3.1), which also satisfies the estimates of Lemma 3.3.1. To summarize,

Theorem 3.3.2. *The symmetric hyperbolic system* (3.3.1) *has a unique solution $u = u(x,t)$. The solution is C^∞-smooth and satisfies the estimates of Lemma* 3.3.1. *The spatially periodic Cauchy problem for a symmetric hyperbolic system is well-posed.*

We shall generalize the result to systems (3.3.1) for which the matrix A is not necessarily Hermitian. For equations with constant coefficients, we defined strong hyperbolicity by the condition that the eigenvalues of A are real and that there is a complete set of eigenvectors. According to Lemma 2.4.2, a matrix A has this property if and only if there is a positive definite Hermitian matrix H with

$$HA = A^* H. \tag{3.3.4}$$

For a variable-coefficient equation we can "almost" use this condition pointwise, and obtain a well-posed problem. More precisely, we make the

Definition. The variable-coefficient system (3.3.1) is called strongly hyperbolic if there exists a smooth 1-periodic "symmetrizer", i.e., a smooth 1-periodic positive definite matrix function $H(x,t)$ such that (3.3.4) is valid for all x, t.

For a strongly hyperbolic system we can proceed in the same way as for a symmetric one, if we replace the L_2-inner product by

$$(u,v)_{H(t)} = \int_0^1 \langle u(x), H(x,t)v(x)\rangle\, dx. \tag{3.3.5}$$

The estimates of the solution and its derivatives follow; see also Section 3.2.5. One obtains

Theorem 3.3.3. *The results of Theorem* 3.3.2 *hold if the system is strongly hyperbolic, i.e., if we can find a smooth* 1*-periodic symmetrizer.*

Note that if the problem is only "pointwise" strongly hyperbolic then a matrix $H(x,t)$ with (3.3.4) exists at each point (x,t) according to Lemma 2.4.2. The only extra assumption we have made is that $H(x,t)$ depends *smoothly* on (x,t). The existence of such a smooth symmetrizer is guaranteed if, for example, for all (x,t) the $n \times n$ matrix $A(x,t)$ has n real *distinct* eigenvalues; in this case the system (3.3.1) is called *strictly hyperbolic*. Therefore, as in the case of constant coefficients, strict hyperbolicity implies strong hyperbolicity.

3.3.2. Properties of Scalar Equations

The simplest case. Let a be a real constant. The simplest hyperbolic differential equation is given by the scalar equation

$$u_t + au_x = 0.$$

Its solution for 1-periodic smooth initial data

$$u(x,0) = f(x) \tag{3.3.6}$$

can be written down explicitly:

$$u(x,t) = f(x - at).$$

Thus the solution is constant along the so-called characteristic lines $x = x_0 + at$. (The important concept of a characteristic is defined below for more general equations.)

This shows that there is no smoothing effect; the solution is as smooth as the initial data. For generalized solutions, whose initial data have discontinuities, this fact is particularly important. Consider, for example, the periodic function

$$f(x) = \begin{cases} 1, & i \le x < i + \frac{1}{2}, \\ 0, & i + \frac{1}{2} \le x < i + 1, \end{cases} \qquad i \in \mathbf{Z}, \tag{3.3.7}$$

which jumps at all integers and half-integers. We approximate f by functions $f_\epsilon \in C^\infty$ by "rounding the corners." For every fixed $\epsilon > 0$ our problem is solved by $u_\epsilon(x,t) = f_\epsilon(x - at)$. Therefore the generalized solution is $u(x,t) = f(x - at)$; in particular, the discontinuities move along the characteristics through i and $i + \frac{1}{2}$, $i \in \mathbf{Z}$.

The case $a = a(x,t)$. Suppose $a = a(x,t)$ is a smooth function, 1-periodic in x. A parametrized line

$$\big(x(t), t\big), \quad 0 \le t < \infty,$$

is called a *characteristic* for the equation

$$u_t + a(x,t)u_x = 0 \tag{3.3.8}$$

if

$$\frac{dx}{dt}(t) = a\big(x(t), t\big), \quad t \ge 0. \tag{3.3.9}$$

Since $a(x,t)$ is bounded in every finite time interval, it is clear that a unique characteristic exists through any given point $(\overline{x}, \overline{t})$. If u solves (3.3.8) and $\big(x(t), t\big)$ is a characteristic, then

$$\begin{aligned} \frac{d}{dt} u\big(x(t), t\big) &= u_t + u_x \frac{dx}{dt} \\ &= u_t + u_x\, a(x,t) = 0. \end{aligned}$$

Thus the solution u of

$$u_t + a(x,t)u_x = 0, \quad u(x,0) = f(x)$$

carries the value $f(x_0)$ along the characteristic $\big(x(t), t\big)$ starting at $\big(x(0), 0\big) = (x_0, 0)$. If the initial function $f(x)$ has a jump discontinuity at $x = x_0$, then the generalized solution $u(x, t)$ will jump when crossing the characteristic

$$\big(x(t), t\big) \quad \text{with} \quad x(0) = x_0.$$

In other words, discontinuities move along the characteristics.

The inhomogeneous equation. We can also solve inhomogeneous problems

$$u_t + a(x,t)u_x = F(x,t), \quad u(x,0) = f(x). \tag{3.3.10}$$

Characteristics are defined as for the case $F \equiv 0$, i.e., by condition (3.3.9). If $\big(x(t), t\big)$ is a characteristic and $u(x, t)$ solves (3.3.10), then

$$\frac{d}{dt}\, u\big(x(t), t\big) = F\big(x(t), t\big),$$

and therefore

$$u\big(x(t), t\big) = f(x_0) + \int_0^t F\big(x(\tau), \tau\big)\, d\tau, \;\; x(0) = x_0. \tag{3.3.11}$$

Suppose F is only piecewise smooth with jump discontinuities along lines which are nowhere tangent to a characteristic. In this case the integral in (3.3.11) mollifies the jumps of F. The generalized solution u is continuous if $f(x)$ is continuous. If f has jumps, these travel along the characteristics; the discontinuities of F do not introduce new jumps in the function u, but – in general – in the derivatives of u.

The general linear case. For equations

$$u_t + a(x,t)u_x = b(x,t)u + F(x,t)$$

the characteristics are again defined by condition (3.3.9). If u solves the above equation, then

$$\frac{d}{dt}\, u\big(x(t), t\big) = b\big(x(t), t\big)\, u\big(x(t), t\big) + F\big(x(t), t\big)$$

along each characteristic. Thus we can obtain $u\big(x(t), t\big)$ by solving a linear ordinary differential equation.

3.3.3. Properties of Hyperbolic Systems

The case A = const, $B = 0$. Consider a strongly hyperbolic system

$$u_t + Au_x = F(x,t) \tag{3.3.12}$$

with a constant coefficient matrix A. There is a nonsingular transformation S such that

$$S^{-1}AS = \Lambda = \text{diag}\,(\lambda_1, \ldots, \lambda_n), \quad \lambda_j \text{ real.}$$

If we introduce the new variables $v = S^{-1}u$, then the equation (3.3.12) transforms to

$$v_t + \Lambda v_x = H(x,t), \quad H = S^{-1}F.$$

Thus we obtain n uncoupled scalar equations

$$v_{jt} + \lambda_j v_{jx} = H_j(x,t), \quad j = 1, \ldots, n,$$

and the considerations of the previous section apply.

The wave equation. An important example is obtained from the wave equation

$$y_{tt} = y_{xx}$$

with 1-periodic initial data

$$y(x,0) = f(x), \quad y_t(x,0) = g(x). \tag{3.3.13}$$

We introduce a new variable v by

$$v_t = y_x.$$

The wave equation then gives us

$$y_{tt} = y_{xx} = v_{tx} = v_{xt},$$

hence integration in t yields

$$y_t = v_x + h(x).$$

For v we have the initial condition

$$v_x(x,0) = g(x) - h(x),$$

and if we choose $h(x) \equiv h_0 = \int_0^1 g(\xi)\,d\xi$, then

$$v(x,0) = \int_0^x \big[g(\xi) - h_0\big]\,d\xi$$

is a 1-periodic initial condition. To summarize, if y solves the wave equation with initial conditions (3.3.13), then

$$u = \begin{pmatrix} y \\ v \end{pmatrix}, \quad v(x,t) = \int_0^t y_x(x,\tau)\,d\tau + \int_0^x \big[g(\xi) - h_0\big]\,d\xi$$

solves the hyperbolic system

$$u_t = \begin{pmatrix} 0 & 1 \\ 1 & 0 \end{pmatrix} u_x + \begin{pmatrix} h_0 \\ 0 \end{pmatrix} =: Au_x + \begin{pmatrix} h_0 \\ 0 \end{pmatrix},$$

$$u(x,0) = \begin{pmatrix} f(x) \\ \int_0^x \left[g(\xi) - h_0\right] d\xi \end{pmatrix}.$$

Conversely, if u solves the above system, then the first component $y = u_1$ solves the original problem. The eigenvalues of A are $\lambda_1 = +1, \quad \lambda_2 = -1$. If

$$S^{-1}AS = \text{diag}\,(\lambda_1, \lambda_2) = \Lambda,$$

then the transformation $w = S^{-1}u$ leads to

$$w_t = \Lambda w_x + S^{-1}\begin{pmatrix} h_0 \\ 0 \end{pmatrix},$$

and we can obtain a solution in closed form. As a result, one finds that

$$y(x,t) = \frac{1}{2}\{f(x+t) + f(x-t)\} + \frac{1}{2}\int_{x-t}^{x+t} g(\xi)\, d\xi.$$

Strongly hyperbolic systems. We will explain here how to obtain the solution of a general strongly hyperbolic system by solving a sequence of scalar equations and going to the limit. The scalar equations themselves can be solved by the method of characteristics, as discussed above. In this way one can study in detail how discontinuities propagate and one can prove that information travels at a finite speed.

Consider a general linear system

$$u_t + A(x,t)u_x = B(x,t)u + F(x,t), \quad u(x,0) = f(x),$$

and assume that there is a smooth transformation $S = S(x,t)$ such that

$$S^{-1}AS = \Lambda(x,t) = \text{ diag }\big(\lambda_1(x,t), \ldots, \lambda_n(x,t)\big), \ \lambda_j(x,t) \text{ real.}$$

Introducing new variables $v = S^{-1}u$, we obtain

$$S_t v + S v_t + AS_x v + AS v_x = BSv + F,$$

and therefore

$$\text{(3.3.14)} \qquad \begin{aligned} &v_t + \Lambda v_x = Cv + H, && v(\cdot, 0) = g = S^{-1}f, \\ &C = S^{-1}BS - S^{-1}S_t - S^{-1}AS_x, && H = S^{-1}F. \end{aligned}$$

Solution by iteration. System (3.3.14) suggests the iteration

$$v_t^{k+1} + \Lambda v_x^{k+1} = G^k, \qquad G^k = Cv^k + H,$$
$$v^{k+1}(x,0) = g(x), \qquad k = 0,1,2,\ldots, \tag{3.3.15}$$

starting with $v^0(x,t) \equiv 0$. For every $k = 0,1,2,\ldots$ the initial value problem (3.3.15) represents n scalar equations which one can solve by the method of characteristics. Accordingly, one makes the

Definition. The lines

$$\big(x_j(t),t\big), \quad t \geq 0, \quad j = 1,\ldots,n,$$

with

$$\frac{dx_j(t)}{dt} = \lambda_j\big(x_j(t),t\big), \quad t \geq 0,$$

are called characteristics for the system (3.3.14) (and also for the given system written in u-variables).

Through any point $(\overline{x},\overline{t})$ there are n characteristics $\big(x_j(t),t\big)$ with

$$x_j(\overline{t}) = \overline{x}, \quad j = 1,\ldots,n.$$

These are distinct if the system is strictly hyperbolic, i.e.,

$$\lambda_1(x,t) < \lambda_2(x,t) < \ldots < \lambda_n(x,t), \tag{3.3.16}$$

but otherwise two characteristics can be identical.

To study the iteration (3.3.15), we fix a point $(\overline{x},\overline{t})$. The characteristics through this point intersect the x-axis in $x_1(0),\ldots,x_n(0)$.

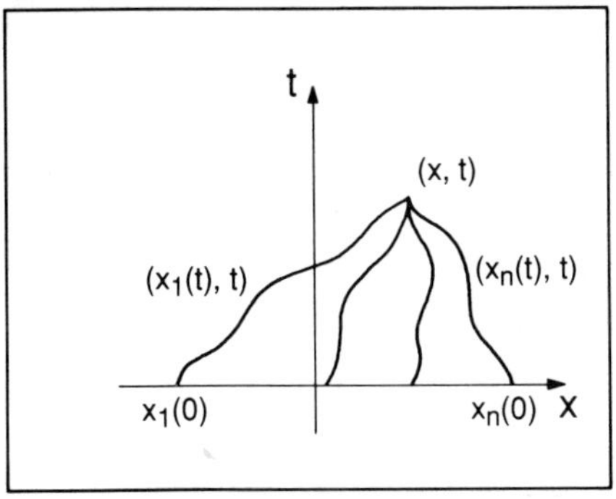

FIGURE 3.3.1. Characteristics through a given point $(\overline{x},\overline{t})$.

Note that the characteristics do not depend on the iteration index k. We obtain from (3.3.15)

$$\frac{d}{dt} v_j^{k+1}\big(x_j(t), t\big) = G_j^k\big(x_j(t), t\big),$$

$$G^k(x,t) = C(x,t)v^k(x,t) + H(x,t).$$

Integration yields

$$v_j^{k+1}(\overline{x}, \overline{t}) = \int_0^{\overline{t}} G_j^k\big(x_j(\tau), \tau\big)\, d\tau + g\big(x_j(0)\big), \quad j = 1, \ldots, n. \tag{3.3.17}$$

Let us show that the sequence v^k converges to the solution v. We will use the following auxiliary result, which we refer to as **Picard's Lemma**.

Lemma 3.3.4. *Let $\eta^k(t)$, $k = 0, 1, \ldots$, denote a sequence of nonnegative continuous functions which satisfy the inequalities*

$$\eta^{k+1}(t) \le a + b \int_0^t \eta^k(\tau)\, d\tau, \quad 0 \le t \le T,$$

with nonnegative constants a, b. Then

$$\eta^k(t) \le a \sum_{\nu=0}^{k-1} \frac{b^\nu t^\nu}{\nu!} + \frac{b^k t^k}{k!} \max_{0 \le \tau \le t} \eta^0(\tau)$$

for $0 \le t \le T$ and $k = 0, 1, \ldots$ In particular, the sequence $\eta^k(t)$, $0 \le t \le T$, is uniformly bounded. If $a = 0$, then the sequence converges uniformly to zero.

Proof. For $k = 0$ the estimate is true. Assume it is valid up to the index k. Then

$$\eta^{k+1}(t) \le a + a \sum_{\nu=1}^{k} \frac{b^\nu t^\nu}{\nu!} + \frac{b^{k+1} t^{k+1}}{(k+1)!} \max_{0 \le \tau \le t} \eta^0(\tau),$$

and the lemma is proved.

We use this to show the following convergence result.

Lemma 3.3.5. *Suppose v solves* (3.3.14)*, and the sequence v^k is defined by* (3.3.15)*. Then*

$$\max\left\{|v(x,t) - v^k(x,t)| : x \in \mathbf{R},\ 0 \le t \le T\right\} \to 0 \text{ as } k \to \infty,$$

for any fixed T. Here it is assumed that all data are smooth and 1-periodic in x.

Proof. The solution v satisfies

$$\frac{d}{dt} v_j\big(x_j(t), t\big) = G_j\big(x_j(t), t\big),$$

$$G(x,t) = C(x,t)v(x,t) + H(x,t).$$

Therefore we obtain, for the difference $w^k = v^k - v$,

$$\frac{d}{dt}\, w_j^{k+1}\big(x_j(t), t\big) = \big\{C\big(x_j(t), t\big) w^k\big(x_j(t), t\big)\big\}_j\,;$$

thus

$$w_j^{k+1}(\overline{x}, \overline{t}) = \int_0^{\overline{t}} \big\{C\big(x_j(\tau), \tau\big) w^k\big(x_j(\tau), \tau\big)\big\}_j\, d\tau, \ j = 1, \ldots, n.$$

If we set

$$\eta^k(t) = \max\big\{|w^k(x,\tau)| : x \in \mathbf{R}, \ \ 0 \le \tau \le t\big\},$$

then, for $0 \le \overline{t} \le T$,

$$|w^{k+1}(\overline{x}, \overline{t})| \le C_T \int_0^{\overline{t}} \eta^k(t)\, dt,$$

$$\eta^{k+1}(\overline{t}) \le C_T \int_0^{\overline{t}} \eta^k(t)\, dt.$$

An application of Picard's Lemma (with $a = 0$) yields

$$\eta^k(t) \le \ \text{const}\ \frac{1}{k!}\ (C_T t)^k.$$

This proves the convergence.

Propagation of discontinuities. We can use (3.3.17) to deduce properties for the sequence $v^k = v^k(x,t)$ and then go to the limit as $k \to \infty$. Consider, for example, the discontinuous initial data

$$v_1(x,0) = f(x), \quad v_2(x,0) = \ldots = v_n(x,0) = 0,$$

where f is given in (3.3.7). For simplicity assume $F \equiv H \equiv 0$ and assume the system is strictly hyperbolic, i.e., (3.3.16) holds. The function $v_1^1(x,t)$ represents the pulses whose edges are moving along the characteristics $\big(x_1(t), t\big)$ starting at $\big(x_1(0), 0\big) = (i, 0)$, $\big(x_1(0), 0\big) = \big(i + \frac{1}{2}, 0\big)$, $i \in \mathbf{Z}$. All other components $v_j^1(x,t) \equiv 0, \ \ j = 2, \ldots, n$.

By our considerations for scalar equations, it follows that $v_1^2(x,t)$ has jumps when crossing the characteristics $\big(x_1(t),t\big)$ mentioned above, whereas $v_j^2(x,t)$, $j=2,\ldots,n$, is continuous. The reason for the continuity of the latter functions is that the characteristics $\big(x_j(t),t\big)$, $j = 2,...,n$, are nowhere tangent to the discontinuity lines $\big(x_1(t),t\big)$ of the forcing. This is the qualitative behavior for all iterates v^k, and therefore of the solution itself. There are no difficulties in discussing the behavior for more general piecewise smooth initial data because we can split these up into a smooth function plus simple jumps. A discontinuity of the initial data at $x = x_0$ propagates along the characteristics which start at $(x,t) = (x_0, 0)$.

The finite speed of propagation. Another important property of hyperbolic differential equations is the so-called "finite speed of propagation" which we explain in the next theorem. Consider again the system (3.3.14) under assumption (3.3.16).

Theorem 3.3.6. *For a given point* $(\overline{x},\overline{t})$ *let* $\big(x_1(t),t\big),\ldots,\big(x_n(t),t\big)$ *denote the characteristics through* $(\overline{x},\overline{t})$*; see Figure* 3.3.1. *Let* Ω *denote the region bounded by the two characteristics*

$$\big(x_1(t),t\big),\quad \big(x_n(t),t\big),\quad 0 \le t \le \overline{t},$$

and the piece of the initial line

$$(x,0),\quad x_1(0) \le x \le x_n(0).$$

Then the restriction of the solution $v(x,t)$ *to the region* Ω *only depends on the restriction of* H *to* Ω *and of* g *to the interval* $x_1(0) \le x \le x_n(0)$. *In other words, we can change* H *and* g *outside these regions without affecting* v *inside* Ω.

Proof. Consider any point $(\tilde{x},\tilde{t}) \in \Omega$. Then the characteristics

$$\big(\tilde{x}_j(t),t\big)\ \text{ for }\ 0 \le t \le \tilde{t},\quad j = 1,\ldots,n,$$

through $(\tilde{x},\tilde{t})$ also belong to Ω. Now consider the iteration (3.3.15). It is clear that the restriction of v^1 to Ω depends only on the corresponding restrictions of H and g. By induction we obtain that all functions v^k have this property, and therefore it is valid for the limit also. This proves the theorem.

In Section 6.2 we shall prove a similar theorem for hyperbolic systems in several space dimensions using a difference approximation.

3.4. Mixed Hyperbolic-Parabolic Systems

Consider a system of the following block form

$$w_t = \begin{pmatrix} A_2 & 0 \\ 0 & 0 \end{pmatrix} w_{xx} + \begin{pmatrix} B_{11} & B_{12} \\ B_{21} & B_{22} \end{pmatrix} w_x + Cw + F,$$

$$w = \begin{pmatrix} u \\ v \end{pmatrix}, \qquad w(x,0) = f(x). \tag{3.4.1}$$

It is assumed that all coefficients and data are C^∞-smooth and 1-periodic in x. Also, let

$$A_2 + A_2^* \geq 2\delta I > 0 \text{ in } 0 \leq t \leq T, \quad \delta = \delta(T) > 0, \quad \text{and } B_{22} = B_{22}^*. \tag{3.4.2}$$

If $B_{12} = C_{12} = 0,\ B_{21} = C_{21} = 0$, then we have a strongly parabolic u-equation and an uncoupled symmetric hyperbolic v-equation. Assuming (3.4.2), we want to show that the spatially periodic Cauchy problem for the mixed system (3.4.1) with coupling terms is well-posed. First consider the parabolic systems

$$w_t = \begin{pmatrix} A_2 & 0 \\ 0 & \epsilon I \end{pmatrix} w_{xx} + \begin{pmatrix} B_{11} & B_{12} \\ B_{21} & B_{22} \end{pmatrix} w_x + Cw + F, \quad \epsilon > 0. \tag{3.4.3}$$

According to Theorem 3.2.7, there is a unique solution $w = w^\epsilon$ of the Cauchy problem for $\epsilon > 0$. We shall show that the solutions w^ϵ and their derivatives can be estimated independently of $\epsilon > 0$; we start with the basic energy estimate.

Lemma 3.4.1. *For any $T > 0$ there is a constant $K(T)$, independent of $\epsilon > 0$, with*

$$\|w^\epsilon(\cdot,t)\|^2 \leq K(T)\left\{\|f\|^2 + \int_0^t \|F(\cdot,\tau)\|^2\, d\tau\right\}, \quad 0 \leq t \leq T.$$

Proof. We write $w = w^\epsilon$. The norms $|\ |_\infty$ below refer to the interval $0 \leq t \leq T$. We have

$$\begin{aligned}\frac{d}{dt}(w,w) &= (w,w_t) + (w_t,w) \\ &= (u,u_t) + (v,v_t) + (u_t,u) + (v_t,v) \\ &\leq -2\epsilon\|v_x\|^2 + \mathrm{I} + \mathrm{II} + \mathrm{III} + \mathrm{IV} + 2\{|C|_\infty\ \|w\|^2 + \|w\|\ \|F\|\},\end{aligned}$$

where

$$\mathrm{I} = (u, A_2 u_{xx}) + (A_2 u_{xx}, u) + (u, B_{11} u_x) + (B_{11} u_x, u),$$

$$\mathrm{II} = (v, B_{22} v_x) + (B_{22} v_x, v),$$

$$\text{III} = (u, B_{12}v_x) + (B_{12}v_x, u),$$

$$\text{IV} = (v, B_{21}u_x) + (B_{21}u_x, v).$$

The term I can be estimated as explained in Section 3.1.1 where we treated parabolic systems:

$$\begin{aligned} \text{I} &\leq -2\delta\|u_x\|^2 + 2\{|A_{2x}|_\infty + |B_{11}|_\infty\}\ \|u\|\ \|u_x\| \\ &\leq -\delta\|u_x\|^2 + c_1\|u\|^2. \end{aligned}$$

The term II can be treated as explained in Section 3.3.1 for the symmetric hyperbolic case:

$$\text{II} \leq |B_{22x}|\ \|v\|^2.$$

Integration by parts gives us for III:

$$\begin{aligned} \text{III} &= -(u_x, B_{12}v) - (u, B_{12x}v) - (B_{12}v, u_x) - (B_{12x}v, u) \\ &\leq 2|B_{12}|_\infty\ \|v\|\ \|u_x\| + 2|B_{12x}|_\infty\ \|u\|\ \|v\| \\ &\leq \frac{\delta}{2}\ \|u_x\|^2 + c_2\{\|u\|^2 + \|v\|^2\}. \end{aligned}$$

Finally,

$$\text{IV} \leq \frac{\delta}{2}\ \|u_x\|^2 + c_3\|v\|^2.$$

Summarizing, we have shown that

$$\frac{d}{dt}\ \|w\|^2 \leq c_4\{\|w\|^2 + \|F\|^2\}$$

where c_4 does not depend on ϵ. This proves the lemma.

It should be remarked that the estimate of Lemma 3.4.1 is also valid for $\epsilon = 0$ if we assume the existence of a smooth solution for this case. Thus *uniqueness* for (3.4.1) is shown. To obtain existence, we estimate all derivatives of w^ϵ independently of ϵ. Differentiation of the system (3.4.3) with respect to x yields an equation for w_x^ϵ with the same structure. Observing that we already have estimates for $\|w^\epsilon\|$, we can bound $\|w_x^\epsilon\|$ in the same way as above. Estimates for higher space derivatives are obtained by repeated differentiation. This proves

Lemma 3.4.2. *For any $T > 0$ and any $p = 1, 2, \ldots$, there is a constant $K(p, T)$, independent of $\epsilon > 0$, with*

$$\|w^\epsilon(\cdot, t)\|_{H^p}^2 \leq K(p, T)\ \Big\{\|f\|_{H^p}^2 + \int_0^t \|F(\cdot, \tau)\|_{H^p}^2\, d\tau\Big\}. \tag{3.4.4}$$

Here $K(p, T)$ depends only on the maximum norm of the coefficients and their derivatives of order $\leq p$.

Time derivatives and mixed derivatives of w^ϵ can always be expressed by space derivatives using the differential equation. Thus we have estimates for all derivatives. Now consider a sequence $\epsilon \to 0$. According to Appendix 4, we can select a uniformly smooth subsequence of w^ϵ which converges along with all its derivatives. The limit solves (3.4.1) and obeys the estimate (3.4.4). This proves

Theorem 3.4.3. *The spatially periodic Cauchy problem for the mixed system* (3.4.1) *is well-posed if* (3.4.2) *is true.*

Thus far, the underlying uncoupled equations were assumed to be strongly parabolic or symmetric hyperbolic. We can generalize our result to the case where instead of (3.4.2) it is assumed that there are smooth symmetrizers:

$$H_1 A_2 + A_2^* H_1 \geq \delta H_1 > 0, \quad H_2 B_{22} = B_{22}^* H_2,$$
$$H_j = H_j^* > 0 \text{ smooth, 1-periodic in } x, \quad j = 1, 2.$$

To obtain the estimates for the more general case, we use the scalar product $(w, \tilde{w})_H = (u, H_1 \tilde{u}) + (v, H_2 \tilde{v})$ instead of the L_2-scalar product $(w, \tilde{w}) = (u, \tilde{u}) + (v, \tilde{v})$.

3.5. The Linearized Navier-Stokes Equations in One Space Dimension

Suppose the variables u, v, w, ρ, p do not depend on y, z. Then the Navier-Stokes equations read

$$\rho \begin{pmatrix} u \\ v \\ w \end{pmatrix}_t + \rho \begin{pmatrix} uu_x \\ uv_x \\ uw_x \end{pmatrix} + \begin{pmatrix} p_x \\ 0 \\ 0 \end{pmatrix}$$
$$= \mu \begin{pmatrix} u \\ v \\ w \end{pmatrix}_{xx} + (\mu + \mu') \begin{pmatrix} u_{xx} \\ 0 \\ 0 \end{pmatrix} + \rho \begin{pmatrix} F_1 \\ F_2 \\ F_3 \end{pmatrix},$$

$$\rho_t + (u\rho)_x = 0, \quad p = r(\rho).$$

This system uncouples into a system for (u, ρ),

$$\begin{aligned} \rho u_t + \rho u u_x + p_x &= (2\mu + \mu')u_{xx} + \rho F_1 \\ \rho_t + \rho u_x + u\rho_x &= 0, \quad p = r(\rho), \end{aligned} \tag{3.5.1}$$

and two scalar equations

$$v_t + u v_x = \frac{\mu}{\rho}\, v_{xx} + F_2,$$

$$w_t + u w_x = \frac{\mu}{\rho}\, w_{xx} + F_3.$$

The equations for v and w are linear if u, ρ are considered as known functions. Clearly, these equations are parabolic in the viscous case ($\mu > 0$), and strongly hyperbolic in the inviscid case ($\mu = 0$). We now linearize the system (3.5.1) at a smooth flow $U = U(x,t), R = R(x,t), P = P(x,t)$ with $P = r(R)$. Substitute $u = U + u'$, $\rho = R + \rho'$, $p = P + p'$ into the equations (3.5.1) and neglect terms quadratic in the corrections. The equation of state gives us

$$p = P + p' = r(R + \rho') \sim r(R) + \kappa\rho', \ \ \kappa = \frac{dr}{d\rho}(R);$$

thus $p' = \kappa\rho'$ to first order. Therefore we can eliminate p' and obtain for u', ρ' a linear system

$$\begin{pmatrix} u' \\ \rho' \end{pmatrix}_t + \begin{pmatrix} U & \kappa/R \\ R & U \end{pmatrix} \begin{pmatrix} u' \\ \rho' \end{pmatrix}_x + C\begin{pmatrix} u' \\ \rho' \end{pmatrix} = (2\mu + \mu') \begin{pmatrix} u'_{xx} \\ 0 \end{pmatrix} + \begin{pmatrix} G_1 \\ G_2 \end{pmatrix}.$$

Here C and G are determined by U, R, P, and F. For $\mu > 0, \mu' \geq 0$, the system is of mixed hyperbolic-parabolic type. In the inviscid case $\mu = \mu' = 0$ we obtain, upon neglecting the zero-order term and the forcing function,

$$q_t + A(x,t)q_x = 0, \quad q = \begin{pmatrix} u' \\ \rho' \end{pmatrix}, \quad A = \begin{pmatrix} U & \kappa/R \\ R & U \end{pmatrix}. \tag{3.5.2}$$

Now suppose that

$$\kappa = \kappa(x,t) = \frac{dr}{d\rho}\,\big(R(x,t)\big) =: a^2(x,t) > 0. \tag{3.5.3}$$

Then the eigenvalues $\lambda_{1,2}$ of A are real and distinct:

$$\lambda_{1,2}(x,t) = U(x,t) \pm a(x,t).$$

The system (3.5.2) is strictly hyperbolic and has the symmetrizer

$$H(x,t) = \begin{pmatrix} 1 & 0 \\ 0 & h \end{pmatrix}, \quad h = \kappa/R^2.$$

This symmetrization simply corresponds to the introduction of a scaled density in (3.5.2); if

$$\tilde{u} = u', \quad \tilde{\rho} = \sqrt{h}\,\rho' = \frac{\sqrt{\kappa}}{R}\,\rho', \text{ i.e.,} \quad \tilde{q} = H^{1/2}q,$$

then the system for $\tilde{q}$ becomes

$$\tilde{q}_t + H^{1/2}\,A\,H^{-1/2}\tilde{q}_x + \{H^{1/2}H_t^{-1/2} + H^{1/2}A\,H_x^{-1/2}\}\tilde{q} = 0.$$

One easily confirms directly that the new system matrix is symmetric,

$$H^{1/2}AH^{-1/2} = \begin{pmatrix} U & \sqrt{\kappa}/R \\ \sqrt{\kappa}/R & U \end{pmatrix}.$$

The positivity assumption (3.5.3), which is crucial for hyperbolicity, is physically reasonable because one expects increasing pressure with increasing density. Furthermore, the interpretation of the eigenvalues $\lambda_{1,2} = U \pm a$ as characteristic speeds (see Section 3.3) makes it plausible that

$$a = \sqrt{\kappa} = \left\{\frac{dr}{d\rho}(R)\right\}^{1/2}$$

is the sound speed corresponding to the base flow with density R. It is the speed of propagation of small disturbances in the base flow.

3.6. The Linearized KdV and the Schrödinger Equations

We have shown the well-posedness of the spatially periodic Cauchy problem for linear variable-coefficient equations in case of a parabolic, a strongly hyperbolic or a mixed system. Basically, integration by parts applied to the differential equation and the differentiated differential equation gave us estimates of any possible solution and its derivatives. For the parabolic case we then used a difference scheme to obtain existence; the strongly hyperbolic and mixed systems were considered as limits of parabolic ones. These techniques can be applied to other equations also, as will be demonstrated here for the linearized KdV and the Schrödinger equations.

3.6.1. The Linearized Korteweg-de Vries Equation

The nonlinear equation

$$w_t + ww_x = \delta w_{xxx}, \quad \delta = \text{ const}, \quad \delta \text{ real},$$

is known as the KdV equation. We linearize about a smooth function $U = U(x,t)$. Upon neglecting quadratic terms in u, the ansatz $w = U + u$ leads to

$$\begin{aligned} u_t &= au_x + bu + \delta u_{xxx} + F, \\ a &= -U, \quad b = -U_x, \quad F = \delta U_{xxx} - UU_x - U. \end{aligned} \tag{3.6.1}$$

We assume that all functions are real, C^∞-smooth, and 1-periodic in x. If u solves (3.6.1) then

$$\begin{aligned} \frac{1}{2}\frac{d}{dt}(u,u) &= (u,u_t) \\ &= (u, au_x) + (u, bu) + \delta(u, u_{xxx}) + (u, F). \end{aligned}$$

Here

$$(u, au_x) = (ua, u_x) = -(u_x a, u) - (ua_x, u),$$

and thus

$$(u, au_x) = -\frac{1}{2}(ua_x, u).$$

Also,

$$(u, u_{xxx}) = -(u_{xxx}, u); \quad \text{thus} \quad (u, u_{xxx}) = 0.$$

If $T > 0$ is fixed and $|\ |_\infty$ refers to the interval $0 \le t \le T$, then

$$\begin{aligned} \frac{1}{2}\frac{d}{dt}(u,u) &\le |b - \frac{1}{2}a_x|_\infty \ \|u\|^2 + \|u\| \ \|F\| \\ &\le \{\frac{1}{2} + |b - \frac{1}{2}a_x|_\infty\} \ \|u\|^2 + \frac{1}{2}\|F\|^2, \quad 0 \le t \le T. \end{aligned}$$

This gives the basic energy estimate

Lemma 3.6.1. *For any $T > 0$ there exists $K(T) > 0$ such that any solution of* (3.6.1) *satisfies the estimate*

$$\|u(\cdot,t)\|^2 \le K(T)\Big\{\|u(\cdot,0)\|^2 + \int_0^t \|F(\cdot,\tau)\|^2\, d\tau\Big\}, \quad 0 \le t \le T.$$

The above lemma implies uniqueness of a solution for given initial data

$$u(x,0) = f(x), \quad f \in C^\infty, \quad f(x) \equiv f(x+1). \tag{3.6.2}$$

To estimate $v = u_x$, we differentiate the equation (3.6.1) and obtain

$$v_t = av_x + (b + a_x)v + \delta v_{xxx} + G, \quad G = F_x + b_x u.$$

Thus we can estimate $v = u_x$ following the same arguments as above. With repeated differentiation one obtains

Lemma 3.6.2. *For any $T > 0$ and any $p = 1, 2, \ldots$ there exists $K(p, T)$ with*

$$\|u(\cdot, t)\|_{H^p}^2 \le K(p, T)\left\{ \|u(\cdot, 0)\|_{H^p}^2 + \int_0^t \|F(\cdot, \tau)\|_{H^p}^2 \, d\tau \right\}, \quad 0 \le t \le T.$$

The constant $K(p, T)$ depends only on bounds for a and b and their derivatives up to order p.

Estimates of time derivatives and mixed derivatives of any solution follow from the differential equation. To obtain existence of a solution of (3.6.1), (3.6.2), we can use the difference scheme

$$\frac{dv_\nu}{dt} = a_\nu \, D_0 v_\nu + b_\nu v + \delta D_0 D_+ D_- v_\nu + F_\nu, \quad v_\nu(0) = f_\nu,$$

and mimic the previous estimates. The Fourier interpolants $w^h(x, t)$ of $v_\nu = v_\nu^h(t)$ w.r.t. x are uniformly smooth. Exactly as in Section 3.2.4, one obtains

$$w^h(x, t) \to u(x, t)$$

for a sequence $h = h_j \to 0$. The limit u solves the Cauchy problem.

Theorem 3.6.3. *The spatially periodic Cauchy problem for the linearized KdV equation is well-posed. The solution satisfies the estimates of Lemma* 3.6.2.

3.6.2. A Schrödinger Equation

Consider the spatially periodic Cauchy problem

(3.6.3) $$u_t = i(pu_x)_x, \quad p \text{ real}, \ p(x, t) \ge p_0 > 0,$$

(3.6.4) $$u(x, 0) = f(x), \quad p(x, t) \equiv p(x + 1, t), \quad f(x) \equiv f(x + 1).$$

The equation is neither parabolic nor hyperbolic. For $\epsilon > 0$, the parabolic problem

(3.6.5) $$u_t = \epsilon u_{xx} + i(pu_x)_x, \quad u(x, 0) = f(x)$$

has a unique solution $u = u^\epsilon$. For $u = u^\epsilon$ it holds that

$$\frac{d}{dt}(u, u) = (u, u_t) + (u_t, u) = -2\epsilon \|u_x\|^2 \le 0,$$

and thus $\|u(\cdot,t)\| \le \|u(\cdot,0)\|$. This estimate is valid for $\epsilon = 0$ also; thus *uniqueness* of a solution of (3.6.3), (3.6.4) follows.

We now show how to bound derivatives of $u = u^\epsilon$. We let $v = u_t$. The differential equation (3.6.5) yields

$$u_{xx} = \frac{1}{\epsilon + ip} v - \frac{ip_x}{\epsilon + ip} u_x,$$

and therefore (see Lemma A.3.2 in Appendix 3)

$$\begin{aligned} \|u_{xx}\| &\le \frac{1}{p_0}\|v\| + \frac{|p_x|_\infty}{p_0}\|u_x\| \\ &\le \frac{1}{p_0}\|v\| + \frac{1}{2}\|u_{xx}\| + C_1\|u\|. \end{aligned}$$

Thus we have the estimates

$$\|u_{xx}\| \le C\{\|v\| + \|u\|\}, \quad \|u_x\| \le C\{\|v\| + \|u\|\}, \quad 0 \le t \le T, \tag{3.6.6}$$

where $C = C(T)$. Differentiation of (3.6.5) w.r.t. t gives us

$$v_t = \epsilon v_{xx} + i(pv_x)_x + i(p_t u_x)_x.$$

By (3.6.6) we find

$$\begin{aligned} \frac{d}{dt}(v,v) &= (v, v_t) + (v_t, v) \\ &\le 2\|v\|\Big\{ |p_t|_\infty \|u_{xx}\| + |p_{tx}|_\infty \|u_x\| \Big\} \\ &\le \text{const}\Big\{ \|v\|^2 + \|u\|^2 \Big\}. \end{aligned}$$

Thus we can bound

$$\|v\| = \|u_t\|, \quad \|u_x\|, \quad \|u_{xx}\|$$

independently of ϵ in any finite time interval $0 \le t \le T$. Higher derivatives can be estimated in the same way by repeated differentiation. The functions u^ϵ are uniformly smooth, and by the same arguments as before, we obtain a solution of the Schrödinger equation for $\epsilon \to 0$.

Notes on Chapter 3

Parallel to the theory of differential equations, the theory of difference approximations was developed; see Osher (1972), Michelson (1983, 1987), Richtmeyer and Morton (1967), Kreiss and Oliger (1973), Kreiss (1968), Gustafsson, Kreiss, and Sundström (1972). It is not new to prove existence of solutions by difference approximations; see, for example, Friedrichs (1954).

Instead of using L_2-estimates, parabolic differential equations are often treated via the construction of fundamental solutions and maximum norm estimates. The classical books are Friedman (1964) and Eidelman (1964). We develop the L_2-theory because we are interested in coupled hyperbolic-parabolic systems. For hyperbolic systems in more than one space dimension, estimates in the maximum norm can, in general, only be obtained under "loss of derivatives"; see Brenner (1966).

A more detailed discussion of the method of characteristics can be found in many books, for example in Petrovskii (1954), Courant and Hilbert (1962), Carrier and Pearson (1976).

4 A Nonlinear Example: Burgers' Equation

The Cauchy problem for the viscous and the inviscid Burgers' equation will be discussed in detail in this chapter. The techniques we apply can mostly be generalized in a straightforward way to more complicated nonlinear parabolic or hyperbolic equations, to systems in one or more space dimensions. These generalizations will be carried out in Chapters 5 and 6. In other words, we will use Burgers' equation as a simple example to illustrate a number of general techniques for the treatment of nonlinear evolutionary systems. Intentionally, we do not apply the celebrated Cole-Hopf transformation, which reduces Burgers' equation to the heat equation. (Though the Cole-Hopf transformation does have interesting generalizations also, these seem to be too limited, at present, to discuss the Navier-Stokes system.) We emphasize in this chapter:

1. local (in time) existence of solutions via a linear iteration;
2. local existence together with *global* (in time) a priori estimates leads to global existence;
3. smoothing properties for the parabolic, i.e., viscous case;
4. breakdown of smooth solutions for the hyperbolic, i.e., inviscid case in finite time.

Concerning these points, our discussion of Burgers' equation is representative for parabolic and hyperbolic systems.

With regard to other aspects, namely *global* a priori estimates, the maximum principle, and our discussion of shocks, the techniques for Burgers' equation do not readily generalize. The difficulties involved will become apparent in Chapters 5 and 6.

4.1. Burgers' Equation: A Priori Estimates and Local Existence

In this section we start our discussion of Burgers' equation

$$\begin{aligned} u_t &= uu_x + \epsilon u_{xx}, \quad \epsilon = \text{ const } > 0, \\ u(x,0) &= f(x), \quad f \in C^\infty, \quad f(x) \equiv f(x+1), \end{aligned} \tag{4.1.1}$$

with periodic boundary conditions. We assume that all functions are real-valued. Burgers' equation is of interest to us, since it shares its mathematical structure

$$u_t = \text{quadratic first order term + diffusion term}$$

with the Navier-Stokes system. After showing the uniqueness of a classical solution, we prove a priori estimates in some time interval $0 \le t \le T$. To obtain these estimates, we intentionally do *not* make use of the maximum principle. Therefore, the proof generalizes to *systems* for which maximum principles are generally not available. Also, we show a priori estimates *which are independent of* ϵ. Thus we can use the same arguments later to discuss the inviscid case ($\epsilon = 0$).

To show *existence* of a solution in some time interval, one can proceed in different ways. A first possibility is to use a difference scheme as in Chapter 3, and to show estimates of its solutions v^h independent of the step-size h. We will write down a suitable difference approximation, but leave the necessary estimates to the reader. These estimates can be obtained along the same lines as the a priori bounds for the solution u. In the same way as in the linear case, one can Fourier interpolate the functions v^h w.r.t. x and obtain uniformly smooth functions $w^h(x,t)$; these converge to a solution $u(x,t)$ as $h \to 0$.

In the text we shall use another approach to show existence, namely the iteration

$$\begin{aligned} u_t^{n+1} &= u^n u_x^{n+1} + \epsilon\, u_{xx}^{n+1}, \\ u^{n+1}(x,0) &= f(x), \qquad n = 0, 1, 2, \dots, \end{aligned} \tag{4.1.2}$$

where $u^0(x,t) \equiv f(x)$. The sequence u^n will be shown to converge, and its limit will solve (4.1.1). Again, we will not apply the maximum principle, and thus the local existence proof generalizes to systems.

4.1.1. Uniqueness

A classical solution of (4.1.1) is a function $u \in C^1(t) \cap C^2(x)$ which satisfies (4.1.1) pointwise. We first show

Lemma 4.1.1. *The* 1*-periodic Cauchy problem* (4.1.1) *has at most one classical solution.*

Proof. Let u and v be solutions. Their difference $w = u - v$ satisfies

$$w_t = \frac{1}{2}(\alpha w)_x + \epsilon w_{xx}, \quad \alpha = u + v, \quad w(x, 0) = 0.$$

From

$$\big(w, (\alpha w)_x\big) = (w, \alpha_x w + \alpha w_x) = (w, \alpha_x w) - \big((\alpha w)_x, w\big)$$

one finds that

$$\big(w, (\alpha w)_x\big) = \frac{1}{2}(w, \alpha_x w), \tag{4.1.3}$$

and therefore

$$\begin{aligned} \frac{1}{2}\frac{d}{dt}(w, w) = (w, w_t) &= \frac{1}{2}\big(w, (\alpha w)_x\big) - \epsilon(w_x, w_x) \\ &= \frac{1}{4}(w, \alpha_x w) - \epsilon(w_x, w_x) \\ &\le \frac{1}{4}|\alpha_x|_\infty (w, w). \end{aligned}$$

The initial condition $w(x, 0) = 0$ implies $w \equiv 0$.

4.1.2. A Priori Estimates

Let u denote a C^∞-solution of (4.1.1) defined for $0 \le t \le T$. First note that

$$\frac{1}{2}\frac{d}{dt}(u, u) = (u, uu_x) - \epsilon\|u_x\|^2.$$

The first term on the right side is zero since

$$(u, uu_x) = (u^2, u_x) = -2(u, uu_x).$$

Thus we find $\frac{d}{dt}(u, u) \le 0$ and obtain the bound

$$\|u(\cdot, t)\| \le \|u(\cdot, 0)\| = \|f\|, \quad 0 \le t \le T. \tag{4.1.4}$$

To show estimates for space derivatives, we use the notation $u_j = \partial^j u/\partial x^j$. The function u_1 satisfies

$$u_{1t} = (uu_x)_1 + \epsilon u_{1xx},$$

and therefore

$$\begin{aligned}\frac{1}{2}\frac{d}{dt}(u_1, u_1) &= \big(u_1, (uu_x)_1\big) - \epsilon\|u_2\|^2 \\ &= \frac{1}{2}(u_1, u_1^2) - \epsilon\|u_2\|^2 \quad \text{(by (4.1.3))} \\ &\le \frac{1}{2}|u_1|_\infty \; \|u_1\|^2 - \epsilon\|u_2\|^2.\end{aligned}$$

From Lemma 3.2.5,

$$|u_1|_\infty^2 \;\le\; \|u_1\|^2 + 2\|u_1\| \; \|u_2\|.$$

Furthermore, the estimate

$$\|u_1\| \le \frac{1}{2\pi}\|u_2\|$$

follows from Fourier expansion. Hence we have

$$|u_1|_\infty \le \|u_2\|. \tag{4.1.5}$$

(For $\epsilon > 0$, we could now derive a bound of $\|u_1(\cdot, t)\|$; the timeinterval where the bound holds would depend on ϵ, however. Therefore we proceed differently and show first an estimate for u_2.) The function u_2 satisfies

$$u_{2t} = (uu_x)_2 + \epsilon u_{2xx},$$

and thus

$$\frac{1}{2}\frac{d}{dt}(u_2, u_2) = \big(u_2, (uu_x)_2\big) - \epsilon\|u_3\|^2 \le (u_2, uu_{2x}) + 3(u_2, u_x u_2).$$

Since

$$\begin{aligned}(u_2, uu_{2x}) &= -\big((u_2 u)_x, u_2\big) \\ &= -(u_{2x}u, u_2) - (u_2 u_x, u_2),\end{aligned}$$

we obtain

$$\begin{aligned}\frac{d}{dt}(u_2, u_2) &\le 5(u_2, u_x u_2) \le 5|u_x|_\infty\|u_2\|^2 \\ &\le 5\;\|u_2\|^3.\end{aligned}$$

Using the abbreviations

$$y(t) = \|u_2(\cdot, t)\|^2, \quad \phi(y) = 5y^{3/2},$$

we have shown the differential inequality

$$y'(t) \le \phi(y(t)), \quad 0 \le t \le T.$$

To estimate $y(t)$, we need a nonlinear extension of Lemma 3.1.1.

Lemma 4.1.2. *Let $\phi \in C^1[0, \infty)$, and let $y(t)$ and $y_0(t)$ denote nonnegative C^1-functions defined for $0 \le t \le T$. If*

$$y'(t) \le \phi(y(t)), \quad y_0'(t) = \phi(y_0(t)), \quad 0 \le t \le T,$$
$$y(0) \le y_0(0),$$

then $y(t) \le y_0(t)$ in $0 \le t \le T$.

Proof. For any two arguments y, y_0 of ϕ we have

$$\begin{aligned}\phi(y) - \phi(y_0) &= \int_{y_0}^{y} \phi'(v)\, dv \\ &= \left\{\int_0^1 \phi'(y_0 + (y - y_0)s)\, ds\right\} (y - y_0).\end{aligned}$$

We set

$$\eta(t) = y(t) - y_0(t), \quad c(t) = \int_0^1 \phi'\Big(y_0(t) + (y(t) - y_0(t))s\Big)\, ds$$

and obtain

$$\eta'(t) \le \phi(y(t)) - \phi(y_0(t)) = c(t)\eta(t).$$

If we introduce

$$z(t) = \exp(-\int_0^t c(\tau)\, d\tau)\, \eta(t),$$

then $z'(t) \le 0$ and $z(0) = \eta(0) \le 0$. Therefore $z(t) \le 0$, and thus $\eta(t) \le 0$. This proves the lemma.

To apply this result, we note that the solution $y_0(t)$ of

$$y_0'(t) = 5y_0^{3/2}(t), \quad y_0(0) = \|f_{xx}\|^2$$

is increasing, and if $y_0(0) > 0$, then there is a time $T_\infty > 0$ with

$$\lim_{t \to T_\infty} y_0(t) = \infty.$$

Therefore, the previous lemma gives us a bound for $\|u_2(\cdot, t)\|$ in any interval $0 \le t \le T < T_\infty$. By (4.1.4) and (4.1.5), $\|u_1(\cdot, t)\|$ and $\|u(\cdot, t)\|$ are also bounded for $0 \le t \le T < T_\infty$. The proof clearly shows that the time T_∞ and the bound do not depend on $\epsilon \ge 0$. The value $\epsilon = 0$ is allowed. We summarize:

Lemma 4.1.3. *Consider* (4.1.1) *with* $\epsilon \ge 0$. *There exists a time* $T > 0$ *and a constant* K_2, *both depending only on* $\|f\|_{H^2}$, *with the following property: If a solution* $u(x, t)$ *is defined for* $0 \le t \le T$ *then*

$$\|u(\cdot, t)\|_{H^2} \le K_2 \quad \text{in} \quad 0 \le t \le T.$$

We proceed by showing estimates for all space derivatives $u_j = \partial^j u/\partial x^j$ *of* u. *Interestingly, the time interval for which the estimates are valid does not depend on* j *nor on* $\epsilon \ge 0$.

Lemma 4.1.4. *Suppose* u *is a* C^∞-*solution of* (4.1.1) *defined for* $0 \le t \le T$, T *being fixed as in the previous lemma. For* $j = 2, 3, \ldots$ *there is a constant* K_j, *depending only on* $\|f\|_{H^j}$, *with*

$$\|u(\cdot, t)\|_{H^j} \le K_j \quad \text{in} \quad 0 \le t \le T.$$

Proof. We use induction on j. The case $j = 2$ has been treated in the previous lemma; thus let $j \ge 3$. The identity

$$u_{jt} = (uu_x)_j + \epsilon u_{jxx}$$

implies that

$$\frac{1}{2}\frac{d}{dt}\big(u_j, u_j\big) = \big(u_j, (uu_x)_j\big) - \epsilon \|u_{j+1}\|^2.$$

Here

$$(u\, u_x)_j = \sum_{\nu=0}^{j} \binom{j}{\nu} u_\nu\, u_{j+1-\nu}.$$

The terms $(u_j,\, u_\nu u_{j+1-\nu})$ can be estimated as follows: For $\nu = 1, \ldots, j-2$,

$$\begin{aligned}(u_j,\, u_\nu u_{j+1-\nu}) &\le |u_\nu|_\infty\, \|u\|_{H^j}^2 \\ &\le \|u\|_{H^{j-1}}\, \|u\|_{H^j}^2.\end{aligned}$$

For $\nu = j-1$ and $\nu = j$,

$$\begin{aligned}(u_j,\, u_\nu u_{j+1-\nu}) &\le |u_{j+1-\nu}|_\infty \, \|u\|_{H^j} \, \|u\|_{H^\nu} \\ &\le \|u\|_{H^{j-1}} \, \|u\|_{H^j}^2 \qquad (\text{ since } j \ge 3).\end{aligned}$$

For $\nu = 0$,

$$\begin{aligned}(u_j,\, u\, u_{j+1}) &= -\left((u_j u)_x, u_j\right) \\ &= -\left(u_{j+1} u, u_j\right) - \left(u_j u_x, u_j\right),\end{aligned}$$

thus

$$\left(u_j, u\, u_{j+1}\right) = -\frac{1}{2}\left(u_j u_x, u_j\right) \le \|u\|_{H^{j-1}} \, \|u\|_{H^j}^2 .$$

If c_j is a suitable numerical constant, we find that

$$\begin{aligned}\frac{d}{dt}\|u_j\|^2 &\le c_j \, \|u\|_{H^{j-1}} \, \|u\|_{H^j}^2 \\ &\le c_j \, \|u\|_{H^{j-1}} \left(\|u_j\|^2 + \|u\|_{H^{j-1}}^2\right) \\ &\le c_j \, K_{j-1} \left(\|u_j\|^2 + K_{j-1}^2\right).\end{aligned}$$

Thus the induction step can be completed, and the lemma is proved.

Since we have bounded all space derivatives of u in $0 \le t \le T$, we can use the differential equation $u_t = uu_x + \epsilon u_{xx}$ to bound all time derivatives and mixed derivatives. Each term

$$\frac{\partial^{p+q}}{\partial x^p \partial t^q} u(x,t)$$

can be written as a sum of products of space derivatives, and hence it can be bounded. The bounds are uniform for $0 \le \epsilon \le \epsilon_0$, $\epsilon_0 > 0$ being fixed.

4.1.3. Existence of Solutions via Difference Approximations

As in Section 3.2, we can discretize in space and approximate (4.1.1) by a system of ordinary differential equations:

$$\begin{aligned}\frac{dv_\nu}{dt} &= \frac{1}{3}\left(v_\nu D_0 v_\nu + D_0 v_\nu^2\right) + \epsilon D_+ D_- v_\nu, \\ v_\nu(0) &= f_\nu.\end{aligned}$$

Here $D_0 = 1/2\left(D_+ + D_-\right)$, and we have approximated $uu_x = 1/3\left(uu_x + (u^2)_x\right)$ by $1/3\left(v_\nu D_0 v_\nu + D_0 v_\nu^2\right)$. This approximation has the technical advantage that

$$(v, vD_0v)_h + (v, D_0v^2)_h = (v^2, D_0v)_h - (D_0v, v^2)_h = 0,$$

and therefore we obtain the basic energy estimate from

$$\frac{1}{2}\frac{d}{dt}(v, v)_h = -\epsilon \|D_+v\|_h^2 \leq 0.$$

Bounds for $\|D_+^p\, d^qv/dt^q\|_h$ independent of h can be obtained along the same lines as the a priori estimates for u in the previous section. Thus we can proceed as in Section 3.2: Fourier interpolate $v = v^h$ with respect to x and send $h \to 0$. We obtain existence in some interval $0 \leq t \leq T$, T depending only on $\|f\|_{H^2}$.

4.1.4. Existence via Iteration

To prove existence via the iteration (4.1.2), we estimate the functions u^n independently of n. Again, our estimates will not take advantage of the diffusion term ϵu_{xx}, and so will not depend on $\epsilon \geq 0$. All of the arguments are without reference to the maximum principle, thus generalize from scalar equations to systems.

To show uniform smoothness of the sequence u^n, we start with the following analogue of Lemma 4.1.3.

Lemma 4.1.5. *For a suitable time $T_1 = T_1\left(\|f\|_{H^2}\right) > 0$ it holds that*

$$\|u^n(\cdot, t)\|_{H^2} \leq 2\|f\|_{H^2} \quad in \quad 0 \leq t \leq T_1.$$

Proof. We use induction on n; the case $n = 0$ is trivial. As abbreviations, let $w = u^n$, $v = u^{n+1}$, thus

$$v_t = w\, v_x + \epsilon v_{xx}.$$

Note that

$$\begin{aligned}
\frac{1}{2}\frac{d}{dt}(v, v) &= (v,\, w\, v_x) - \epsilon(v_x, v_x) \\
&= -\frac{1}{2}\,(v, w_x v) - \epsilon\|v_x\|^2 \\
&\leq \frac{1}{2}\,|w_x|_\infty\, \|v\|^2 \\
&\leq \frac{1}{2}\,\|w\|_{H^2}\, \|v\|^2 \quad \text{(by(4.1.5))}.
\end{aligned}$$

For the derivative $v_x = v_1$ it holds that

$$v_{1\,t} = (wv_x)_x + \epsilon v_{1\,xx},$$

and therefore

$$\begin{aligned}\frac{1}{2}\frac{d}{dt}(v_1, v_1) &= (v_1, (wv_x)_x) - \epsilon\,\|v_{1x}\|^2\\ &= \frac{1}{2}(v_1,\, w_x\, v_1) - \epsilon\,\|v_{1x}\|^2 \quad \text{(by (4.1.3))}\\ &\le \frac{1}{2}\,\|w\|_{H^2}\,\|v_1\|^2.\end{aligned}$$

Finally,

$$v_{2t} = (wv_1)_{xx} + \epsilon v_{2xx}$$

implies that

$$\begin{aligned}\frac{1}{2}\frac{d}{dt}(v_2, v_2) &= (v_2, (wv_1)_{xx}) - \epsilon\,\|v_{2x}\|^2\\ &= (v_2,\, wv_{2x}) + 2(v_2,\, w_x\, v_2) + (v_2,\, w_2\, v_1) - \epsilon\,\|v_{2x}\|^2\\ &\le \frac{3}{2}\,\|w\|_{H^2}\,\|v_2\|^2 + |v_1|_\infty\,\|v_2\|\,\|w_2\| \le \frac{5}{2}\,\|w\|_{H^2}\,\|v_2\|^2.\end{aligned}$$

Adding the three inequalities, we obtain from the induction assumption that

$$\frac{d}{dt}\|v\|_{H^2}^2 \le 5\,\|w\|_{H^2}\,\|v\|_{H^2}^2 \le 10\,\|f\|_{H^2}\,\|v\|_{H^2}^2,$$

and therefore

$$\|v(\cdot,t)\|_{H^2}^2 \le \exp\left(10\|f\|_{H^2}\,t\right)\,\|v(\cdot,0)\|_{H^2}^2.$$

Thus we can complete the induction step if $T_1 > 0$ is so small that

$$\exp\left(10\|f\|_{H^2}T_1\right) \le 4.$$

This proves the lemma.

We show now that *all* space derivatives of u^n can be estimated in the same interval $0 \le t \le T_1$.

Lemma 4.1.6. *For each $j = 2, 3, \ldots$ there exists K_j with*

$$\|u^n(\cdot,t)\|_{H^j} \le K_j \quad \text{in} \quad 0 \le t \le T_1.$$

The constant K_j depends on $\|f\|_{H^j}$, but is independent of n and $\epsilon \ge 0$.

Proof. As previously, let $w = u^n$, $v = u^{n+1}$. We can assume $j \ge 3$ and

$$\|w\|_{H^{j-1}} \le K_{j-1}, \qquad \|v\|_{H^{j-1}} \le K_{j-1}.$$

From

$$v_{jt} = (wv_x)_j + \epsilon v_{jxx}$$

it follows that

$$\frac{1}{2}\frac{d}{dt}(v_j, v_j) = \big(v_j, (wv_x)_j\big) - \epsilon\,\|v_{j+1}\|^2.$$

Here

$$(wv_x)_j = \sum_{\nu=0}^{j}\binom{j}{\nu} w_\nu v_{j+1-\nu}.$$

For $\nu = 1, \ldots, j-2$ it holds that

$$|(v_j,\, w_\nu\, v_{j+1-\nu})| \le \|v_j\|\,|w_\nu|_\infty\,\|v_{j+1-\nu}\| \le K_{j-1}\,\|v_j\|^2.$$

(Note that $\|v_l\| \le \|v_j\|$ for $l \le j$, by Fourier expansion and Parseval's relation.) For $\nu = j-1,\, j$ it holds that (see (4.1.5))

$$|(v_j,\, w_{j-1}\, v_2)| \le K_{j-1}\|v_j\|\,|v_2|_\infty \le K_{j-1}\|v_j\|^2,$$

$$|(v_j,\, w_j\, v_1)| \le |v_1|_\infty\,\|v_j\|\,\|w_j\| \le K_{j-1}(\|v_j\|^2 + \|w_j\|^2).$$

For $\nu = 0$, integration by parts gives us

$$|(v_j,\, wv_{j+1})| \le \frac{1}{2}\,K_{j-1}\,\|v_j\|^2.$$

Thus, there is a numerical constant $c = c_j$ with

$$\frac{d}{dt}\,\|v_j\|^2 \le c\Big(\|v_j\|^2 + \|w_j\|^2\Big), \quad 0 \le t \le T_1.$$

By Gronwall's Lemma 3.1.1,

$$\|u_j^{n+1}(\cdot,t)\|^2 \le K\Big\{\|f_j\|^2 + \int_0^t \|u_j^n(\cdot,\tau)\|^2\,d\tau\Big\}, \quad 0 \le t \le T_1.$$

Now the desired estimate follows from Picard's Lemma 3.3.4.

Summarizing, we have estimated the L_2-norms of all space derivatives of the sequence u^n in $0 \le t \le T_1$, $T_1 = T_1\left(\|f\|_{H^2}\right)$. By the Sobolev inequality stated in Lemma 3.2.5, the functions $\partial^j u^n/\partial x^j$ are also bounded in maximum norm. Since we can always use the differential equation (4.1.2) to replace time derivatives by space derivatives, it follows that

$$\left|\frac{\partial^{p+q}u^n}{\partial x^p \partial t^q}(x,t)\right| \le C(p,q), \qquad 0 \le t \le T_1. \tag{4.1.6}$$

Here $C(p, q)$ is independent of n and $0 \le \epsilon \le \epsilon_0$, where ϵ_0 is arbitrary but fixed.

We finally show

Theorem 4.1.7. *Let* $T_1 = T_1\left(\|f\|_{H^2}\right)$ *be determined as in Lemma* 4.1.5. *For any* $\epsilon \ge 0$ *Burgers' equation has a* C^∞*-solution* $u(x, t)$ *defined for* $0 \le t \le T_1$.

Proof. Consider the sequence u^n defined by the iteration (4.1.2) and abbreviate

$$v = u^{n+1} - u^n, \qquad w = u^n - u^{n-1}.$$

Then

$$v_t = u^n v_x + u_x^n w + \epsilon v_{xx}, \qquad v(x, 0) = 0$$

implies that

$$\begin{aligned}\frac{1}{2}\frac{d}{dt}\|v\|^2 &\le (v, u^n v_x) + (v, u_x^n w) \\ &= -\frac{1}{2}(v,\, u_x^n\, v) + (v,\, u_x^n w) \\ &\le \text{const}\left(\|v\|^2 + \|w\|^2\right), \qquad 0 \le t \le T_1.\end{aligned}$$

Therefore (by Gronwall's Lemma 3.1.1),

$$\left\|u^{n+1}(\cdot, t) - u^n(\cdot, t)\right\|^2 \le K \int_0^t \left\|u^n(\cdot, \tau) - u^{n-1}(\cdot, \tau)\right\|^2 d\tau,$$

where K is independent of n.

Thus, by Lemma 3.3.4, the sequence $u^n(\cdot, t)$ converges to a function $u(\cdot, t) \in L_2$. The smoothness estimates (4.1.6) imply that $u \in C^\infty$, and the convergence $u^n \to u$ holds pointwise and also holds for all derivatives. (See Appendix 4.) Then the equation (4.1.2) implies that u solves Burgers' equation.

4.2. Global Existence for the Viscous Burgers' Equation

In this section we use the Cauchy problem for Burgers' equation to illustrate an important principle. Loosely formulated it says: *Local existence together with global a priori estimates leads to global existence.* We will formulate the a priori estimate in the following theorem. Throughout this section, $\epsilon > 0$ is arbitrary but fixed.

Theorem 4.2.1. *Let $f = f(x)$ denote 1-periodic C^∞ initial data, and let $u \in C^\infty$ denote a solution of* (4.1.1) *defined for $0 \le t \le T$. There is a constant K, depending on $\|f\|_{H^2}$ (and ϵ) but independent of T, with*

$$\|u(\cdot, t)\|_{H^2} \le K \quad \text{for} \quad 0 \le t \le T. \tag{4.2.1}$$

(The smoothness assumption $f \in C^\infty$ is unnecessarily restrictive and will be relaxed later.)

Assuming this result is proven, we can show all-time existence for the problem (4.1.1) as follows: According to the Local Existence Theorem 4.1.7, there is a time $T_1 > 0$ with a C^∞-solution u defined for $0 \le t \le T_1$. By (4.2.1) we have

$$\|u(\cdot, T_1)\|_{H^2} \le K.$$

We can use the function $x \to u(x, T_1)$ as new initial data and apply the Local Existence Theorem 4.1.7 again.

The solution starting with the initial data $u(\cdot, T_1)$ exists in a time interval $0 \le t \le T_2$, T_2 depending only on K (and ϵ).

Clearly, putting the two solutions together, we have a C^∞-solution u of (4.1.1) defined for $0 \le t \le T_1 + T_2$. The important point is that the a priori estimate (4.2.1) implies that

$$\|u(\cdot, T_1 + T_2)\|_{H^2} \le K$$

with the same constant K as before. Thus, using the Local Existence Theorem, we can again extend the solution for a time interval of length T_2, etc. This shows

Theorem 4.2.2. *Let $f \in C^\infty$ and $\epsilon > 0$. The 1-periodic Cauchy problem* (4.1.1) *has a unique solution $u \in C^\infty$ defined for $0 \le t < \infty$.*

It remains to prove the a priori estimate. The proof is based on the **maximum principle**.

Lemma 4.2.3. *Let $u \in C^\infty$ solve* (4.1.1) *in $0 \le t \le T$. Then $|u(\cdot, t)|_\infty \le |u(\cdot, 0)|_\infty$ for $0 \le t \le T$.*

Proof. Let $\alpha > 0$ be a constant and set $v(x, t) = e^{-\alpha t} u(x, t)$. The function v satisfies

$$v_t = u v_x + \epsilon v_{xx} - \alpha v. \tag{4.2.2}$$

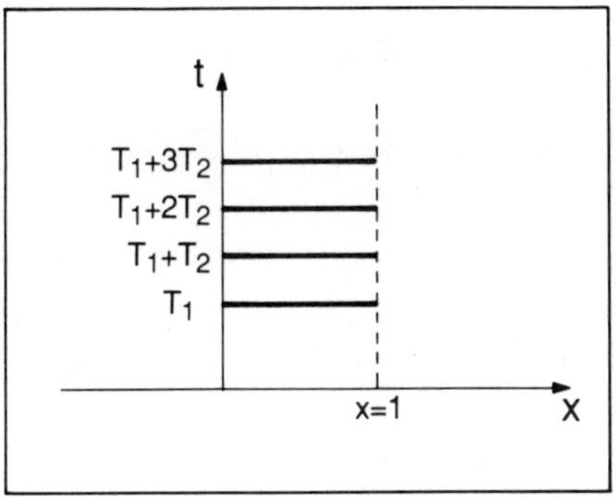

FIGURE 4.2.1. Restarting process.

Fix $T > 0$ and assume that $v(x,t)$ attains its maximum over

$$x \in \mathbf{R}, \qquad 0 \le t \le T$$

at (x_0, t_0) with $v(x_0, t_0) > 0$. If $0 < t_0 \le T$ then

$$v_t(x_0, t_0) \ge 0, \qquad v_x(x_0, t_0) = 0, \qquad v_{xx}(x_0, t_0) \le 0,$$

in contradiction to (4.2.2). Thus $t_0 = 0$, i.e., a positive maximum can only be attained at $t = 0$. Similarly, a negative minimum can only occur at $t = 0$. Therefore,

$$|v(\cdot, t)|_\infty \le |v(\cdot, 0)|_\infty \quad \text{for} \quad t \ge 0,$$

and if we let $\alpha \to 0$, then the corresponding estimate for u follows.

***Proof of Theorem* 4.2.1.** The basic energy inequality

$$\frac{d}{dt}\|u\| = -2\epsilon\|u_1\| \le 0$$

implies that

$$\|u(\cdot,t)\|^2 \le \|u(\cdot,0)\|^2, \quad 2\epsilon \int_0^T \|u_1(\cdot,\tau)\|^2\, d\tau \le \|u(\cdot,0)\|^2. \tag{4.2.3}$$

For the function $u_1 = u_x$, the equation

$$u_{1t} = (uu_1)_x + \epsilon u_{1xx}$$

gives us

$$\frac{d}{dt}\|u_1\|^2 \leq -2(u_{1x}, uu_1) - 2\epsilon\|u_{1x}\|^2$$
$$\leq \frac{1}{\epsilon}\,|u|_\infty^2\|u_1\|^2 - \epsilon\|u_{1x}\|^2.$$

Therefore,

$$\|u_1(\cdot,t)\|^2 \leq \|u_1(\cdot,0)\|^2 + \frac{|f|_\infty^2}{\epsilon}\int_0^t \|u_1(\cdot,\tau)\|^2\,d\tau - \epsilon\int_0^t \|u_2(\cdot,\tau)\|^2\,d\tau.$$

Thus, by (4.2.3), we obtain an estimate for

$$\text{(4.2.4)} \qquad \max_{0\leq t\leq T}\|u_1(\cdot,t)\|^2 \quad \text{and} \quad \epsilon\int_0^T \|u_2(\cdot,\tau)\|^2\,d\tau$$

independent of T. For u_2 the equation

$$u_{2t} = (uu_1)_{xx} + \epsilon u_{2xx}$$

implies that

$$\frac{1}{2}\frac{d}{dt}\|u_2\|^2 = \big(u_2,(uu_1)_{xx}\big) - \epsilon\|u_3\|^2 = (u_2,uu_3) + 3(u_2,u_2u_1) - \epsilon\|u_3\|^2$$
$$= -5(u_2,uu_3) - \epsilon\|u_3\|^2$$
$$\leq 5|f|_\infty\,\|u_2\|\,\|u_3\| - \epsilon\|u_3\|^2$$
$$\leq \frac{25}{2\epsilon}\,|f|_\infty^2\,\|u_2\|^2 - \frac{\epsilon}{2}\,\|u_3\|^3.$$

Therefore, (4.2.4) gives us bounds for

$$\text{(4.2.5)} \qquad \max_{0\leq t\leq T}\|u_2(\cdot,t)\|^2 \quad \text{and} \quad \epsilon\int_0^T \|u_3(\cdot,\tau)\|^2\,d\tau$$

independent of T. This finishes the proof of Theorem 4.2.1.

Remark. To obtain all-time existence, we only needed an a priori estimate for $\|u\|_{H^2}$. The reason is that the time interval of the Local Existence Theorem 4.1.7 depends on $\|f\|_{H^2}$ only, and does not depend on bounds of higher derivatives of f. Nevertheless, one can ask for a priori estimates of higher derivatives $\|u_j\|$, $j \geq 3$, in terms of $\|f\|_{H^j}$. These can indeed be derived. Just note that

$$u_{jt} = (uu_x)_j + \epsilon u_{jxx},$$

and thus

$$\frac{d}{dt}(u_j, u_j) = -2\left(u_{j+1}, (uu_x)_{j-1}\right) - 2\epsilon \left\|u_{j+1}\right\|^2$$
$$\leq -\epsilon \left\|u_{j+1}\right\|^2 + \frac{1}{\epsilon} \left\|(uu_x)_{j-1}\right\|^2.$$

We can make an induction argument using bounds for $\|u_{j-1}\|^2$ and $\epsilon \int_0^T \|u_j\|^2 \, d\tau$.

4.3. Generalized Solutions for Burgers' Equation and Smoothing*

We have shown that the 1-periodic Cauchy problem

$$\begin{aligned} u_t &= uu_x + \epsilon\, u_{xx}, \quad \epsilon > 0, \\ u(x,0) &= f(x), \quad f \in C^\infty, \end{aligned} \tag{4.3.1}$$

has a unique C^∞-solution $u(x,t)$ existing for all times $t \geq 0$. In this section we first define a *generalized solution* for any initial function

$$f \in L_1, \quad f(x) \equiv f(x+1).$$

For example, f could be piecewise continuous with finitely many jumps. Afterwards, we show that the generalized solution is always C^∞-smooth for $t > 0$ if $f \in L_\infty$.

4.3.1. Construction of Generalized Solutions

Let f, g denote 1-periodic C^∞-functions. We estimate the difference of the solutions u and v of

$$u_t = uu_x + \epsilon u_{xx}, \quad u(x,0) = f(x),$$
$$v_t = vv_x + \epsilon v_{xx}, \quad v(x,0) = g(x),$$

in terms of $f - g$. It is convenient to work with the L_1-norm

$$|h|_1 = \int_0^1 |h(x)| \, dx.$$

The difference $w = u - v$ satisfies the linear equation

$$w_t = \frac{1}{2}\left((u+v)w\right)_x + \epsilon w_{xx}, \quad w(x,0) = f(x) - g(x).$$

*This section might be omitted on the first reading.

The key result is

Lemma 4.3.1. *Let $a = a(x,t)$, $w = w(x,t)$ denote 1-periodic C^∞-functions with*

$$w_t = (aw)_x + \epsilon w_{xx}.$$

Then $|w(\cdot,t)|_1 \le |w(\cdot,0)|_1$ *for all* $t \ge 0$.

Proof. 1) To obtain smooth approximations to the sign function, we choose a fixed function Sgn $\in C^1$ with

$$\text{Sgn}(y) = -1 \quad \text{for} \quad y \le -1, \qquad \text{Sgn}(y) = +1 \quad \text{for} \quad y \ge 1,$$
$$\text{Sgn}(0) = 0, \qquad \text{Sgn}'(y) \ge 0 \quad \text{for all } y.$$

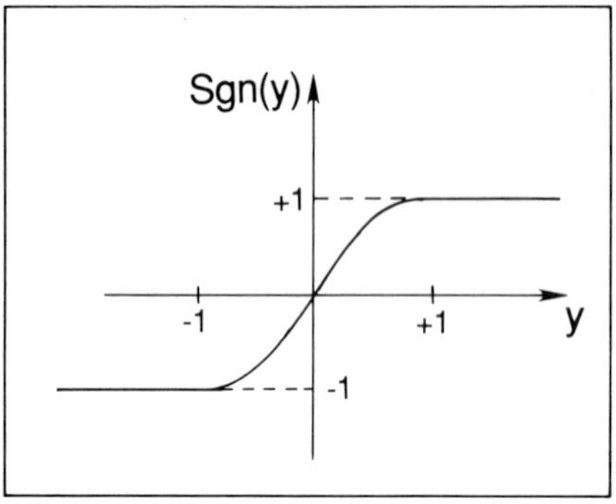

FIGURE 4.3.1. Approximation for the sign function.

Then we set

$$\text{sgn}_\delta(y) = \text{Sgn}\,(y/\delta), \quad y \in \mathbf{R}, \quad \delta > 0.$$

For $\delta \to 0$ we have that

$$\text{sgn}_\delta(y) \to \text{sgn}\, y = \begin{cases} 1 & \text{for } y > 0, \\ -1 & \text{for } y < 0, \\ 0 & \text{for } y = 0, \end{cases}$$

and thus

$$\int_0^1 \text{sgn}_\delta(w) w \, dx \to |w|_1 \quad \text{as} \quad \delta \to 0.$$

Therefore, we have expressed $|w|_1$ approximately as an integral of a smooth function. This is the main technical point of the proof.
2) For $\delta > 0$ we find that

$$\frac{d}{dt}\int_0^1 \operatorname{sgn}_\delta(w)w\,dx = \int_0^1 \operatorname{sgn}'_\delta(w)w_t w\,dx + \int_0^1 \operatorname{sgn}_\delta(w)w_t\,dx,$$

where

$$\begin{aligned}\int_0^1 \operatorname{sgn}_\delta(w)w_t\,dx &= \int_0^1 \operatorname{sgn}_\delta(w)(aw)_x\,dx + \epsilon\int_0^1 \operatorname{sgn}_\delta(w)w_{xx}\,dx\\ &= -\int_0^1 \operatorname{sgn}'_\delta(w)w_x aw\,dx - \epsilon\int_0^1 \operatorname{sgn}'_\delta(w)(w_x)^2\,dx\\ &\le -\int_0^1 \operatorname{sgn}'_\delta(w)w_x aw\,dx.\end{aligned}$$

Therefore

$$\frac{d}{dt}\int_0^1 \operatorname{sgn}_\delta(w)w\,dx \le \int_0^1 \operatorname{sgn}'_\delta(w)wb\,dx =: R_\delta.$$

Here $b = w_t - aw_x$ is independent of δ. Integration from 0 to t yields

$$\int_0^1 \operatorname{sgn}_\delta\big(w(x,t)\big)w(x,t)\,dx - \int_0^1 \operatorname{sgn}_\delta\big(w(x,0)\big)w(x,0)\,dx \le \int_0^t R_\delta(\tau)\,d\tau,$$

and we send $\delta \to 0$. The left-hand side converges to

$$|w(\cdot,t)|_1 - |w(\cdot,0)|_1.$$

The integrand $R_\delta(\tau)$ of the right-hand side is

$$R_\delta(\tau) = \int_0^1 \operatorname{sgn}'_\delta\big(w(x,\tau)\big)w(x,\tau)b(x,\tau)\,dx. \tag{4.3.2}$$

If $w(x,\tau) \neq 0$ then $\operatorname{sgn}'_\delta\big(w(x,\tau)\big) \to 0$ as $\delta \to 0$. Thus the integrand of (4.3.2) converges pointwise to zero as $\delta \to 0$. Also, $\operatorname{sgn}'_\delta(w)w = \operatorname{Sgn}'\,\big(w/\delta\big)\,(w/\delta)$ is bounded independently of δ and x, because the (fixed) function $y \to \operatorname{Sgn}'(y)\,y$ is bounded. Therefore, Lebesgue's Dominated-Convergence Theorem yields $R_\delta(\tau) \to 0$ as $\delta \to 0$ for each τ. The convergence

$$\int_0^t R_\delta(\tau)\,d\tau \to 0 \quad \text{as} \quad \delta \to 0$$

follows by another application of Lebesgue's theorem and finishes the proof of the lemma.

Remark. A different proof could be given, which uses a monotonic difference scheme. This proof would not require Lebesgue's theorem.

The estimate shown in Lemma 4.3.1 allows one to obtain generalized solutions of

$$u_t = uu_x + \epsilon u_{xx}, \quad u(x,0) = f(x), \tag{4.3.3}$$

for nonsmooth initial functions f by the usual approximation process: Let

$$f \in L_1, \quad f(x) \equiv f(x+1),$$

and approximate f by 1-periodic C^∞-functions $f^{(\nu)}$:

$$|f^{(\nu)} - f|_1 \to 0.$$

If $u^{(\nu)}(x,t)$ denotes the solution for initial data $u^{(\nu)}(x,0) = f^{(\nu)}(x)$ we have

$$|u^{(\nu)}(\cdot,t) - u^{(\mu)}(\cdot,t)|_1 \;\le\; |f^{(\nu)} - f^{(\mu)}|_1 \;\to\; 0 \;\text{ as }\; \nu,\ \mu \to \infty.$$

Thus, $u^{(\nu)}(\cdot,t) \to u(\cdot,t)$ with respect to $|\cdot|_1$; the limit function u is independent of the specific choice of the approximating sequence $f^{(\nu)}$. By definition, u is the generalized solution of (4.3.3).

Remark. Let us note another application of Lemma 4.3.1. Suppose $u = u_\epsilon$ solves (4.3.1), and set $w = u_x$. Differentiation of (4.3.1) yields

$$w_t = (uw)_x + \epsilon w_{xx}.$$

Hence, we can apply Lemma 4.3.1 and obtain that

$$|u_x(\cdot,t)|_1 \le |u_x(\cdot,0)|_1 = |f_x|_1, \quad t \ge 0.$$

This ϵ-independent bound of the L_1-norm of $u_x = u_{\epsilon x}$ is the key estimate to obtain a weak solution of the inviscid ($\epsilon = 0$) equation by a compactness argument. We will not present the details, however.

4.3.2. Smoothing for Burgers' Equation

We want to show that the generalized solution $u(x,t)$ defined above is a C^∞-function for $t > 0$ if $f \in L_\infty$. This is remarkable and not self-evident from linear theory. To show smoothing for parabolic *linear* equations, we used the smoothness of coefficients. In the present case we could think of u as solution of the linear equation

$$u_t = au_x + \epsilon u_{xx}$$

with $a(x,t) = u(x,t)$, but the coefficient $a(x,t)$ is rough initially. To obtain smoothness of the generalized solutions, we first show estimates of *derivatives* of u in terms of the *maximum* norm of the initial data $u(\cdot,0) = f$ where $f \in C^\infty$. We remind the reader of the maximum principle

$$|u(\cdot,t)|_\infty \leq |f|_\infty, \quad t \geq 0,$$

and the notation $u_j = \partial^j u/\partial x^j$.

Lemma 4.3.2. *For any interval* $0 \leq t \leq T$ *and any* $j = 0, 1, 2, \ldots$ *there exists a constant* $C = C(j, \epsilon, |f|_\infty, T)$ *with*

$$t^j \|u_j(\cdot,t)\|^2 \leq C, \quad 0 \leq t \leq T. \tag{4.3.4}$$

The constant C *does not depend on bounds for derivatives of* f.

Proof. Recall the basic energy inequality

$$\frac{d}{dt}(u,u) = 2(u,u_t) = -2\epsilon\|u_1\|^2 \leq 0,$$

which implies that

$$\|u(\cdot,t)\|^2 \leq \|f\|^2 \leq |f|_\infty^2, \quad 2\epsilon \int_0^T \|u_1\|^2\, d\tau \leq \|f\|^2. \tag{4.3.5}$$

For u_1 we have that

$$\begin{aligned}\frac{d}{dt}\{t(u_1,u_1)\} &= (u_1,u_1) + 2t(u_1,u_{1t})\\ &= \|u_1\|^2 + 2t\{-(u_2,uu_1) - \epsilon\|u_2\|^2\}\\ &\leq \|u_1\|^2 + \frac{t}{\epsilon}|f|_\infty^2\,\|u_1\|^2 - t\epsilon\|u_2\|^2.\end{aligned}$$

Integration yields

$$t\|u_1(\cdot,t)\|^2 \leq \int_0^t \left(1 + \frac{\tau}{\epsilon}|f|_\infty^2\right)\|u_1(\cdot,\tau)\|^2\, d\tau - \epsilon\int_0^t \tau\|u_2(\cdot,\tau)\|^2\, d\tau.$$

By (4.3.5) it follows that

$$\max_{0\leq t\leq T} t\|u_1(\cdot,t)\|^2 \leq \text{const}, \quad \epsilon\int_0^T \tau\|u_2(\cdot,\tau)\|^2 d\tau \leq \text{const}. \tag{4.3.6}$$

Similarly, (see the computations in the proof of Theorem 4.2.1)

$$\frac{d}{dt}\{t^2\|u_2\|^2\} \leq 2t\|u_2\|^2 + c_1 t^2\,\|u_2\|^2 - \epsilon t^2\,\|u_3\|^2,$$

and therefore,

$$t^2\|u_2(\cdot,t)\| \le c_2 \int_0^t \tau\|u_2(\cdot,\tau)\|^2\,d\tau - \epsilon \int_0^t \tau^2\|u_3(\cdot,\tau)\|^2\,d\tau.$$

Using (4.3.6), we obtain bounds for

$$\max_{0\le t\le T} t^2\|u_2(\cdot,t)\|^2 \quad \text{and} \quad \epsilon \int_0^T \tau^2\|u_3(\cdot,\tau)\|^2\,d\tau.$$

Clearly, this process can be continued and the lemma follows.

We have established bounds for the L_2-norms of *all* derivatives $u_j = \partial^j u/\partial x^j$ in terms of $|f|_\infty$. Applying the Sobolev inequality of Lemma 3.2.5, we find bounds for u_j in maximum norm, and – using the differential equation to express time derivatives by space derivatives – we obtain that

$$\left|\frac{\partial^{p+q}u}{\partial x^p \partial t^q}(x,t)\right| \le \text{const}, \qquad 0 < \tau \le t \le T.$$

The constant does not depend on derivatives of f, but only on $|f|_\infty$. Thus far we have assumed that $f \in C^\infty$. Now let

$$f \in L_\infty, \quad f(x+1) \equiv f(x),$$

and approximate f by a sequence $f^{(\nu)} \in C^\infty$, $f^{(\nu)}(x) \equiv f^{(\nu)}(x+1)$,

$$|f^{(\nu)}|_\infty \le |f|_\infty + 1, \qquad |f^{(\nu)} - f|_1 \to 0.$$

All derivatives $\left|\partial^{p+q}u^{(\nu)}/\partial x^p \partial t^q\right|$ are bounded independently of ν in $0 < \tau \le t \le T$. This shows that the limit function u is C^∞-smooth for $t > 0$. Also, the derivatives of $u^{(\nu)}$ converge pointwise to the corresponding derivatives of u for $t > 0$. We summarize:

Theorem 4.3.3. *Burgers' equation* (4.3.3) *has a* 1*-periodic generalized solution* $u = u(x,t)$ *for each initial function*

$$f \in L_\infty, \quad f(x) \equiv f(x+1).$$

For $t > 0$, the solution u is C^∞-smooth and satisfies the differential equation pointwise.

4.4. The Inviscid Burgers' Equation: A First Study of Shocks

The viscous equations*

$$(4.4.1)\qquad \begin{aligned} &u_t + uu_x = \epsilon u_{xx}, \quad t \geq 0, \quad \epsilon > 0, \\ &u(x,0) = f(x), \quad f \in C^\infty, \quad f(x+1) \equiv f(x), \end{aligned}$$

have unique C^∞-solutions $u = u_\epsilon$, existing for all time $t \geq 0$. The corresponding inviscid equation

$$(4.4.2)\qquad u_t + uu_x = 0, \quad u(x,0) = f(x)$$

shall be treated as the limit of (4.4.1) as $\epsilon \to 0$. The viscous solutions $u_\epsilon,\ \epsilon > 0$, are uniformly smooth in some interval $0 \leq t \leq T, T > 0$ independent of ϵ, as shown in Section 4.1. Thus we can send $\epsilon \to 0$ and obtain a smooth solution u of the inviscid equation in the same time interval. (Indeed, if $u = u_{\epsilon_1},\ v = u_{\epsilon_2}$, then we can argue as in the proof of Lemma 4.1.1 and show that

$$\|w(\cdot,t)\| = O(\epsilon_1 + \epsilon_2), \quad w = u_{\epsilon_1} - u_{\epsilon_2}, \quad 0 \leq t \leq T.$$

Then, by Appendix 4, convergence $u_\epsilon \to u$ follows.) According to Lemma 4.1.1, this limit u is the *unique* classical solution in $0 \leq t \leq T$. It is natural to ask whether the functions $u_\epsilon(x,t)$ converge for all $t \geq 0$ and whether we can extend the smooth solution $u(x,t)$ for all times. Both questions have indeed affirmative answers. However, the smoothness of the solution u breaks down, in general, at a certain time T_b, at which one or several shocks form. This will be proved below using the method of characteristics.

In gas dynamics one treats systems of equations which exhibit a similar mathematical structure as the inviscid Burgers' equation. In view of these applications, it is desirable to extend the solution u beyond the time T_b, but this is possible only as a nonsmooth function. Of course, the concept of a nonsmooth solution of a differential equation requires some explanation. Concerning this question, we will restrict ourselves to Burgers' equation, and will first treat the special case of limits of traveling waves. Afterwards, in Section 4.4.3, we discuss more general piecewise smooth weak solutions.

4.4.1. A Solution Formula via Characteristics

Suppose $u = u(x,t)$ is a smooth function solving

*We alter here the sign of the uu_x-term; this corresponds simply to the transformation $x \to -x$. The nonlinear term appears now with the same sign as the convection term in the Navier-Stokes equations.

(4.4.3) $$u_t + uu_x = 0, \quad u(x,0) = f(x)$$

in $0 \le t \le T$. We can consider $u(x,t) = a(x,t)$ as a coefficient in (4.4.3) and u as the solution of $u_t + au_x = 0$. This suggests applying the idea of characteristics discussed for linear equations in Section 3.3. Accordingly, we call

$$\big(x(t), t\big), \quad 0 \le t \le T,$$

a characteristic line of the equation $u_t + uu_x = 0$ for the specific solution u, if

(4.4.4) $$\frac{dx}{dt}(t) = u\big(x(t), t\big), \quad 0 \le t \le T.$$

This agrees completely with the definition for the linear equation $u_t + au_x = 0$ if $a = u$. (In the nonlinear case, the characteristics depend on the solution under consideration, however.) As in the linear case, it follows that u is constant along each characteristic:

$$\frac{d}{dt}\, u\big(x(t), t\big) = u_x \frac{dx}{dt} + u_t = u_x u + u_t = 0;$$

i.e., $u\big(x(t), t\big) = f(x_0)$ if $x(0) = x_0$. Now (4.4.4) implies that the characteristic $\big(x(t), t\big)$ is a straight line:

$$x(t) = x_0 + tf(x_0).$$

We have shown

Lemma 4.4.1. *Suppose u solves*

(4.4.5) $$u_t + uu_x = 0, \quad u(x,0) = f(x).$$

Then

(4.4.6) $$u\big(x_0 + t\, f(x_0), t\big) = f(x_0), \quad x_0 \in \mathbf{R},$$

in any t-interval $0 \le t \le T$ where u is smooth.

Note that this result is completely independent of spatial periodicity. Henceforth we drop the periodicity assumption and consider (4.4.5), where $f \in C^\infty$ is a given function with

$$|f|_\infty = \sup |f(x)| < \infty, \quad |f'|_\infty = \sup |f'(x)| < \infty.$$

We can use the formula (4.4.6) to obtain a solution $u = u(x,t)$ as follows: Suppose $x \in \mathbf{R}$, $t \ge 0$ are given and suppose that there is a *unique* $x_0 \in \mathbf{R}$ with

(4.4.7) $$x_0 + tf(x_0) = x.$$

Then *define* $u(x,t) = f(x_0)$. Let us explain why u solves (4.4.5). First note that, since f is bounded, the equation (4.4.7) always has at least one solution x_0. If there are two different solutions x_0 and x_1, then

$$t = -\frac{x_1 - x_0}{f(x_1) - f(x_0)} = -\frac{1}{f'(\xi)} \quad \text{for some } \xi.$$

Thus, if we let

$$T_b = \begin{cases} \infty & \text{if } f'(\xi) \geq 0 \text{ for all } \xi, \\ -\frac{1}{\inf f'(\xi)} & \text{otherwise,} \end{cases}$$

then (4.4.7) has a *unique* solution $x_0 = x_0(x,t)$ for given $x \in \mathbf{R},\ 0 \leq t < T_b$, and consequently the function

$$u(x,t) = f(x_0(x,t)), \quad x \in \mathbf{R}, \quad 0 \leq t < T_b, \tag{4.4.8}$$

is well-defined. By the Implicit Function Theorem (see below), it follows that $x_0(x,t)$ is C^∞-smooth. Assuming this is known, we prove now that (4.4.8) solves the inviscid Burgers' equation.

Lemma 4.4.2. *The C^∞-function $u = u(x,t)$ defined by* (4.4.8) *is the unique classical solution of* (4.4.5).

Proof. It is clear that $x_0(x,0) = x$, and thus $u(x,0) = f(x)$. To show that $u_t + uu_x = 0$, we differentiate

$$x_0 + tf(x_0) = x$$

with respect to x and t, and find that

$$x_{0x} + tf'x_{0x} = 1, \quad x_{0t} + f + tf'x_{0t} = 0.$$

Hence $x_{0x}(1 + tf') = 1,\ x_{0t}(1 + tf') = -f$, and therefore,

$$x_{0t} + fx_{0x} = 0 \quad (\text{note that } 1 + tf' \neq 0).$$

Thus the definition $u(x,t) = f(x_0(x,t))$ gives us

$$u_t + uu_x = f'x_{0t} + ff'x_{0x} = 0.$$

For completeness we state the

Implicit Function Theorem. *Let $U \subset \mathbf{R}^n$, $V \subset \mathbf{R}^m$ be open, and let $F : U \times V \to \mathbf{R}^m$ be a C^p-function, $p \geq 1$. Let $a_0 \in U$, $b_0 \in V$ and assume that*

$$F(a_0, b_0) = 0 \quad \textit{and} \quad F_b(a_0, b_0) \quad \textit{is nonsingular}.$$

(Here F_b is the Jacobian matrix of F w.r.t. the b-variables.) Then there exists an open neighborhood U_0 of a_0, and a unique C^p-function $g : U_0 \to \mathbf{R}^m$ with

$$g(a_0) = b_0 \quad \textit{and} \quad F(a, g(a)) = 0 \quad \textit{for all } a \in U_0.$$

Roughly speaking, the equation $F(a, b) = 0$ defines b *implicitly* as a function of a, $b = g(a)$, and g is (at least) as smooth as F.

This result applies to the previous situation with $n = 2$, $m = 1$, and

$$F(x, t, b) = b + tf(b) - x, \quad a = (x, t), \quad -\epsilon < t < T_b, \quad x \in \mathbf{R},$$

in the neighborhood of any point

$$(a_0, b_0) = (x, t, x_0(x, t)),$$

and shows that x_0 depends smoothly on (x, t). Here

$$F_b(x, t, b) = 1 + tf'(b) \neq 0, \quad -\epsilon < t < T_b$$

is essential.

Let us illustrate the solution formula (4.4.8): If $f'(\xi) \geq 0$ for all ξ, then $x_0 < x_1$ implies $f(x_0) \leq f(x_1)$; thus the characteristics

$$(4.4.9) \qquad \big(x_0 + tf(x_0), t\big), \quad \big(x_1 + tf(x_1), t\big), t \geq 0$$

never cross. We have $T_b = \infty$, and a smooth solution exists for all time $t \geq 0$. On the other hand, if $f'(\xi) < 0$ for some ξ, then there are values

$$x_0 < x_1 \quad \text{with} \quad f(x_0) > f(x_1)$$

and the characteristics (4.4.9) cross. The solution steepens up with time. The time T_b is finite. We cannot extend the solution u beyond T_b as a smooth solution. See figures 4.4.1 to 4.4.3.

4.4.2. Traveling Wave-Solutions for Burgers' Equation

Before we discuss the extension of the inviscid solution u as a nonsmooth function, let us study explicitly some simple solutions of the viscous equation. We start with the steady state problem

$$(4.4.10) \qquad -\epsilon\phi_{xx} + \phi\phi_x = 0, \qquad -\infty < x < \infty,$$

under boundary conditions

$$(4.4.11) \qquad \lim_{x \to -\infty} \phi(x) = a, \qquad \lim_{x \to \infty} \phi(x) = -a.$$

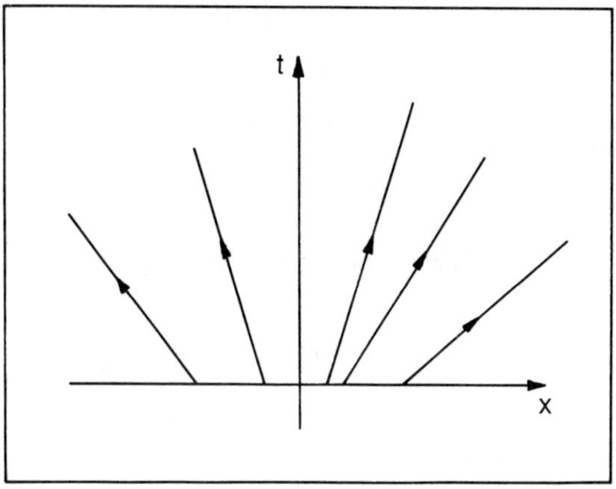

FIGURE 4.4.1. Characteristics for $f(x)$ increasing.

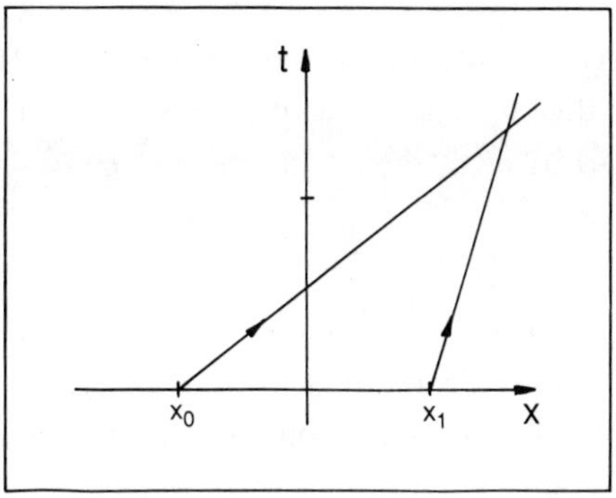

FIGURE 4.4.2. Crossing characterictics.

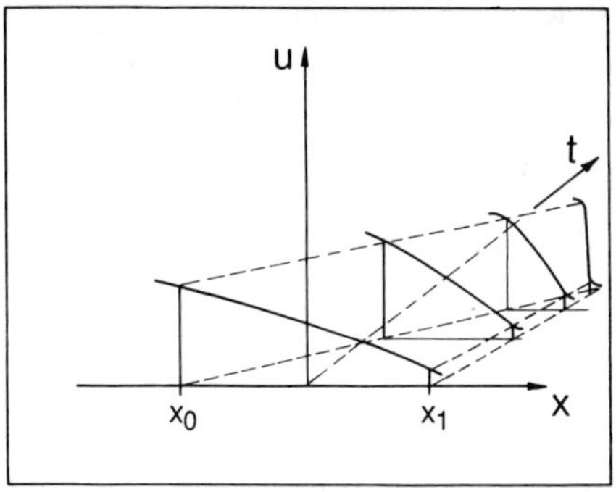

FIGURE 4.4.3. Solution steepens up.

If $\phi(x)$ is a solution, then so is the translated function $\phi(x+c)$ for any real constant c. To fix the constant, we require that

$$\phi(0) = 0 \tag{4.4.12}$$

and prove

Theorem 4.4.3. *If $a \geq 0$ the problem* (4.4.10)–(4.4.12) *has a unique solution given by*

$$\phi(x) = a\frac{1 - e^{ax/\epsilon}}{1 + e^{ax/\epsilon}}. \tag{4.4.13}$$

If $a < 0$ there is no solution.

Proof. First suppose ϕ is a solution. Integration of (4.4.10) yields

$$-\epsilon\phi_x + \frac{1}{2}\phi^2 = \text{const.} \tag{4.4.14}$$

It follows that either ϕ is constant or $\phi_x(x) \neq 0$ for all x. The monotonicity of ϕ and (4.4.11) imply that ϕ_x tends to zero as $x \to \pm\infty$. Thus we can replace the constant in (4.4.14) by $a^2/2$:

$$\epsilon\phi_x = \frac{1}{2}\left(\phi^2 - a^2\right). \tag{4.4.15}$$

Clearly, if $a \neq 0$, then $\phi_x(0) < 0$, and therefore ϕ is decreasing. The boundary conditions (4.4.11) can only be fulfilled for $a > 0$. The o.d.e. (4.4.15) with initial condition (4.4.12) has a unique solution, and by separation of variables one arrives at formula (4.4.13).

The stationary profiles (4.4.13) are shown in figure 4.4.4. In the limit as $\epsilon \to 0$, the functions $\phi(x)$ approach a jump discontinuity.

Now consider the time dependent problem

$$u_t + uu_x = \epsilon u_{xx} \tag{4.4.16}$$

under boundary conditions

$$\lim_{x\to-\infty} u(x,t) = b > c = \lim_{x\to\infty} u(x,t). \tag{4.4.17}$$

We use an ansatz

$$u(x,t) = \phi(x - st) + \frac{1}{2}(b + c)$$

with a constant wave-speed s. It leads to

$$u_x = \phi', \qquad u_{xx} = \phi'', \qquad u_t = -s\phi',$$

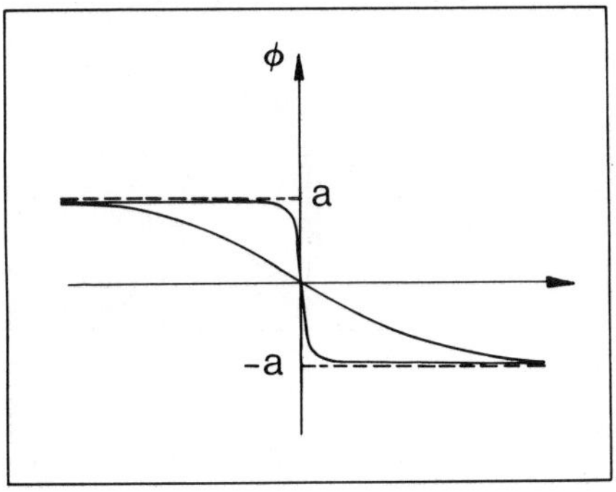

FIGURE 4.4.4. Stationary profile.

and the differential equation (4.4.16) requires

$$-s\phi' + \left(\phi + \frac{1}{2}(b+c)\right)\phi' = \epsilon\phi''.$$

If we set the wave-speed

$$s = \frac{1}{2}(b+c),$$

then we obtain the previous stationary differential equation for ϕ. The boundary conditions for ϕ are those given in (4.4.11) with $a = (b-c)/2$.

We have shown

Theorem 4.4.4. *The time dependent problem* (4.4.16), (4.4.17) *has traveling wave-solutions*

$$u(x,t) = \phi(x - st) + \frac{1}{2}(b+c), \qquad s = \frac{1}{2}(b+c).$$

For $\epsilon \to 0$ *these solutions converge to the piecewise constant function*

$$u_0(x,t) = \begin{cases} b & \text{for } x < st, \\ c & \text{for } x > st. \end{cases}$$

4.4.3. Weak Solutions of the Inviscid Burgers' Equation

The inviscid equation

$$u_t + uu_x = 0, \qquad u(x,0) = f(x), \tag{4.4.18}$$

$$f \in C^\infty, \qquad |f|_\infty < \infty, \qquad |f'|_\infty < \infty,$$

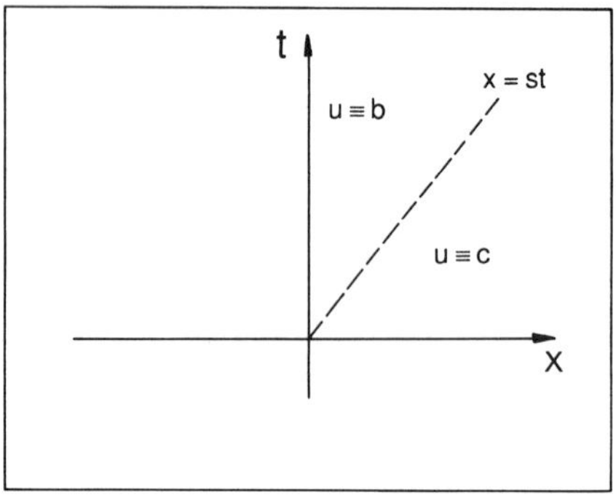

FIGURE 4.4.5. Limit of traveling waves.

has a unique smooth solution $u = u(x,t)$ only in $0 \le t < T_b$. Here T_b is finite unless $f'(\xi) \ge 0$ for all ξ. This result has been shown in Section 4.4.1. We now broaden the solution concept and first define weak solutions. To this end, let C_0^∞ denote the set of all test functions $\phi = \phi(x,t)$, $(x,t) \in \mathbf{R}^2$:

$$C_0^\infty = \{\phi \in C^\infty : \text{there exists } K = K_\phi \text{ with } \phi(x,t) = 0 \text{ for } |x| + |t| > K\}.$$

Suppose u is a classical solution of (4.4.18) and $\phi \in C_0^\infty$. If we integrate

$$u_t\phi + \frac{1}{2}(u^2)_x\phi = 0$$

over $-\infty < x < \infty$, $0 \le t < \infty$, then integration by parts gives us

$$\int_0^\infty \int_{-\infty}^\infty (u\phi_t + \frac{1}{2}u^2\phi_x)\,dx\,dt + \int_{-\infty}^\infty f(x)\phi(x,0)\,dx = 0. \tag{4.4.19}$$

The double integral makes sense even if u is not smooth. We make the

Definition. A bounded measurable function $u = u(x,t)$ defined for $x \in \mathbf{R}$, $t \ge 0$, is called a *weak solution* of (4.4.18) if equation (4.4.19) holds for all $\phi \in C_0^\infty$.

Clearly, we have broadened the solution concept. Using the solutions u_ϵ of the viscous equation and sending $\epsilon \to 0$, one can show that there always exists a weak solution. It turns out, however, that weak solutions are not uniquely determined by their initial data.

Before discussing these issues further, we ask for conditions characterizing a *piecewise smooth* function as a weak solution. First note: If u is a weak solution, which is smooth in a neighborhood U of a point (x_0, t_0), then u satisfies the differential equation $u_t + uu_x = 0$ in the usual sense at (x_0, t_0). To show this, we take an arbitrary $\phi \in C_0^\infty$ with support in U and find that

$$0 = \int_0^\infty \int_{-\infty}^\infty (u\phi_t + \frac{1}{2}u^2\phi_x)\,dx\,dt = -\int_U (u_t + uu_x)\phi\,dx\,dt;$$

thus $u_t + uu_x = 0$ in (x_0, t_0), since ϕ is arbitrary.

Next assume there is a smooth line (parametrized over time)

$$\Gamma : \big(s(t), t\big), \qquad t_0 \le t \le t_1, \tag{4.4.20}$$

and that the weak solution u has a jump discontinuity along Γ. The left and right limits are assumed to exist:

$$u_l\big(s(t), t\big) := \lim_{x \to s(t)-} u(x,t), \quad u_r\big(s(t), t\big) := \lim_{x \to s(t)+} u(x,t);$$

and we suppose that these limits depend smoothly on t. Except for this jump, we assume that u is smooth in a neighborhood of Γ. Taking a test function ϕ with support in a neighborhood U along Γ we find that

$$\begin{aligned}
0 &= \int_U (u\phi_t + \frac{1}{2}u^2\phi_x)\,dx\,dt = \int_{U_l} \cdots + \int_{U_r} \cdots \\
&= -\int_{U_l} \left\{u_t + \frac{1}{2}(u^2)_x\right\} \phi\,dx\,dt - \int_\Gamma \left(u_l\phi\,dx - \frac{1}{2}u_l^2\phi\,dt\right) \\
&\quad - \int_{U_r} \left\{u_t + \frac{1}{2}(u^2)_x\right\} \phi\,dx\,dt + \int_\Gamma \left(u_r\phi\,dx - \frac{1}{2}u_r^2\phi\,dt\right).
\end{aligned}$$

The integrals over U_l and U_r on the right-hand side vanish, and therefore

$$\begin{aligned}
0 &= \int_\Gamma \phi \left\{(u_l - u_r)\,dx - \frac{1}{2}(u_l^2 - u_r^2)\right\} dt \\
&= \int_{t_0}^{t_1} \phi \left\{(u_l - u_r)s'(t) - \frac{1}{2}(u_l^2 - u_r^2)\right\} dt.
\end{aligned}$$

Since ϕ can take arbitrary values along Γ, it follows that

$$(u_l - u_r)s'(t) = \frac{1}{2}(u_l^2 - u_r^2). \tag{4.4.21}$$

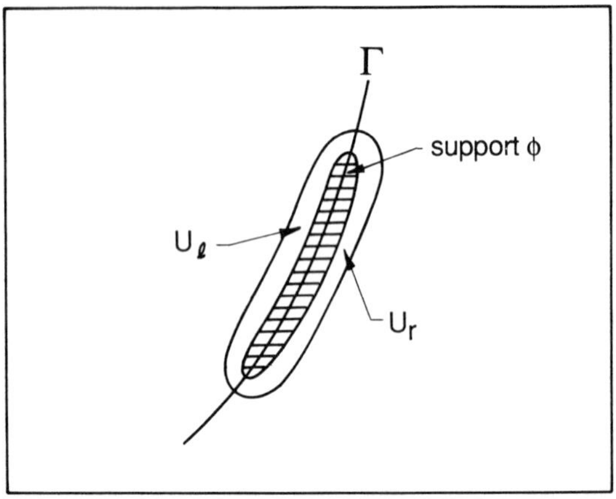

FIGURE 4.4.6. Discontinuity curve.

This is a special case of the so-called **Rankine-Hugoniot jump condition**, which relates the shock speed $s'(t)$ to the states u_l and u_r of u along the shock line:

$$s'(t) = \frac{1}{2}(u_l + u_r)\big(s(t), t\big). \tag{4.4.22}$$

A particular case of this condition occurred already in Theorem 4.4.4 for the limit of traveling waves. In that example the shock speed s was given by $s = (1/2)(b + c)$.

Thus far, we assumed that u is a piecewise smooth weak solution and that the initial function $f(x)$ is smooth. However, if f is only piecewise smooth, we obtain the same result. Conversely, reversing the arguments given above, one can show that a piecewise smooth function u is a weak solution of (4.4.1) if

(a) the initial condition $u(x, 0) = f(x)$ is satisfied at the points of continuity of f;
(b) the differential equation is satisfied in the pointwise sense in all regions of smoothness of u;
(c) the jump condition (4.4.22) is satisfied along the lines of discontinuity of u.

We summarize:

Theorem 4.4.5. *Suppose u is a piecewise smooth function with jump discontinuities along finitely many smooth lines. Then u is a weak solution of* (4.4.18) *if and only if the conditions* $(a), (b), (c)$ *above are met.*

Example 1. (A propagating shock.)

$$f(x) = 1 \text{ for } x < 0, \qquad f(x) = 0 \text{ for } x > 0.$$

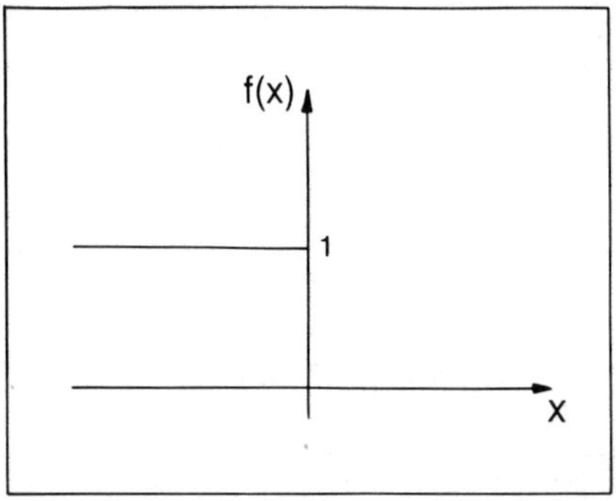

FIGURE 4.4.7. Discontinuous initial function.

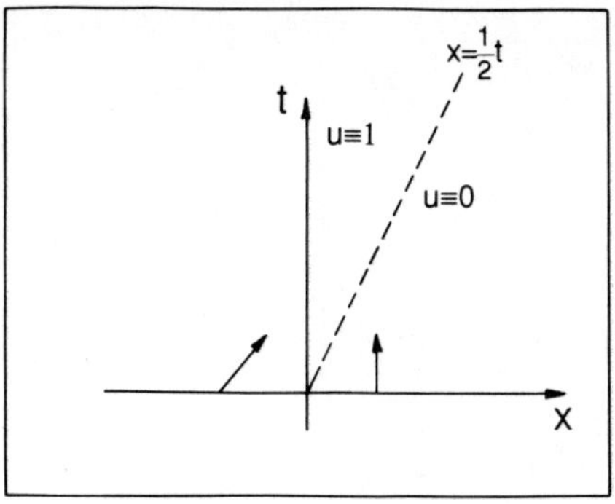

FIGURE 4.4.8. Propagating shock.

Here

$$u(x,t) = \begin{cases} 1 & \text{for } x < \dfrac{1}{2}t, \\ 0 & \text{for } x > \dfrac{1}{2}t, \end{cases}$$

is a weak solution. The shock speed $s = 1/2 = (1/2)(u_l + u_r)$ fulfills the jump condition (4.4.22). This solution is a limit of traveling waves; compare Theorem 4.4.4.

Example 2. (A rarefaction.)

$$f(x) = 0 \text{ for } x < 0, \qquad f(x) = 1 \text{ for } x > 0.$$

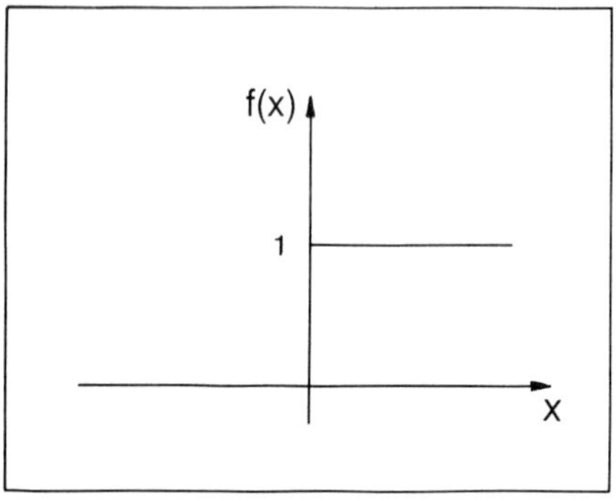

FIGURE 4.4.9. Discontinuous initial function.

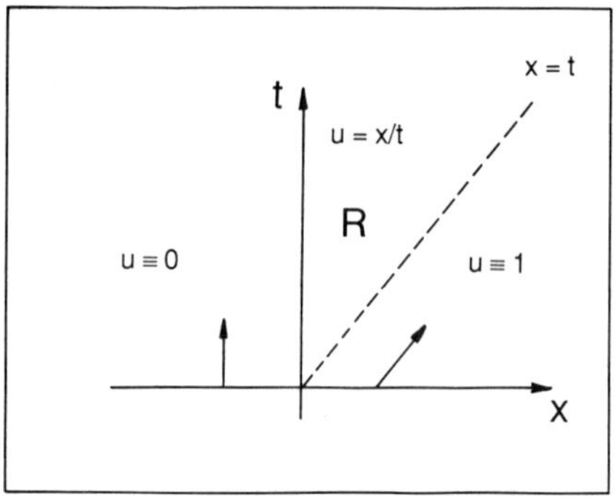

FIGURE 4.4.10. Rarefaction wave.

Here the characteristics starting at $t = 0$ do not enter the region

$$R = \left\{(x,t) : 0 < \frac{x}{t} < 1\right\}.$$

If one uses an ansatz

$$u(x,t) = r\left(\frac{x}{t}\right), \qquad 0 < \frac{x}{t} < 1,$$

then the differential equation requires that

$$0 = u_t + uu_x = \frac{1}{t}r'\left(\frac{x}{t}\right)\left\{-\frac{x}{t} + r\left(\frac{x}{t}\right)\right\},$$

and one obtains

$$u(x,t) = \frac{x}{t}, \qquad 0 < \frac{x}{t} < 1,$$

as a smooth solution in R, which connects the constant states

$$u = 0 \text{ for } x \le 0, \qquad u = 1 \text{ for } x \ge t.$$

This solution is called a *rarefaction wave*. The discontinuity at $t = 0$ is resolved instantaneously.

Example 3. (A pulse.)

$$f(x) = 1 \text{ for } 0 < x < 1, \qquad f(x) = 0 \text{ otherwise.}$$

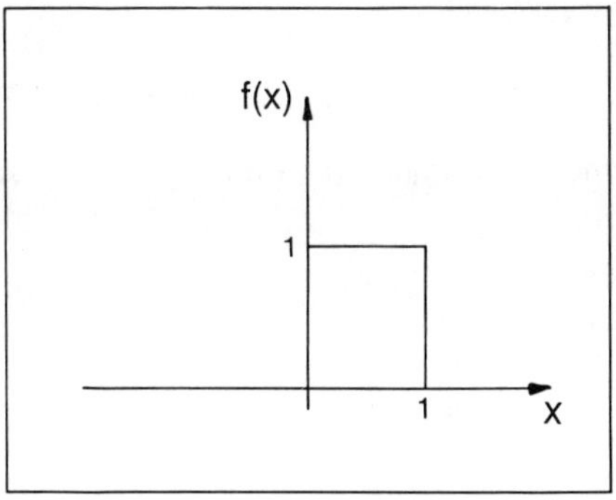

FIGURE 4.4.11. Initial pulse.

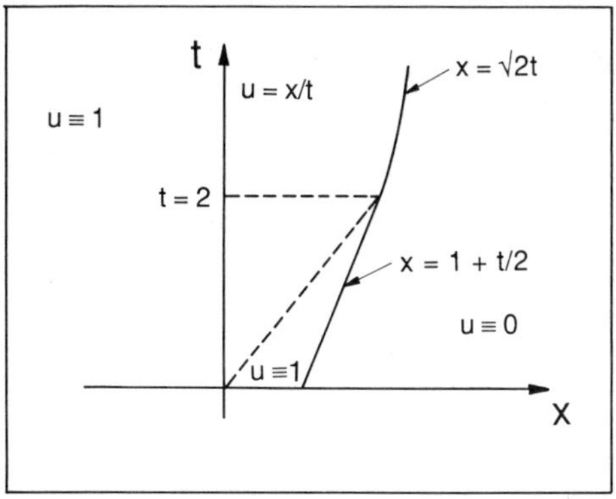

FIGURE 4.4.12. Rarefaction and shock.

At time $t = 2$ the rarefaction wave x/t starting at $x = 0$ catches up with the shock wave starting at $x = 1$. We can use the jump condition (4.4.22) to determine the location $s(t)$ of the shock for $t \geq 2$: The equation

$$s'(t) = \frac{1}{2}\frac{s(t)}{t}, \qquad s(2) = 2$$

is solved by

$$s(t) = \sqrt{2}\,t^{1/2}.$$

In these examples we have used the characteristics, the jump condition, and the form of a rarefaction to construct weak solutions. Unfortunately weak solutions are not unique. This follows from

Example 4. (A non-physical shock.)

$$f(x) = 0 \text{ for } x < 0, \qquad f(x) = 1 \text{ for } x > 0.$$

Thus we take the same initial condition which led to the rarefaction wave of Example 2. The function

$$u(x,t) = \begin{cases} 0 & x < \frac{1}{2}t, \\ 1 & x > \frac{1}{2}t, \end{cases}$$

is a piecewise smooth weak solution since the conditions of Theorem 4.4.5 are met.

Examples 2 and 4 describe two weak solutions to the same initial function.

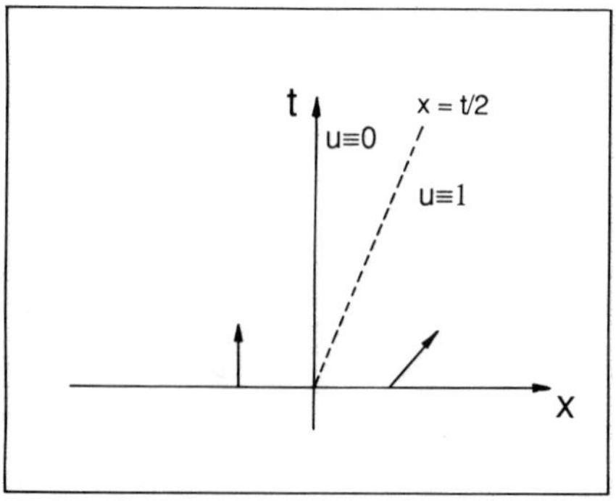

FIGURE 4.4.13. Non-physical shock.

Which solution should be preferred? There are two reasons why the shock solution of Example 4 is unphysical:

1) It is *not* true that two characteristics meet at the shock.
2) The solutions u_ϵ of the corresponding viscous problem converge to the rarefaction wave and do not converge to the shock solution of Example 4.

We do not prove the second statement here. The result is plausible, however, since we have shown in Section 4.4.2 that there are no traveling waves for $\epsilon > 0$ which are *increasing* in x; the shock solution of Example 4, however, corresponds to such a wave for $\epsilon = 0$.

Let us now outline the results for the general case. The viscous problems

$$u_t + uu_x = \epsilon u_{xx}, \qquad u(x, 0) = f(x),$$
$$f \in C^\infty, \qquad |f|_\infty < \infty, \qquad |f'|_\infty < \infty$$

have unique smooth solutions u_ϵ for all $\epsilon > 0$. These exist for all time. (We have shown this for periodic f. The present case can be treated as the limit where the period tends to infinity.) If ϕ is any test function, it follows that

(4.4.23)
$$\epsilon \int_0^\infty \int_{-\infty}^\infty u_\epsilon \phi_{xx}\, dx\, dt = -\int_0^\infty \int_{-\infty}^\infty (u_\epsilon \phi_t + \frac{1}{2} u_\epsilon^2 \phi_x)\, dx\, dt - \int_{-\infty}^\infty f(x)\phi(x, 0)\, dx.$$

Using a compactness argument for u_ϵ, one can show that there is a sequence $\epsilon = \epsilon_k$ and a bounded measurable function u with

$$u_\epsilon(x, t) \to u(x, t) \quad \text{as} \quad \epsilon = \epsilon_k \to 0.$$

(Compare the remark at the end of Section 4.3.1.) The convergence holds pointwise almost everywhere. By the maximum principle, the functions u_ϵ are uniformly bounded. Thus the integrals in (4.4.23) converge as $\epsilon = \epsilon_k \to 0$, and one obtains (4.4.19); therefore the limit u is a weak solution.

This shows the *existence* of weak solutions. As we have seen, weak solutions are not unique. However, if one considers only piecewise smooth weak solutions and requires that

$$u_l(x, t) > u_r(x, t) \qquad (4.4.24)$$

at each discontinuity, then – in this restricted class – weak solutions turn out to be unique. Condition (4.4.24) is a plausible requirement at each shock: The characteristic speed should be larger to the left of the shock than to the right. For practical purposes, one can restrict attention to piecewise smooth weak solutions. The *physically relevant* one is characterized by (4.4.24). This weak solution is the limit (almost everywhere) of the viscous solutions u_ϵ as $\epsilon \to 0$.

Roughly speaking, one has the following picture: The physically relevant weak solution u of the inviscid equation is a piecewise smooth function. The shock speeds are governed by the Rankine-Hugoniot condition (4.4.22). The function u jumps downwards if one crosses a shock from left to right. For small $\epsilon > 0$, the viscous solutions u_ϵ are uniformly close to u away from shocks. The convergence is of order $O(\epsilon)$ as $\epsilon \to \infty$. In a small neighborhood of the shocks, the viscous solutions behave like traveling waves connecting the state $u_l(x,t)$ with $u_r(x,t)$.

Notes on Chapter 4

Burgers' equation has been discussed in many books; for example in Whitham (1974). We have discussed the estimates in detail because the same procedure can be used – as we will see – in very general situations (systems, many space dimensions). One can make the global estimates more precise and prove $|\partial^p u/\partial x^p|_\infty = O(\epsilon^{-p})$. E. Hopf (1950) discussed the properties of the solutions of Burgers' equation by transforming the equation into the heat equation (Cole-Hopf transformation). In particular, he proved that the solutions converge for $\epsilon \to 0$ to a weak solution of the limit equation.

The solutions of parabolic equations

$$w_t = aw_{xx} + bw_x + cw, \quad a > 0,$$

have the property that the number of sign changes does not increase with time; see Matano (1982), Protter and Weinberger (1967). This can be used to obtain very precise information about the solutions of Burgers' equation

$$u_t + uu_x = \epsilon u_{xx}, \quad u(x,0) = f(x).$$

The derivative $v = u_x$ satisfies

$$v_t + uv_x + v^2 = \epsilon v_{xx}, \quad v(x,0) = f_x(x).$$

Assume first $f_x \geq 0$; then $v \geq 0$ for all times, and

$$\frac{d}{dt}(\max_x v) \leq -(\max_x v)^2.$$

Thus the solution u "becomes flat", i.e., we have a rarefaction. If $f_x \leq 0$ then

$$\frac{d}{dt}(\max_x |v|) \simeq (\max_x |v|)^2,$$

provided we can neglect the viscosity term. This is the case if $|v| \ll \epsilon^{-1}$. Therefore, the solution u "steepens up". When $|v| \simeq \epsilon^{-1}$, the viscosity term stops the growth of $|v|$, and we obtain a traveling wave. If f_x has p sign changes then – after some time – the solution u will show at most p traveling waves, separated by intervals where the solution is smooth. All these results can be generalized to equations

$$u_t + (f(u))_x = \epsilon u_{xx}.$$

In particular, Oleinik (1957, 1959) proved convergence of u for $\epsilon \to 0$ to the unique weak solution of

$$u_t + (f(u))_x = 0$$

which satisfies the so-called entropy condition

$$a(u_l(x,t)) > a(u_r(x,t)), \quad a(u) = \frac{df}{du}(u). \tag{4.4.25}$$

This condition generalizes (4.4.24).

There are no difficulties to prove local existence for the solutions of the Korteweg-de Vries equation

$$u_t + uu_x = \delta u_{xxx}.$$

In fact, one can show global existence (Sjöberg (1973)) because of conservation laws which yield estimates of the derivatives for all times. There is a very large body of literature on the KdV equation. An interesting result is the existence of traveling waves, so-called solitons, which can pass through each other without interacting. This is very surprising because of the nonlinear nature of these waves. For details, see Whitham (1974).

5 Nonlinear Systems in One Space Dimension

For Burgers' equation we used an iteration solving linear equations to obtain a local existence result. This technique carries over to first-order hyperbolic, to second-order parabolic and to mixed systems. The important point is again to obtain *a priori* estimates of the solution and its derivatives; these lead in turn to uniform estimates for the iteration sequence mentioned above.

Not to obscure the underlying principles, we shall not consider the most general situation, however, but restrict ourselves mainly to equations

$$\begin{aligned} u_t + A(u)u_x &= \epsilon u_{xx}, \quad \epsilon \geq 0 \\ u(x,0) &= f(x), \quad f \in C^\infty, \quad f(x) \equiv f(x+1) \end{aligned} \tag{5.1.1}$$

where $A = A(u)$ is a given smooth matrix function. As before, all (vector- and matrix-) functions are assumed to be *real*, for simplicity, and 1-periodic in x.

In Section 5.1 we treat the case where $A(u)$ and all its derivatives are globally bounded. For $\epsilon > 0$, one obtains existence for all time. If the system

$$u_t + A(u)u_x = 0 \tag{5.1.2}$$

is hyperbolic, one obtains short time existence for $\epsilon = 0$.

The assumption of global boundedness of $A(u)$ is, in general, not fulfilled in applications. In Section 5.2 we use a simple cut-off technique to treat more general cases and obtain short time existence results if $A(u)$ is smooth in a neighborhood of the initial data.

If a finite time $T_0 > 0$ is given (T_0 is not necessarily small), how can one decide whether a solution of (5.1.1) exists in $0 \le t \le T_0$? As for ordinary differential equations, this problem is often of a quantitative nature. We will sketch in Section 5.3, how asymptotic expansions and numerical calculations can be used to answer the finite time existence question, at least in principle.

Finally, in Section 5.4, we discuss a global existence result under more specific assumptions.

5.1. The Case of Bounded Coefficients

We first settle the uniqueness problem.

Lemma 5.1.1. *If $\epsilon > 0$ or if ($\epsilon = 0$ and $A(u) = A^*(u)$) then the problem* (5.1.1) *has at most one classical solution.*

Proof. Suppose u, v are solutions and $w = u - v$. We write

$$A(u)\ u_x - A(v)\ v_x = A(u)w_x + \big(A(u) - A(v)\big)v_x$$

and note that

$$|A(u) - A(v)| \le C_1\ |w|.$$

Therefore,

$$\begin{aligned}\frac{1}{2}\frac{d}{dt}(w, w) = (w, w_t) &\le -\big(w, A(u)w_x\big) + C_1|v_x|_\infty\ \|w\|^2 - \epsilon\|w_x\|^2 \\ &\le C_2\|w\|\ \|w_x\| + C_1|v_x|_\infty\ \|w\|^2 - \epsilon\|w_x\|^2 \\ &\le C_3(\epsilon)\ \|w\|^2 \quad \text{if } \epsilon > 0.\end{aligned}$$

If $\epsilon = 0$, $A = A^*$, then we have

$$-\big(w, A(u)w_x\big) = \frac{1}{2}\big(w, A(u)_x\ w\big),$$

and again it follows that

$$\frac{d}{dt}(w, w) \le C_4\ \|w\|^2.$$

In both cases the initial condition $w(x, 0) = 0$ implies $w \equiv 0$.

A priori estimates. We assume in this section that $A = A(u)$ and all its derivatives are globally bounded:

$$\sum_{|\nu|=p} |D^{\nu}\ A(u)| \leq K_p \ \text{ for all } \ u \in \mathbf{R}^n, \quad p = 0, 1, 2, \ldots \tag{5.1.3}$$

(This assumption, which is very restrictive for applications, will be relaxed below.) As in the uniqueness lemma, the case $\epsilon = 0$ requires an additional assumption, namely hyperbolicity.

Definition. *The first-order system*

$$u_t + A(u)u_x = 0 \tag{5.1.4}$$

is called hyperbolic if there is a smooth symmetrizer $H = H(u)$, i.e., a smooth matrix function $H(u)$ with

$$H(u) = H^*(u) > 0, \quad H(u)A(u) = A^*(u)H(u), \quad u \in \mathbf{R}^n.$$

To simplify the proof below, we will restrict ourselves to symmetric hyperbolic systems where

$$A(u) \equiv A^*(u), \quad H \equiv I\ . \tag{5.1.5}$$

Let $u = u(x,t)$ denote a smooth solution of (5.1.1), and let $u_j = \partial^j u/\partial\, x^j$. We first consider the parabolic case $\epsilon > 0$ and estimate u_j in terms of the initial data and its derivatives.

Lemma 5.1.2. *Assume the solution $u = u(x,t)$ of* (5.1.1) *is defined for $0 \leq t < T$, and the global bounds* (5.1.3) *hold. For each $\epsilon > 0$ there exists a constant $C_j = C_j(T, \epsilon)$ with*

$$\|u_j(\cdot, t)\| \leq C_j \ \text{ in } \ 0 \leq t < T. \tag{5.1.6}$$

The constant C_j depends on $\|f_j\|$.

Proof. Equation (5.1.1) implies that

$$\begin{aligned} \frac{d}{dt}(u,u) = 2(u, u_t) &= -2\big(u, A(u)u_1\big) - 2\epsilon\|u_1\|^2 \\ &\leq 2K_0\|u\|\ \|u_1\| - 2\epsilon\ \|u_1\|^2 \\ &\leq \frac{1}{\epsilon}\ K_0^2\|u\|^2. \end{aligned}$$

As usual, this yields

$$\|u(\cdot,t)\|^2 \leq \exp\big(\frac{1}{\epsilon}K_0^2 t\big)\ \|f\|^2.$$

Thus we have proved (5.1.6) for $j = 0$. Now assume (5.1.6) is proved up to $j - 1$. Clearly,

$$\frac{1}{2}\frac{d}{dt}(u_j, u_j) = (u_j, u_{jt}) = -\left(u_j, \left(A(u)u_1\right)_j\right) - \epsilon\|u_{j+1}\|^2$$

$$= \left(u_{j+1}, \left(A(u)u_1\right)_{j-1}\right) - \epsilon\|u_{j+1}\|^2.$$

Here*

$$\left(A(u)u_1\right)_1 = A(u)u_2 + A'(u)u_1u_1,$$

$$\left(A(u)u_1\right)_2 = A(u)u_3 + 3A'(u)u_1u_2 + A''(u)u_1u_1u_1,$$

etc. In general,

$$\left(A(u)u_1\right)_{j-1} = A(u)u_j + Q_j$$

where a bound

$$\|Q_j\| \le c_j$$

follows from (5.1.3), the Sobolev inequality

$$|u_k|_\infty^2 \le 2\left(\|u_{k+1}\|^2 + \|u_k\|^2\right),$$

and the induction hypothesis. Therefore,

$$\frac{1}{2}\frac{d}{dt}(u_j, u_j) \le \left(u_{j+1}, A(u)u_j\right) + c_j\|u_{j+1}\| - \epsilon\|u_{j+1}\|^2$$

$$\le K_0\|u_j\|\;\|u_{j+1}\| + \frac{1}{2\epsilon}\,c_j^2 - \frac{\epsilon}{2}\|u_{j+1}\|^2$$

$$\le \frac{1}{2\epsilon}\,K_0^2\|u_j\|^2 + \frac{1}{2\epsilon}\,c_j^2.$$

Now we can complete the induction step, and the lemma is proved.

Under the global assumption (5.1.3), we have shown *a priori* estimates of all derivatives of a solution for any given interval $0 \le t < T$. The estimates depend on $\epsilon > 0$, however.

Existence results. Existence of a solution can be shown by either using the difference approximation

$$dv_\nu/dt + A(v_\nu)D_+v_\nu = \epsilon D_+D_-v_\nu$$

*Formally, $A'(u)$ is a bilinear map acting an two vectors, $A''(u)$ is a trilinear map acting on three vectors, etc. This calculus is not essential here, however. It is only important that each matrix element of $A(u)$ has bounded partial derivatives of all orders.

or by considering the iteration

$$u_t^{k+1} + A(u^k)u_x^{k+1} = \epsilon\, u_{xx}^{k+1}, \qquad k = 0, 1, 2, \dots ,$$

$$u^{k+1}(x,0) = f(x), \qquad u^0(x,t) \equiv f(x).$$

Since the time interval $0 \le t < T$ for the *a priori* estimate was arbitrary, one obtains global existence. We summarize:

Theorem 5.1.3. *Given the global bounds* (5.1.3), *the parabolic system* (5.1.1) *has a unique smooth solution* $u_\epsilon = u_\epsilon(x,t)$ *defined for* $0 \le t < \infty$. *Here* $\epsilon > 0$ *is arbitrary.*

To treat the first-order system

$$u_t + A(u)u_x = 0, \; u(x,0) = f(x), \; f \in C^\infty, \; f(x) = f(x+1) \tag{5.1.7}$$

we assume symmetric hyperbolicity (5.1.5) and the boundedness assumption (5.1.3). We will show ϵ-independent bounds for the solutions u_ϵ of the parabolic equation in some time interval $0 \le t \le T$. Sending $\epsilon \to 0$, we obtain a solution of (5.1.7) in this interval.

Theorem 5.1.4. *Suppose that* (5.1.3) *and* (5.1.5) *hold. There exists a time* $T > 0$ *and a constant* $C > 0$, *both depending on* $\|f\|_{H^2}$ *but independent of* $\epsilon > 0$, *with*

$$\|u_\epsilon(\cdot,t)\|_{H^2} \le C \quad \text{in } 0 \le t \le T. \tag{5.1.8}$$

The symmetric hyperbolic system (5.1.7) *has a smooth solution* $u = u(x,t)$ *in* $0 \le t \le T$.

Proof. 1) We fix $\epsilon > 0$ and write $u = u_\epsilon$ for the solution of (5.1.1). Equation (5.1.1) gives us

$$\begin{aligned}\frac{d}{dt}(u,u) &= 2(u,u_t) = -2\big(u,\; A(u)u_1\big) - \epsilon\|u_1\|^2\\ &\le 2K_0\|u\|\;\|u_1\|.\end{aligned}$$

For $u_1 = \partial u/\partial x$ we have

$$u_{1t} = -\big(A(u)u_1\big)_1 + \epsilon u_3$$

and therefore

$$\begin{aligned}\frac{d}{dt}(u_1,u_1) &= 2(u_1,u_{1t}) = 2\big(u_2,\; A(u)u_1\big) - \epsilon\|u_2\|^2\\ &\le 2K_0\|u_1\|\;\|u_2\|.\end{aligned}$$

We now estimate u_2 to close the system. For u_2 it holds that

$$u_{2t} = -\big(A(u)u_1\big)_2 + \epsilon u_4,$$

and therefore,

$$\begin{aligned}\frac{d}{dt}\,(u_2, u_2) &\le 2\Big(u_3,\ \big(A(u)u_1\big)_1\Big)\\ &= \Big(u_3,\ \big(A(u)u_1\big)_1\Big) - \Big(u_2,\ \big(A(u)u_1\big)_2\Big).\end{aligned}$$

Here

$$\begin{aligned}\big(A(u)u_1\big)_1 &= A(u)u_2 + A'(u)u_1u_1\ ,\\ \big(A(u)u_1\big)_2 &= A(u)u_3 + 3A'(u)u_1u_2 + A''(u)u_1u_1u_1.\end{aligned}$$

The assumption $A = A^*$ implies cancellation of the terms

$$\big(u_3, A(u)u_2\big) \quad\text{and}\quad \big(u_2, A(u)u_3\big).$$

One finds that

$$\begin{aligned}\frac{d}{dt}\,(u_2, u_2) &\le \big(u_3,\ A'(u)u_1u_1\big) - 3\big(u_2,\ A'(u)u_1u_2\big) - \big(u_2,\ A''(u)u_1u_1u_1\big)\\ &= -5\big(u_2,\ A'(u)u_1u_2\big) - 2\big(u_2,\ A''(u)u_1u_1u_1\big)\\ &\le\ \text{const.}\ \big(|u_1|_\infty\ \|u_2\|^2 + |u_1|_\infty^2\ \|u_1\|\ \|u_2\|\big).\end{aligned}$$

Using Sobolev's inequality to bound $|u_1|_\infty$ by $\|u_2\| + \|u_1\|$, we arrive at a closed system of differential inequalities for $\|u\|^2$, $\|u_1\|^2$, $\|u_2\|^2$. Therefore the estimate (5.1.8) follows with T and C independent of ϵ.

2) Estimates of higher derivatives follow in the same way from

$$\begin{aligned}u_{jt} &= -\big(A(u)u_1\big)_j + \epsilon u_{j+2},\\ \frac{d}{dt}(u_j, u_j) &= 2(u_j, u_{jt}) \le 2\Big(u_{j+1},\ \big(A(u)u_1\big)_{j-1}\Big)\\ &= \Big(u_{j+1},\ \big(A(u)u_1\big)_{j-1}\Big) - \Big(u_j,\ \big(A(u)u_1\big)_j\Big).\end{aligned}$$

The time interval $0 \le t \le T$ determined in part 1 of the proof does not have to be diminished. This shows that the solutions u_ϵ of the parabolic problem are uniformly smooth in $0 \le t \le T$. The existence result for the hyperbolic case follows if we let $\epsilon \to 0$.

5.2. Local Existence Theorems

The assumption of global boundedness (5.1.3) is too restrictive for applications; for example, it is even violated for Burgers' equation. In this section we show how to localize the assumptions to a neighborhood of the initial data and how to obtain existence in a small time interval. The basic idea is simple, namely the solution *stays close to the initial data*, at least for a short time.

For a given initial function

$$f = f(x) \in C^\infty, \quad f(x) \equiv f(x+1)$$

let

$$U_\eta = \{u \in \mathbf{R}^n : |u - f(x)| < \eta \text{ for some } x\}$$

denote the η-*neighborhood* of the initial data. We choose a scalar cut-off function $\phi \in C^\infty(\mathbf{R}^n)$ with

$$\phi(u) = 1 \text{ for } u \in U_{\eta/4}, \ \phi(u) = 0 \text{ for } u \notin U_{\eta/2}.$$

If $A = A(u), A \in C^\infty(U_\eta)$, is the given coefficient matrix in (5.1.1), then set

$$\tilde{A}(u) = \begin{cases} \phi(u)A(u) & u \in U_\eta, \\ 0 & u \notin U_\eta. \end{cases}$$

The altered systems

(5.2.1) $$\tilde{u}_t + \tilde{A}(\tilde{u})\tilde{u}_x = \epsilon\tilde{u}_{xx}, \quad \tilde{u}(x,0) = f(x)$$

satisfy the boundedness assumption (5.1.3). First consider the case $\epsilon > 0$ without structural assumptions on A. By continuity, the solution $\tilde{u}_\epsilon$ of (5.2.1) stays in $U_{\eta/4}$ for some interval $0 \le t \le T_\epsilon$. Here $\tilde{u}_\epsilon$ solves the original system. This shows

Theorem 5.2.1. *If $A = A(u)$ is a C^∞-function in a neighborhood of the initial data, then the parabolic systems* (5.1.1) *have unique smooth solutions $u = u_\epsilon$ in some time interval $0 \le t \le T_\epsilon$. The time $T_\epsilon > 0$ depends on $\epsilon > 0$.*

We add now the (local) assumption of symmetric hyperbolicity:

(5.2.2) $$A(u) = A^*(u) \text{ for all } u \in U_\eta.$$

Clearly $\tilde{A}(u) = \tilde{A}^*(u)$ holds; thus the ϵ-independent estimates of Theorem 5.1.4 apply to $\tilde{u}_\epsilon$. There is a time $\tilde{T} > 0$ and a constant $C > 0$, both independent of ϵ, with

$$|\tilde{u}_{\epsilon t}(x,t)| \le C \text{ for } 0 \le t \le \tilde{T},$$

and thus

$$|\tilde{u}_\epsilon(x,t) - f(x)| = |\tilde{u}_\epsilon(x,t) - \tilde{u}_\epsilon(x,0)| \le Ct.$$

Therefore, in some time interval $0 \le t \le T,\ T > 0$, independent of $\epsilon > 0$, the functions $\tilde{u}_\epsilon$ solve the original system (5.1.1) and are uniformly smooth. For $\epsilon \to 0$, one obtains a solution of the hyperbolic problem. We summarize:

Theorem 5.2.2. *If $A(u)$ satisfies the symmetry condition (5.2.2) in a neighborhood of the initial data, then the problems (5.1.1) have unique smooth solutions $u_\epsilon = u_\epsilon(x,t)$ for $\epsilon \ge 0$ in some time interval $0 \le t \le T$. The time $T > 0$ can be chosen independently of ϵ.*

Generalizations. It is not difficult to generalize the local Existence Theorem 5.2.1 to parabolic equations

$$u_t = A_2(u,x,t,)\ u_{xx} + A_1(u,x,t)\ u_x + C(u,x,t), \tag{5.2.3}$$

where

$$A_2(u,x,t) + A_2^*(u,x,t) \ge 2\delta I, \quad \delta > 0,$$

in a neighborhood of the initial data. Also, more general first-order hyperbolic systems

$$\begin{aligned} v_t &= A(v,x,t)v_x + C(v,x,t), \\ H(v,x,t)A(v,x,t) &= A^*(v,x,t)H(v,x,t), \quad H > 0, \end{aligned} \tag{5.2.4}$$

can be treated as in Theorem 5.2.2. Finally, we can proceed as in Section 3.4 and obtain results for mixed hyperbolic-parabolic systems

$$\begin{pmatrix} u \\ v \end{pmatrix}_t = \begin{pmatrix} A_2 & 0 \\ 0 & 0 \end{pmatrix} \begin{pmatrix} u \\ v \end{pmatrix}_{xx} + \begin{pmatrix} B_{11} & B_{12} \\ B_{21} & B_{22} \end{pmatrix} \begin{pmatrix} u \\ v \end{pmatrix}_x + C. \tag{5.2.5}$$

Here $A_2,\ B_{ij},\ C$ are smooth functions of u, v, x, t and

$$A_2 + A_2^* \ge 2\delta I, \quad \delta > 0, \quad HB_{22} = B_{22}^* H, \quad H > 0.$$

One obtains

Theorem 5.2.3. *Consider the 1-periodic Cauchy problem for a parabolic system (5.2.3), a hyperbolic system (5.2.4) or a mixed hyperbolic-parabolic system (5.2.5). A unique solution exists in a sufficiently small time interval $0 \le t \le T$. The time T depends on the initial data.*

Application to Navier-Stokes. An example of a mixed system (5.2.5) is given by the one-dimensional viscous compressible Navier-Stokes equations:

$$\rho\left(\frac{\partial}{\partial t}+u\,\frac{\partial}{\partial x}\right)\begin{pmatrix}u\\ v\\ w\end{pmatrix}+\begin{pmatrix}p_x\\ 0\\ 0\end{pmatrix}=\begin{pmatrix}(2\mu+\mu')u_{xx}\\ \mu\, v_{xx}\\ \mu\, w_{xx}\end{pmatrix},$$

$$\rho_t+(\rho\, u)_x=0,\quad p=r(\rho).$$

The system reads in matrix form:

$$\frac{\partial}{\partial t}\begin{pmatrix}u\\ v\\ w\\ \rho\end{pmatrix}+\begin{pmatrix}u&0&0&\frac{1}{\rho}r'(\rho)\\ 0&u&0&0\\ 0&0&u&0\\ \rho&0&0&u\end{pmatrix}\frac{\partial}{\partial x}\begin{pmatrix}u\\ v\\ w\\ \rho\end{pmatrix}$$

$$=\frac{1}{\rho}\begin{pmatrix}2\mu+\mu'&0&0&0\\ 0&\mu&0&0\\ 0&0&\mu&0\\ 0&0&0&0\end{pmatrix}\frac{\partial^2}{\partial x^2}\begin{pmatrix}u\\ v\\ w\\ \rho\end{pmatrix}.$$

Theorem 5.2.3 applies if $\mu>0,\ \mu'\geq 0$ and if the initial density $\rho(x,0)$ is positive. Under the assumption $r'(\rho)>0$, the matrix

$$A(\mathbf{u},\rho)=\begin{pmatrix}u&0&0&\frac{1}{\rho}r'(\rho)\\ 0&u&0&0\\ 0&0&u&0\\ \rho&0&0&u\end{pmatrix}$$

has the symmetrizer

$$H(\mathbf{u},\rho)=\begin{pmatrix}1&0&0&0\\ 0&1&0&0\\ 0&0&1&0\\ 0&0&0&h\end{pmatrix},\quad h=\frac{r'(\rho)}{\rho^2}.$$

Thus one obtains a hyperbolic system for $\mu=\mu'=0,\ r'(\rho)>0$, and Theorem 5.2.3 applies.

5.3. Finite-Time Existence and Asymptotic Expansions

In most of the previous results we could obtain short-time existence only; however, in principle we were able to quantify the length of a guaranteed interval of existence. If a good approximate solution u_0 of our problem is known, then we can rewrite the equation as a problem for the difference $u-u_0$, and can employ the previous techniques to the rewritten equation. In this way, the guaranteed existence interval can often be improved considerably.

Example. Let us demonstrate the foregoing by the following example:

$$(5.3.1a) \quad u_t + uu_x = \alpha u, \quad u(x,0) = f(x), \quad f \in C^\infty, \quad f(x) \equiv f(x+1),$$

where $\alpha \in \mathbf{R}$ is a given constant. The general techniques yield short-time existence in an interval $0 \le t \le T$, $T = T(\|f\|_{H^2})$. Now suppose

$$(5.3.1b) \qquad f(x) = 1 + \epsilon g(x), \quad 0 < \epsilon \ll 1.$$

For $\epsilon = 0$, the problem has the solution

$$u_0(x,t) = e^{\alpha t},$$

existing for all time; this suggests the ansatz

$$u(x,t) = u_0(x,t) + \epsilon w(x,t).$$

For w one obtains that

$$(5.3.2) \qquad w_t + u_0 w_x + \epsilon w w_x = \alpha w, \quad w(x,0) = g(x).$$

It should be emphasized that no terms have been neglected, and therefore, if we obtain existence for w in $0 \le t \le T$, we also can solve the original problem in $0 \le t \le T$. Clearly, the nonlinear term in (5.3.2) is multiplied by the small parameter ϵ. (In generalizations of this example, the value of ϵ is a measure of the accuracy to which u_0 solves the given problem.)

Let us discuss (5.3.2). Proceeding as in Section 4.1.2, we obtain

$$\begin{aligned}
\frac{d}{dt}\|w_{xx}(\cdot,t)\|^2 &= -2\epsilon(w_{xx}, (ww_x)_{xx}) + 2\alpha\|w_{xx}\|^2 \\
&= -5\epsilon(w_{xx}, w_x w_{xx}) + 2\alpha\|w_{xx}\|^2 \\
&\le 5\epsilon|w_x|_\infty\|w_{xx}\|^2 + 2\alpha\|w_{xx}\|^2 \\
&\le 5\epsilon\|w_{xx}\|^3 + 2\alpha\|w_{xx}\|^2 \quad \text{(observe (4.1.5))}.
\end{aligned}$$

Consequently, if $y(t)$ is the solution of

$$\frac{d}{dt}y(t) = 2\alpha y(t) + 5\epsilon y^{3/2}(t), \quad y(0) = \|g_{xx}\|^2,$$

then w exists in any interval $0 \le t \le T$ where $y(t)$ is bounded. It is, of course, important to note that the nonlinear term $y^{3/2}$ – which may cause a singularity in y – appears multiplied by ϵ. Explicit discussion shows the following:

Case 1. $\alpha < 0$. If

$$\|g_{xx}\| < \frac{2|\alpha|}{5\epsilon},$$

then $y(t)$ decays exponentially, and thus w exists for all time.

Case 2. $\alpha = 0$. Now $y(t)$ has a singularity at a time

$$T_B = O(\frac{1}{\epsilon}) \quad (\text{if } \|g_{xx}\| = O(1)).$$

Thus we have long-time existence for small ϵ.

Case 3. $\alpha > 0$. In this case, $y(t)$ grows exponentially and has a pole at

$$T_B = O\big(\log \frac{1}{\epsilon\|g_{xx}\|}\big).$$

Again, if $\|g_{xx}\| = O(1)$, the blow-up time $T_B \to \infty$ as $\epsilon \to 0$, but the increase of T_B with decreasing ϵ is very slow. (In order to obtain long-time existence in this case, the perturbation term $\epsilon\|g_{xx}\|$ has to be "exponentially" small:

$$\epsilon\|g_{xx}\| \sim e^{-cT_B}.)$$

The behaviour of the blow-up times T_B as functions of ϵ can be predicted on basis of the *linearized* equation. Linearization of (5.3.1) at u_0 gives us

(5.3.3) $$\phi_{0t} + u_0\phi_{0x} = \alpha\phi_0.$$

If $\alpha < 0$, then ϕ_0 decays exponentially; the solution $u(x,t)$ and the approximation $u_0(x,t)$ stay close for all time if the pertubation ϵg is sufficiently restricted. If $\alpha = 0$, then the solution ϕ_0 stays bounded; the perturbation ϵg will, in general, grow over a long time interval. If $\alpha > 0$, then ϕ_0 grows exponentially; after a short time, the differences between u and u_0 can be $O(1)$ unless the perturbation is exponentially small.

Asymptotic expansion. By solving linear problems, one can increase the *power* of ϵ which multiplies the nonlinearity. Again, we illustrate this for the foregoing example. If ϕ_0 solves

$$\phi_{0t} + u_0\phi_{0x} = \alpha\phi_0, \quad \phi_0(x,0) = g(x),$$

and we write the solution w of (5.3.2) in the form

$$w = \phi_0 + \epsilon w_1,$$

then we obtain for w_1:

$$w_{1t} + (u_0 + \epsilon\phi_0)\phi_{1x} + \epsilon^2 w_1 w_{1x} = (\alpha - \epsilon\phi_{0x})w_1 - \phi_0\phi_{0x},$$
$$w_1(x, 0) = 0.$$

Obviously, the nonlinearity is now multiplied by ϵ^2. Repeating this process, we introduce

$$w = \phi_0 + \epsilon\phi_1 + \dots + \epsilon^{p-1}\phi_{p-1} + \epsilon^p w_p$$

into (5.3.2). Equating terms of order $\epsilon^0, \epsilon^1, \epsilon^2, \dots, \epsilon^{p-1}$ successivly, we obtain linear equations for $\phi_0, \dots, \phi_{p-1}$. If these are solved, the nonlinear equation for w_p can be written down:

$$w_{pt} + g_p w_{px} + \epsilon^{p+1} w_p w_{px} = h_p w_p + F_p$$

where

$$g_p = u_0 + \epsilon\phi_0 + \epsilon^2\phi_1 + \dots + \epsilon^p\phi_{p-1},$$
$$h_p = \alpha - \epsilon\phi_{0x} - \epsilon^2\phi_{1x} - \dots - \epsilon^p\phi_{p-1,x},$$

and F_p is also known in terms of $\phi_0, \dots, \phi_{p-1}$. In this way, we have reduced the nonlinearity to order ϵ^{p+1}; the solution u has the asymptotic expansion

$$u = u_0 + \epsilon\phi_0 + \epsilon^2\phi_1 + \dots + \epsilon^p\phi_{p-1} + \epsilon^{p+1}w_p.$$

Generalization. Without going into formal details, we want to indicate generalizations of the previous example. Consider a system

$$u_t = P(x, t, u, \partial/\partial x)u + F, \quad u(x, 0) = f(x), \tag{5.3.4}$$

where P is a differential operator of any of the forms treated in Sections 5.1, 5.2, and assume the coefficients of P are polynominals in u, for simplicity. Now suppose $u_0(x, t)$ is a known approximate solution in some time interval $0 \le t \le T$; more precisely, if we substitute u_0 into (5.3.4), the equations will be satisfied with small defects:

$$\begin{aligned} u_{0t} &= P(x, t, u_0, \partial/\partial x)u_0 + \tilde{F}, \qquad & F &= \tilde{F} + \epsilon G, \\ u_0(x, 0) &= \tilde{f}(x), & f &= \tilde{f} + \epsilon g. \end{aligned} \tag{5.3.5}$$

(Formally, we can *define* the terms ϵG, ϵg by the above equations. The parameter ϵ is introduced artificially in such a way that $G = O(1)$, $g = O(1)$.) Then we can view the given system (5.3.4) as a pertubation of (5.3.5), which suggests an ansatz

$$u(x, t) = u_0(x, t) + \epsilon w(x, t).$$

For w one obtains:

$$w_t = P_1(x,t,\partial/\partial x)w + \epsilon P_2(x,t,w,\epsilon,\partial/\partial x)w + G,$$

$$w(x,0) = g(x).$$

Here P_1 and P_2 are differential operators whose coefficients also depend on u_0; however, since u_0 is assumed to be known, this is suppressed in the notation. The operator P_1 is linear, and the nonlinear part is multiplied by ϵ.

If the equation for w is of the type discussed in Sections 5.1, 5.2, we can derive a priori estimates for w. As in the example, the growth behaviour of the solution $\tilde{w}$ of the linear problem

(5.3.6) $$\tilde{w}_t = P_1(x,t,\partial/\partial x)\tilde{w}$$

is crucial. If the solutions of (5.3.6) satisfy

$$\frac{d}{dt}\|\tilde{w}\|_{H^2}^2 \le 2\alpha\|\tilde{w}\|_{H^2}^2,$$

then one can obtain a priori estimates for w of the form

$$\frac{d}{dt}\|w\|_{H^2}^2 \le 2(\alpha + O(\epsilon))\|w\|_{H^2}^2 + \text{const}|\epsilon|\|w\|_{H^2}^{2+q}.$$

Therefore, if $|\epsilon|$ is sufficiently small, the solution w – and thus u – will exist in $0 \le t \le T$. Also, if ϕ_0 solves the linearized equation

$$\phi_{0t} = P_1(x,t,\partial/\partial x)\phi_0 + G, \quad \phi_0(x,0) = g(x),$$

then $u = u_0 + \epsilon\phi_0 + O(\epsilon^2)$. Higher order asymptotic expansions

$$u = u_0 + \epsilon\phi_0 + \epsilon^2\phi_1 + \ldots + \epsilon^p\phi_{p-1} + O(\epsilon^{p+1})$$

can be derived as in the example; the functions $\phi_0, \ldots, \phi_{p-1}$ solve linear problems.

Remark on numerical calculations. The approximate solution u_0 can also be obtained through interpolation of numerically computed data v^h. It is possible to bound the defects ϵG, ϵg defined in (5.3.5) in terms of divided differences of v^h; the latter can be computed. For this reason, if a numerical solution v^h is computed in $0 \le t \le T$, and sufficiently many divided differences stay within reasonable bounds, one has an indication that the partial differential equation has a solution in $0 \le t \le T$. A better test is obtained if one repeats the computation with step-sizes

$$h_0 > h_1 > h_2 > \ldots.$$

If the bounds for the divided differences do not grow for decreasing step-sizes, the defects of the interpolant $u_0 = u_0^h$ will finally be small enough to ensure existence of a solution of the p.d.e. On the other hand, if the divided differences do grow for decreasing h, one has an indication for blow-up at a time $T_B < T$.

5.4. On Global Existence for Parabolic and Mixed Systems

Global (in time) existence theorems for systems

$$(5.4.1) \qquad \begin{aligned} u_t + A(u)u_x &= \epsilon u_{xx}, \quad \epsilon > 0, \\ u(x,0) &= f(x), \quad f,\, A \in C^\infty, \quad f(x) \equiv f(x+1) \end{aligned}$$

can only be derived under rather special assumptions. However, since we have a local existence and uniqueness result, we can always piece local (in time) solutions together and obtain a half-open maximal interval of existence $0 \le t < T_0$. Here T_0 is finite or $T_0 = \infty$. If T_0 is finite, then

$$\sup\bigl\{|u(\cdot,t)|_\infty : 0 \le t < T_0\bigr\}$$

cannot be bounded. Otherwise we could apply the cut-off technique described in Section 5.2 and use Theorem 5.1.3 to extend the solution beyond T_0. For the systems (5.4.1) we thus have the alternative:

Lemma 5.4.1. *Either* (i) *a smooth solution u exists for $0 \le t < \infty$, or* (ii) *there is a blow-up time $T_0 < \infty$ with*

$$\sup\bigl\{|u(\cdot,t)|_\infty : 0 \le t < T_0\bigr\} = \infty.$$

Phrased concisely, the lemma states that “the solution exists as long as it stays bounded”.

We want to describe conditions which exclude possibility (ii); thus these conditions imply all-time existence. First consider a specific example:

$$(5.4.2) \qquad \begin{aligned} v_t + vv_x + p_x &= \epsilon v_{xx}, \\ p_t + v_x &= \epsilon p_{xx}, \quad \epsilon > 0. \end{aligned}$$

Suppose $u = (v,p)$ is a smooth solution. The basic energy estimate follows from

$$\begin{aligned} \frac{1}{2}\frac{d}{dt}\bigl\{\|v\|^2 + \|p\|^2\bigr\} &= -(v, vv_x) - (v, p_x) - (p, v_x) - \epsilon\bigl\{\|v_x\|^2 + \|p_x\|^2\bigr\} \\ &= -\epsilon\bigl\{\|v_x\|^2 + \|p_x\|^2\bigr\}. \end{aligned}$$

Integration from 0 to t yields

$$\|v(\cdot,t)\|^2 + \|p(\cdot,t)\|^2 \le \|v(\cdot,0)\|^2 + \|p(\cdot,0)\|^2 \tag{5.4.3}$$

and

$$2\epsilon \int_0^t \left\{\|v_x(\cdot,\tau)\|^2 + \|p_x(\cdot,\tau)\|^2\right\} d\tau \le \|v(\cdot,0)\|^2 + \|p(\cdot,0)\|^2. \tag{5.4.4}$$

The coefficient matrix

$$A(u) = A(v,p) = \begin{pmatrix} v & 1 \\ 1 & 0 \end{pmatrix}$$

of the example (5.4.2) clearly satisfies an estimate of the form

$$|A(u)| \le K_1\big(1 + |u|\big). \tag{5.4.5}$$

We will show in the following theorem that estimates like (5.4.3), (5.4.4), (5.4.5) exclude possibility (ii) of Lemma 5.4.1; thus they imply all-time existence.

Theorem 5.4.2. *Suppose that* (5.4.5) *holds, and a solution* u *of* (5.4.1) *is defined for* $0 \le t < T < \infty$ *and satisfies*

$$\|u(\cdot,t)\| \le K_2, \quad 0 \le t < T, \tag{5.4.6}$$

$$\int_0^t \|u_x(\cdot,\tau)\|^2 \, d\tau \le K_3, \quad 0 \le t < T. \tag{5.4.7}$$

Then

$$\sup_{0\le t<T} \|u(\cdot,t)\|_{H^1} \quad \textit{and thus} \quad \sup_{0\le t<T} |u(\cdot,t)|_\infty$$

are finite; thus u *can be extended beyond* T. *Furthermore, there is a bound*

$$\|u(\cdot,t)\|_{H^1} \le K_4, \quad 0 \le t < T,$$

where K_4 *only depends on* ϵ, K_2, K_3, *and* $\|u(\cdot,0)\|_{H^1}$, *but does not depend on* T.

Proof. Differentiation of (5.4.1) yields

$$\begin{aligned}
\frac{1}{2}\frac{d}{dt}\|u_1\|^2 &= -\Big(u_1,\ \big(A(u)u_1\big)_1\Big) - \epsilon\|u_2\|^2 \\
&= \big(u_2, A(u)u_1\big) - \epsilon\|u_2\|^2 \\
&\le K_1(1 + |u|_\infty)\ \|u_1\|\ \|u_2\| - \epsilon\ \|u_2\|^2 \\
&\le \frac{1}{2\epsilon}\ K_1^2\ (1 + |u|_\infty)^2\ \|u_1\|^2 \le \frac{1}{\epsilon}\ K_1^2\ (1 + |u|_\infty^2)\ \|u_1\|^2.
\end{aligned}$$

To estimate $|u|_\infty$, we use Lemma 3.2.5,

$$|u|_\infty^2 \leq 2\|u\|\,\|u_1\| + \|u\|^2,$$

and find that

$$\begin{aligned}\frac{d}{dt}\,\|u_1\|^2 &\leq \frac{2}{\epsilon}\,K_1^2\big(1 + 2\|u\|\,\|u_1\| + \|u\|^2\big)\,\|u_1\|^2\\ &\leq 2\alpha\|u_1\|^2 + 2\beta\|u_1\|^3, \quad \alpha = \frac{1}{\epsilon}\,K_1^2(1+K_2^2), \quad \beta = \frac{2}{\epsilon}\,K_1^2\,K_2.\end{aligned}$$

If we set

$$y(t) = \|u_1(\cdot,t)\|, \quad 0 \leq t < T,$$

then we have shown that

$$\frac{d}{dt}y^2(t) = 2y(t)y'(t) \leq 2\alpha y^2(t) + 2\beta y^3(t);$$

hence

$$y'(t) \leq \alpha y(t) + \beta y^2(t) \quad \text{if} \quad y(t) \neq 0. \tag{5.4.8}$$

To bound $y(t)$, we fix a time $0 \leq t < T$ with $y(t) > 1$.

Case 1. There is $t_0 < t$ with $y(t_0) = 1$. Let T_0 denote the largest of these t_0-values, thus $y(\tau) > 1$ for $T_0 < \tau \leq t$. Integration of (5.4.8) gives us

$$y(t) - y(T_0) \leq (\alpha+\beta)\int_{T_0}^t y^2(\tau)\,d\tau,$$

and thus $y(t) \leq 1 + (\alpha+\beta)K_3$.

Case 2. For $0 \leq \tau < t$ we have $y(\tau) > 1$. Integration of (5.4.8) yields

$$y(t) \leq y(0) + (\alpha+\beta)K_3,$$

and the theorem is proved.

In the example (5.4.2), the estimates (5.4.6), (5.4.7) are valid *a priori* in any interval $0 \leq t < T$ where the solution exists. Therefore, according to the previous theorem and Lemma 5.4.1, the solution exists for all time. Furthermore, since the constants K_2 and K_3 do not depend on T, the bound K_4 derived in the theorem is uniform for $0 \leq t < \infty$. In particular, the values of the solution stay bounded for all time:

$$|u(\cdot,t)|_\infty \leq K_5, \quad 0 \leq t < \infty.$$

A special mixed system. We shall now show all-time existence for the mixed hyperbolic-parabolic system

$$v_t + vv_x + p_x = \epsilon v_{xx}, \quad \epsilon > 0,$$

$$p_t + v_x = 0.$$

Integration by parts gives us

$$\frac{d}{dt}\{\|v\|^2 + \|p\|^2\} = -2\epsilon\|v_x\|^2,$$

$$\|v(\cdot,t)\|^2 + \|p(\cdot,t)\|^2 \leq \|v(\cdot,0)\|^2 + \|p(\cdot,0)\|^2,$$

$$2\epsilon\int_0^T \|v_x\|^2\,dt \leq \|v(\cdot,0)\|^2 + \|p(\cdot,0)\|^2.$$

Therefore, by the Sobolev inequality, Lemma 3.2.5,

(5.4.9) $$\int_0^T |v|_\infty^2\,dt \leq \text{const}.$$

For the first x-derivatives we obtain

$$\begin{aligned}\frac{d}{dt}\{\|v_1\|^2 + \|p_1\|^2\} &= -2\big(v_1,(vv_1)_x\big) - 2\epsilon\|v_2\|^2\\ &\leq 2|v|_\infty\,\|v_1\|\,\|v_2\| - 2\epsilon\|v_2\|^2\\ &\leq \frac{1}{\epsilon}\,|v|_\infty^2\,\|v_1\|^2 - \epsilon\|v_2\|^2\\ &\leq \frac{1}{\epsilon}\,|v|_\infty^2\,\{\|v_1\|^2 + \|p_1\|^2\} - \epsilon\|v_2\|^2.\end{aligned}$$

Thus, using (5.4.9), we obtain a bound for

$$\sup_{0\leq t<T}\{\|v_1(\cdot,t)\|^2 + \|p_1(\cdot,t)\|^2\}, \quad \int_0^T \|v_2\|^2\,dt, \quad \int_0^T |v_1|_\infty^2\,dt.$$

This process can be continued, and existence follows.

Notes on Chapter 5

For local existence theorems there is no essential difference between linear and nonlinear equations. Global existence requires global bounds, however. One such result for parabolic systems is given in Theorem 5.4.2. Others can be found in Smoller (1983) and in Ladyzhenskaya, Solonnikov, and Uralceva (1968). Much less is known for the inviscid case $\epsilon = 0$. We cannot expect global existence. Classical solutions cease to exist beyond a certain blow-up time; see John (1974), Lax (1964), Glimm and Lax (1970). If one can write the equations in conservation form

$$u_t + (g(u))_x = 0, \quad (g(u))_x = A(u)u_x,$$

then one can introduce again weak solutions. A (not necessarily smooth) function v is a weak solution if

$$\int_0^{\infty}\int_{-\infty}^{+\infty} (\phi_t v + \phi_x g(v))\, dx\, dt + \int_{-\infty}^{+\infty} \phi(x,0)v(x,0)\, dx = 0$$

for all $\phi \in C_0^{\infty}$. Weak solutions are not unique. Generalizing (4.4.25), Lax (1957) introduced entropy conditions for systems. Glimm (1965) proved that there is always a global weak solution which satisfies these conditions, provided the initial data are sufficiently close to a constant vector. However, it is not known whether the solution is unique. For later results we refer to DiPerna (1985), Liu (1983), and Majda (1984). An important example is the system of conservation laws for gasdynamics:

$$\begin{aligned} \rho_t + (\rho u)_x &= 0 \\ (\rho u)_t + (\rho u^2 + p)_x &= 0 \\ (\rho E)_t + (\rho E u + p u)_x &= 0, \quad E = e + \frac{1}{2}u^2. \end{aligned}$$

Here ρ, u, p, e denote the density, velocity, pressure, and internal energy, respectively. Also, an equation of state which connects e, ρ, p has to be specified. Many people have studied these equations for the last 150 years using less mathematical machinery. For an account see Courant and Friedrichs (1948), Whitham (1974), and Landau and Lifshitz (1959). Little is known about the behavior of the solutions of (5.4.1) as $\epsilon \to 0$. However, Gilbarg (1951) proved that the viscous gasdynamical equations allow traveling waves which converge for $\epsilon \to 0$ to a weak solution of the limit equation. For other results we refer to Smoller (1983).

6 The Cauchy Problem for Systems in Several Dimensions

The results of Chapter 3 about *linear* parabolic, strongly hyperbolic, and mixed systems can mostly be generalized from one space dimension to an arbitrary number of space dimensions. These generalizations are carried out in the first three sections of this chapter. In particular, we show well-posedness of the Cauchy problem for the linearized compressible Navier-Stokes and Euler equations. Section 6.4 treats short-time existence for *nonlinear* systems; we carry out the details for the symmetric hyperbolic case and sketch generalizations to systems of different type. For a special class of nonlinear parabolic systems in two space dimensions, we will prove all-time existence in Section 6.5.

6.1. Linear Parabolic Systems

We consider second-order parabolic systems of the general form

$$u_t = \sum_{i,j=1}^{s} D_i(A_{ij}D_ju) + \sum_{i=1}^{s} B_iD_iu + Cu + F. \tag{6.1.1}$$

Here s is the number of space dimensions and D_i is the operator $D_i = \partial/\partial x_i$. The coefficient matrices $A_{ij} = A_{ij}(x,t)$, $B_i = B_i(x,t)$, $C = C(x,t)$ and the forcing function $F = F(x,t)$ are assumed to be C^∞-smooth, real, and 1-periodic in each component $x_i, i = 1, \ldots, s$. The concept of parabolicity will be defined below. For the function $u(x, t)$ an initial condition

$$u(x,0) = f(x), \quad f \in C^\infty, \quad f \text{ 1-periodic in each } x_i \tag{6.1.2}$$

is given, where f takes values in $\mathbf{R}^n$. As before, we seek a smooth solution $u = u(x, t)$ taking values in $\mathbf{R}^n$ which is 1-periodic in each x_i, $i = 1, \ldots, s$.

In Section 6.1.1 the concept of strong parabolicity will be defined; it leads to a simple derivation of solution-estimates. Existence of a solution can again be shown by proving the analogous estimates for a difference scheme. This will be sketched in Section 6.1.2. Existence, uniqueness, and estimates can be generalized from strongly parabolic systems to systems for which a pointwise definition of parabolicity is used. For our applications this generalization is of minor importance, however, and we only sketch the proof.

6.1.1. Estimates for Strongly Parabolic Systems

The following definition generalizes our previous concept of strong parabolicity to systems in any number of space dimensions.

Definition. The system (6.1.1) is called *strongly parabolic* in $0 \le t \le T$ if there exists a constant $\delta > 0$ such that the estimate

$$\sum_{i,j=1}^{s} \langle y_i, A_{ij}(x,t) y_j \rangle \ge \delta \sum_{i=1}^{s} |y_i|^2$$

holds for all $x \in \mathbf{R}^s$, $0 \le t \le T$, and all vectors $y_i \in \mathbf{R}^n$, $i = 1, \ldots, s$.

The basic result about existence, uniqueness, and estimates is

Theorem 6.1.1. *If* (6.1.1) *is strongly parabolic in* $0 \le t \le T$, *then the Cauchy problem* (6.1.1), (6.1.2) *has a unique classical solution* $u = u(x, t)$ *in* $0 \le t \le T$. *The solution* u *is a* C^∞*-function. For each* $p = 0, 1, 2, \ldots$, *there is a constant* K_p *with*

$$\|u(\cdot,t)\|_{H_p}^2 \le K_p \left(\|f\|_{H_p}^2 + \int_0^t \|F(\cdot,\tau)\|_{H_p}^2 \, d\tau \right), \quad 0 \le t \le T. \tag{6.1.3}$$

The constant K_p *depends on* δ *and the maximum norm of the coefficients of* (6.1.1) *and their derivatives of order* $\le p$, *but not on* F, f.

Proof. We start with the estimates. Integration by parts yields

$$\begin{aligned}\frac{1}{2}\frac{d}{dt}(u,u) &= (u,u_t)\\ &= -\sum_{i,j=1}^{s}(D_i u,\ A_{ij}D_j u) + \sum_{i=1}^{s}(u, B_i D_i u) + (u, Cu) + (u, F)\\ &\le \sum_{i=1}^{s}\left(-\delta\|D_i u\|^2 + |B_i|_\infty\ \|D_i u\|\ \|u\|\right) + |C|_\infty\ \|u\|^2 + \|u\|\ \|F\|\\ &\le -\frac{\delta}{2}\sum_{i=1}^{s}\|D_i u\|^2 + \text{ const }\|u\|^2 + \frac{1}{2}\|F\|^2.\end{aligned}$$

Therefore, (6.1.3) follows for $p = 0$ if we apply Lemma 3.1.1. This estimate also shows the uniqueness of a classical solution.

To estimate first derivatives, we differentiate the equation (6.1.1) with respect to x_i, $i = 1, \dots, s$, and obtain a system of similar type for the vector

$$(D_1 u, D_2 u, \dots,\ D_s u).$$

The coefficients of this system depend on A_{ij}, B_i, C and the first derivatives of these matrices. The desired estimate follows as above. Estimates for higher derivatives are obtained by repeated differentiation.

In the next section we sketch the proof of the existence of u.

6.1.2. Existence via Difference Approximations

Let $h = 1/N$ denote a gridlength, N a natural number. For any multi-index

$$\nu = (\nu_1, \dots, \nu_s), \quad \nu_j \in \mathbf{Z},$$

let $x_\nu = (h\nu_1, \dots, h\nu_s)$ denote the corresponding meshpoint. We consider gridfunctions $v_\nu = v(x_\nu)$ which are assumed to be 1-periodic in each coordinate direction. By E_j we denote the translation operator in the j-th coordinate:

$$E_j v_\nu = E_j v(x_\nu) = v(x_\nu + he_j),$$

$$e_j = (0, \dots, 0, 1, 0, \dots, 0) \ \text{ with 1 at position } j.$$

The powers of E_j are

$$E_j^p v_\nu = v(x_\nu + phe_j), \quad p \in \mathbf{Z},$$

and $E_j^0 = I$ is the identity. The forward, backward, and centered difference operators in the j-th coordinate direction are

$$D_{+j} = h^{-1}(E_j - I), \qquad D_{-j} = h^{-1}(I - E_j^{-1}),$$
$$D_{0j} = \frac{1}{2}(D_{+j} + D_{-j}), \qquad j = 1, \ldots, s.$$

Using these notations, we approximate (6.1.1) by the system of ordinary differential equations

$$\frac{dv_\nu}{dt} = \sum_{i,j=1}^{s} D_{-i}\big(A_{ij}(x_\nu, t) D_{+j} v_\nu\big) + \sum_{i=1}^{s} B_i(x_\nu, t) D_{0i} v_\nu + C(x_\nu, t) v_\nu + F(x_\nu, t). \tag{6.1.4}$$

If we introduce a discrete scalar product and norm for gridfunctions by

$$(v, w)_h = h^s \sum_{0 \le \nu_j < N} \langle v_\nu,\ w_\nu \rangle, \quad \|v\|_h = (v, v)_h^{1/2},$$

then all computational rules of Section 3.2.2 carry over in an obvious way to the multi-dimensional case. Therefore, we can estimate the solutions of (6.1.4) and their divided differences independently of h. We just mimic the continuous process. As in the one dimensional case, we can interpolate $v_\nu(t) = v_\nu^h(t)$ with respect to the x-variables by Fourier polynomials $w^h(x, t)$. (See Appendix 2.) These are uniformly (with respect to h) smooth; in the limit $h \to 0$ we obtain a solution $u = u(x, t)$ of the Cauchy problem. There is no essential difference from the one dimensional case.

6.1.3. General Linear Second-Order Parabolic Systems

In Section 2.4.3 we have defined parabolicity for problems with constant coefficients. Applying this concept to all frozen-coefficient problems, we are lead to:

Definition. *The system* (6.1.1) *is called* parabolic *in* $0 \le t \le T$ *if there is a constant* $\delta > 0$ *such that for all* $x \in \mathbf{R}^s$, $0 \le t \le T$, *and all* $\omega = (\omega_1, \ldots, \omega_s) \in \mathbf{R}^s$, *the eigenvalues* κ *of the matrix*

$$\sum_{i,j=1}^{s} \omega_i\, \omega_j\, A_{ij}(x, t)$$

satisfy

$$\operatorname{Re} \kappa \ge \delta |\omega|^2.$$

One can prove

Theorem 6.1.2. *The statements of Theorem* 6.1.1 *are valid if the assumption of strong parabolicity is replaced by the assumption of parabolicity.*

We only sketch the proof of this result. The essential point is to obtain the *a priori* estimates (6.1.3); then existence of a solution follows as before by use of a difference scheme.

To obtain the estimate (6.1.3) for $p = 0$, first fix a point (x_0, t_0) and consider the constant-coefficient parabolic operator

$$P_2(x_0, t_0, D) = \sum_{i,j=1}^{s} D_i \; A_{ij}(x_0, t_0) \; D_j.$$

If we proceed as in Section 2.7 – but replace Fourier transforms by Fourier expansions – we can construct an inner product

$$(u, v)_{H(x_0,t_0)} = \sum_{\omega} \langle \hat{u}(\omega), \; H(x_0, t_0, \omega)\hat{v}(\omega) \rangle$$

with

$$\big(u, P_2(x_0, t_0, D)u\big)_{H(x_0,t_0)} + \big(P_2(x_0, t_0, D)u, u\big)_{H(x_0,t_0)} \leq -\delta \sum_{i=1}^{s} \|D_i u\|^2$$

for all smooth functions $u = u(x)$. Using the corresponding norms and a partition of unity argument as in Section 3.2.5, we obtain the basic energy estimate as before. Estimates for derivatives can be obtained by differentiating the differential equation.

As noted in Section 3.2.6, a basic property of parabolic differential operators is their smoothing ability. This smoothing property is also valid for equations in more than one space dimension and can be proved as in the one dimensional case.

6.2. Linear Hyperbolic Systems

A first-order system

(6.2.1a) $$u_t = P(x, t, \; \partial/\partial x)u + C(x, t)u + F(x, t)$$

is called *symmetric hyperbolic* if $P(x, t, \; \partial/\partial x)$ has the form

(6.2.1b) $$P(x, t, \; \partial/\partial x)u = \frac{1}{2} \sum_{j=1}^{s} \left\{ B_j(x, t) D_j u + D_j \big(B_j(x, t) u \big) \right\},$$

and $B_j(x,t) = B_j^*(x,t)$ for all x, t. As before, the coefficients B_j, C, and the forcing function F are assumed to be real, C^∞-smooth, and 1-periodic in each x_j.

To obtain existence for the Cauchy problem (6.2.1), (6.1.2), we consider first the parabolic problems

$$u_t = \epsilon\,\Delta u + P(x,t,\,\partial/\partial x)u + Cu + F, \quad \epsilon > 0.$$

The existence result of Theorem 6.1.1 applies, and the estimates are independent of $\epsilon > 0$. For $\epsilon \to 0$, one obtains a solution of the symmetric hyperbolic problem.

A basic property of hyperbolic equations is the "finite speed of propagation". We formulate the result in the following theorem; the proof uses the convergence of an explicit difference scheme.

Theorem 6.2.1. *Define*

$$a = \max\left\{\sum_{j=1}^{s} \left|B_j(x,t)\right| : x \in \mathbf{R}^s,\ \ 0 \le t \le t_0\right\}.$$

The solution-value $u(x_0,t_0)$ depends only on those values of $F(x,t)$ for (x,t) lying in the cone

$$|x_0 - x| \le a(t_0 - t), \quad 0 \le t \le t_0,$$

and only on those values of $f(x)$ for x lying in the interval $|x_0 - x| \le at_0$.

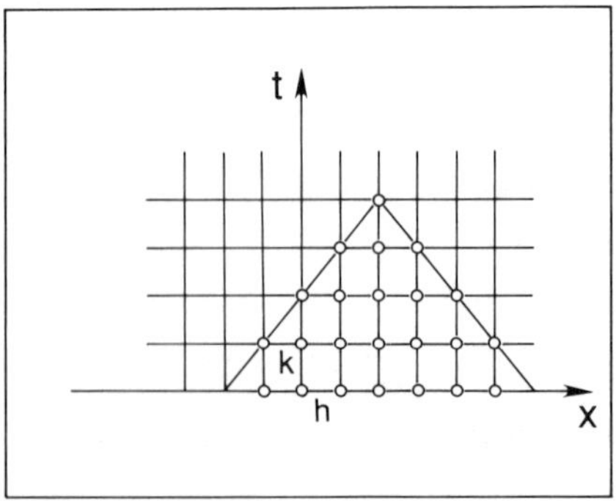

FIGURE 6.2.1. Domain of dependence and mesh.

Proof. 1. As in Section 6.2.1, we introduce meshpoints x_ν, $\nu \in \mathbf{Z}^s$, but discretize in time also. Let $k > 0$ denote the time step-size. We approximate (6.2.1) by the difference approximation

$$\begin{aligned}\big(I - kC(x_\nu, t)\big)\, v_\nu(t+k) &= \big(I + kC(x_\nu, t)\big)\, v_\nu(t-k)\\ &\quad + 2kQ(x_\nu, t)\, v_\nu(t) + 2kF(x_\nu, t),\\ &\quad t = k,\ 2k,\ 3k, \ldots,\end{aligned} \tag{6.2.2}$$

$$Q(x_\nu, t)\, v_\nu(t) = \frac{1}{2}\sum_{j=1}^{s}\left\{B_j(x_\nu, t)D_{0j}v_\nu(t) + D_{0j}\big(B_j(x_\nu, t)v_\nu(t)\big)\right\}.$$

Here

$$D_{0j} = \frac{1}{2}(D_{+j} + D_{-j}) = \frac{1}{2h}(E_j - E_j^{-1})$$

denotes the usual centered difference operator. The three level difference scheme (6.2.2) is supplemented by the starting conditions

(6.2.3)
$$v_\nu(0) = f(x_\nu),\ v_\nu(k) = f(x_\nu) + k\big\{Q(x_\nu, 0) + C(x_\nu, 0)\big\}f(x_\nu) + k\ F(x_\nu, 0).$$

We fix a relation $k = h/(a+\eta)$, $\eta > 0$ arbitrary, and send $h \to 0$. The difference equations (6.2.2), (6.2.3) define an approximation $v = v^h$ for the exact solution u, and we will prove the convergence

$$\lim_{h\to 0}\ \max_{0\le t=\mu k\le t_0}\ \|u(x_\nu, t) - v_\nu(t)\|_h = 0. \tag{6.2.4}$$

The domain of dependence of v^h belongs to the cone

$$|x_0 - x| \le (a+\eta)(t_0 - t), \quad 0 \le t \le t_0.$$

Since $\eta > 0$ is arbitrary and u is smooth, the assertion of the theorem follows.
2. It remains to show (6.2.4). First note that the operators D_{0j} are antisymmetric:

$$\begin{aligned}(p, D_{0j}q)_h &= \frac{1}{2}(p, D_{+j}q)_h + \frac{1}{2}(p, D_{-j}q)_h\\ &= -\frac{1}{2}(D_{-j}p, q)_h - \frac{1}{2}(D_{+j}p, q)_h\\ &= -(D_{0j}p, q)_h.\end{aligned}$$

Therefore, also the operators $Q(t)$ defined in (6.2.2) are antisymmetric:

$$\big(p, Q(t)q\big)_h = -\big(Q(t)p, q\big)_h.$$

We give a simple bound for $Q(t)$:
The estimate

$$\|D_{0j}w\|_h \leq \frac{1}{2h}\{\|E_j w\|_h + \|E_j^{-1}w\|_h\} = \frac{1}{h}\|w\|_h$$

implies that

$$\begin{aligned}\|Q(t)w\|_h &\leq \frac{1}{2}\sum_{j=1}^{s}\{\|B_j\ D_{0j}w\|_h + \|D_{0j}\ B_j w\|_h\}\\ &\leq \frac{1}{h}\ \sum_{j=1}^{s}|B_j(\cdot,t)|_\infty\ \|w\|_h.\end{aligned}$$

The smooth solution $u = u(x,t)$ of the differential problem satisfies

$$\begin{aligned}\frac{1}{2k}\{u_\nu(t+k) - u_\nu(t-k)\} &= u_t(x_\nu,t) + O(k^2),\\ D_{0j}\ u_\nu(t) &= D_j u(x_\nu,t) + O(h^2),\\ \frac{1}{2}C(x_\nu,t)\{u_\nu(t+k) + u_\nu(t-k)\} &= C(x_\nu,t)u(x_\nu,t) + O(k^2).\end{aligned}$$

Thus we can consider $u_\nu(t)$ as a solution of perturbed difference equations:

$$\begin{aligned}\big(I - kC(x_\nu,t)\big)u_\nu(t+k) &= \big(I + kC(x_\nu,t)\big)u_\nu(t-k)\\ &\quad + 2kQ(x_\nu,t)u_\nu(t) + 2kF(x_\nu,t)\\ &\quad + k(k^2+h^2)\psi_\nu(t),\\ u_\nu(0) = f(x_\nu),&\quad u_\nu(k) = v_\nu(k) + k^2 g_\nu.\end{aligned}$$

Here $\psi_\nu(t),\ g_\nu$ are bounded independently of h and k. The difference $w = u - v$ solves

$$\begin{aligned}w_\nu(t+k) - w_\nu(t-k) &= kC(x_\nu,t)\big(w_\nu(t+k) + w_\nu(t-k)\big)\\ &\quad + 2kQ(x_\nu,t)w_\nu(t) + k(k^2+h^2)\psi_\nu(t),\\ w_\nu(0) = 0,&\quad w_\nu(k) = k^2 g_\nu.\end{aligned}$$

Taking the inner product with $w_\nu(t+k) + w_\nu(t-k)$ and summing over ν, we find that

$$\begin{aligned}\|w(t+k)\|_h^2 - \|w(t-k)\|_h^2 \le\ & kc_1\big\{\|w(t+k)\|_h^2 + \|w(t-h)\|_h^2\big\} \\ & - 2k\big(Q(t)w(t+k), w(t)\big)_h \\ & + 2k\big(Q(t)w(t), w(t-k)\big)_h \\ & + k(k^2+h^2)\ \big(w(t+k)+w(t-k),\ \psi(t)\big)_h.\end{aligned}$$

If one defines the auxiliary quantities

$$L(t) = \|w(t)\|_h^2 + \|w(t-k)\|_h^2 + 2k\big(Q(t)w(t),\ w(t-k)\big)_h,$$

then one obtains the recursive estimate

$$\begin{aligned}L(t+k) &\le L(t) + kc_1\big\{\|w(t+k)\|_h^2 + \|w(t-h)\|_h^2\big\} \\ &\quad + 2k\Big(\big(Q(t+k)-Q(t)\big)w(t+k), w(t)\Big)_h \\ &\quad + k(k^2+h^2)\big(w(t+k)+w(t-k), \psi(t)\big)_h \\ &\le L(t) + kc_2\big\{\|w(t+k)\|_h^2 + \|w(t)\|_h^2 + \|w(t-h)\|_h^2\big\} \\ &\quad + k(k^2+h^2)^2\ \|\psi(t)\|_h^2.\end{aligned}$$

It is important to observe that

$$\begin{aligned}\Big|2k\big(Q(t)w(t), w(t-k)\big)_h\Big| &\le \frac{2k}{h}\sum_{j=1}^{s}|B_j(\cdot,t)|_\infty\ \|w(t)\|_h\ \|w(t-h)\|_h \\ &\le \frac{a}{a+\eta}\ \big\{\|w(t)\|_h^2 + \|w(t-h)\|_h^2\big\}.\end{aligned}$$

This implies a lower bound for $L(t)$,

$$L(t) \ge \frac{\eta}{a+\eta}\ \big\{\|w(t)\|_h^2 + \|w(t-h)\|_h^2\big\},$$

and therefore the above recursive estimate gives us that

$$L(t+k) \le L(t) + kc_3L(t+k) + kc_3L(t) + k(k^2+h^2)^2c_4.$$

Hence the auxiliary quantities $L(t)$ also satisfy

$$L(t+k) \le (1+kc_5)L(t) + k(k^2+h^2)^2c_6.$$

Initially, at $t = k$, we have $L(k) = \|w(k)\|_h^2 = O(k^4)$, and by a discrete analog of Gronwall's Lemma (Lemma 3.1.1) it follows that

$$L(t) = O\big((k^2+h^2)^2\big), \quad 0 \le t \le t_0.$$

Since $L(t)$ bounds $\|w(t)\|_h^2$, the convergence (6.2.4) is proved.

The existence result mentioned above for symmetric hyperbolic systems can be extended to more general hyperbolic equations. Consider the first-order system (6.2.1a) where the operator $P(x, t, \partial/\partial x)$ has the form (6.2.1b), but the matrices $B_j(x, t)$ are not necessarily symmetric. We call

$$P(x, t, i\omega) = i \sum_{j=1}^{s} B_j(x, t)\omega_j, \quad \omega \in \mathbf{R}^s, \quad |\omega| = 1,$$

the symbol of the differential operator P and make the

Definition. The system (6.2.1) is called *strongly hyperbolic* if there exists a smooth 1-periodic symmetrizer; i.e., there is a 1-periodic positive definite Hermitian matrix function

$$H(x, t, \omega), \quad x \in \mathbf{R}^s, \quad t \geq 0, \quad \omega \in \mathbf{R}^s, \quad |\omega| = 1,$$

depending smoothly on all variables, with

$$H(x, t, \omega)P(x, t, i\omega) + P^*(x, t, i\omega)H(x, t, \omega) = 0.$$

This definition generalizes the conditions discussed in Section 2.4.1 for the multi-dimensional constant-coefficient case and the conditions in Section 3.3.1 for the one dimensional variable-coefficient case, but it requires in addition *smooth* dependence on (x, t, ω). (In the constant-coefficient case smooth dependence on ω was not essential.)

If we require that all frozen-coefficient problems obtained from (6.2.1) be strongly hyperbolic in the sense of Section 2.4.1, then we can use the matrices

$$S(x, t, \omega), \quad \omega \in \mathbf{R}^s, \quad |\omega| = 1 \ ,$$

of Theorem 2.4.1 and obtain the symmetrizer $H = S^*S$; compare Lemma 2.4.2. The only extra assumption in the above definition is the requirement of periodicity and smoothness of $H = H(x, t, \omega)$. The matrix $S^{-1}(x, t, \omega)$ contains in its columns the eigenvectors of $P(x, t, i\omega)$. Thus, if the eigenvalues of $P(x, t, i\omega)$ are all purely imaginary and if linearly independent eigenvectors can be chosen as smooth periodic functions of (x, t, ω), then the system is strongly hyperbolic. This is the case if the system is strictly hyperbolic, i.e., the eigenvalues of $P(x, t, i\omega)$ are purely imaginary and always *distinct*.

One can prove

Theorem 6.2.2. *If the system* (6.2.1) *is strongly hyperbolic, then the Cauchy problem is well-posed. For every set of smooth data* $F = F(x, t)$, $f = f(x)$

there exists a unique smooth solution. Estimates as in the one dimensional case are valid; i.e., the solution is as smooth as the data.

Sketch of the proof. We use the theory of pseudodifferential operators and refer to Eskin (1981, Chapter 5) or Nirenberg (1973). To conform with the usual framework, we do not assume that the data and coefficients are 1-periodic. Instead, we assume that $F,\ f \in C^\infty$ and that the coefficients are constant outside a bounded set. Because of the finite speed of propagation, this is no restriction.

Let $\phi_j = \phi_j(x)$ denote a partition of unity where the support of ϕ_j has diameter d_j and $\sup_j d_j$ is sufficiently small. Also, at every point x at most p functions ϕ_j are different from zero, p independent of x. We define the pseudodifferential operator $\mathbf{H}(t)$ by

$$\mathbf{H}(t)v(x) = \int_{\mathbf{R}^s} H(x,t,\omega/|\omega|)e^{2\pi i\langle\omega,x\rangle}\hat{v}(\omega)\,d\omega,$$

$$\hat{v}(\omega) = \int_{\mathbf{R}^s} e^{-2\pi i\langle\omega,x\rangle}v(x)\,dx.$$

For sufficiently small d_j

$$(u,v)_{H(t)} = \sum_j \operatorname{Re}\Big(u_j, \mathbf{H}(t)v_j\Big), \qquad u_j = \phi_j u, \quad v_j = \phi_j v,$$

defines a scalar product which is equivalent to the usual L_2-scalar product. To prove this, we define

$$\mathbf{H}_j(t)v(x) = \int_{\mathbf{R}^s} H(x_j,t,\omega/|\omega|)e^{2\pi i\langle\omega,x\rangle}\hat{v}(\omega)\,d\omega, \qquad x_j \in \text{ support } \phi_j,$$

and note that

$$\|\mathbf{H}(t)v_j - \mathbf{H}_j(t)v_j\| = O(d_j)\|v_j\|.$$

This ensures that

$$\|v\|_{H(t)} = \big(v,v\big)^{1/2}_{H(t)}$$

can be estimated from below by const $\|v\|$.

Let $\phi = \phi(|\omega|),\ \phi \in C^\infty$, denote a "cut-off" function with

$$\phi(|\omega|) = \begin{cases} 0 & \text{for } 0 \le |\omega| \le 1/2, \\ 1 & \text{for } |\omega| \ge 3/4. \end{cases}$$

We write (at each t)

$$\mathbf{H} = \bar{\mathbf{H}} + \tilde{\mathbf{H}},$$

where

$$\bar{\mathbf{H}}v(x) = \int_{\mathbf{R}^s} \phi(|\omega|)\, H(x,t,\omega/|\omega|)\, e^{2\pi i\langle\omega,x\rangle}\, \hat{v}(\omega)\, d\omega,$$

$$\tilde{\mathbf{H}}v(x) = \int_{\mathbf{R}^s} \big(1-\phi(|\omega|)\big)\, H(x,t,\omega/|\omega|)\, e^{2\pi i\langle\omega,x\rangle}\, \hat{v}(\omega)\, d\omega.$$

With these notations,

$$\begin{aligned}\operatorname{Re}(u_j, \mathbf{H}Pu_j) &= \operatorname{Re}(u_j, \bar{\mathbf{H}}Pu_j) + \operatorname{Re}(u_j, \tilde{\mathbf{H}}Pu_j)\\ &= \frac{1}{2}\Big(u_j, (\bar{\mathbf{H}}P + P^*\bar{\mathbf{H}}^*)u_j\Big) + \operatorname{Re}(u_j, \tilde{\mathbf{H}}Pu_j).\end{aligned}$$

Again, $\bar{\mathbf{H}}P + P^*\bar{\mathbf{H}}^*$ is a pseudodifferential operator. The calculus of pseudodifferential operators implies that its symbol has the form

$$\phi(|\omega|)\Big(H(x,t,\omega/|\omega|)P(x,t,i\omega) + P^*(x,t,i\omega)H(x,t,\omega/|\omega|)\Big) + S(x,t,\omega).$$

The part multiplied by ϕ vanishes, and $S(x,t,\omega)$ is bounded. Hence, $\bar{\mathbf{H}}P + P^*\bar{\mathbf{H}}^*$ is a bounded operator. Because the factor $1-\phi(|\omega|)$ in the definition of $\tilde{\mathbf{H}}$ cuts off high frequencies, the operator $\tilde{\mathbf{H}}P$ is also bounded.

Therefore one obtains that

$$\frac{d}{dt}(u,u)_{H(t)} \leq \text{const}\,(u,u)_{H(t)} + (u,F)_{H(t)},$$

and the basic energy estimate follows.

6.3. Mixed Hyperbolic-Parabolic Systems and the Linearized Navier-Stokes Equations

There are no problems in obtaining existence and uniqueness results, and estimates of smooth solutions for mixed hyperbolic-parabolic systems of the form

$$\begin{aligned}u_t &= P_2(x,t,\partial/\partial x)u + R_{12}(x,t,\partial/\partial x)v + G_2,\\ v_t &= P_1(x,t,\partial/\partial x)v + R_{21}(x,t,\partial/\partial x)u + G_1.\end{aligned} \tag{6.3.1}$$

Here we assume that the uncoupled systems

$$\begin{aligned}\overline{u}_t &= P_2(x,t,\partial/\partial x)\overline{u} + G,\\ \overline{v}_t &= P_1(x,t,\partial/\partial x)\overline{v} + F\end{aligned} \tag{6.3.2}$$

are second-order strongly parabolic and first-order symmetric hyperbolic, respectively. The coupling terms R_{12} and R_{21} are general first-order operators.

The proof proceeds as in the one dimensional case; we refer to Section 3.4.1. The solutions are as smooth as the data.

One can also generalize this result to the case where the uncoupled systems (6.3.2) are second-order parabolic and first-order strongly hyperbolic, respectively. The proof for this more general case requires that one modify the L_2-norm to obtain the estimates.

We shall now discuss how these results apply to the *linearized* Navier-Stokes equations. The (nonlinear) viscous compressible system reads

$$\rho \frac{D}{Dt}\mathbf{u} + \text{ grad } p = \mu\Delta\mathbf{u} + (\mu+\mu') \text{ grad}\,(\text{div}\,\mathbf{u}) + \rho\mathbf{F},$$

(6.3.3) $$\frac{D}{Dt}\,\rho + \rho \text{ div}\,\mathbf{u} = 0, \quad p = r(\rho),$$

$$\frac{D}{Dt} = \frac{\partial}{\partial t} + u\,\frac{\partial}{\partial x} + v\,\frac{\partial}{\partial y} + w\,\frac{\partial}{\partial z}.$$

Suppose $\mathbf{U}(\mathbf{x},t)$, $P(\mathbf{x},t)$, $R(\mathbf{x},t)$ are smooth functions with $R > 0$ and $P = r(R)$; these functions are considered as an exact or approximate solution of (6.3.3). We substitute

$$\mathbf{u} = \mathbf{U} + \mathbf{u}', \quad p = P + p', \quad \rho = R + \rho'$$

into (6.3.3) and neglect all terms which are quadratic in the corrections $\mathbf{u}'$, p', ρ'. If we drop the $'$ sign in our notation, we obtain the linear equations

$$R\frac{\tilde{D}}{\tilde{D}t}\mathbf{u} + \kappa \text{ grad } \rho + R(\mathbf{u}\cdot \text{ grad})\mathbf{U} + \rho\,\frac{\tilde{D}}{\tilde{D}t}\,\mathbf{U}$$
$$= \mu\Delta\mathbf{u} + (\mu+\mu') \text{ grad}\,(\text{div }\mathbf{u}) + \mathbf{G}_1$$

(6.3.4) $$\frac{\tilde{D}}{\tilde{D}t} = \frac{\partial}{\partial t} + U\,\frac{\partial}{\partial x} + V\,\frac{\partial}{\partial y} + W\,\frac{\partial}{\partial z},$$

$$\kappa = \frac{dr}{d\rho}(R),$$

$$\frac{\tilde{D}}{\tilde{D}t}\,\rho + R \text{ div }\mathbf{u} + \mathbf{u}\cdot \text{ grad } R + \rho \text{ div }\mathbf{U} = G_2.$$

The inhomogeneous terms $\mathbf{G}_1$, G_2 are determined by $\mathbf{U}$, P, R and vanish if $\mathbf{U}$, P, R solve (6.3.3) exactly. Clearly, the equations (6.3.4) are of the general form (6.3.1) with $u = \mathbf{u}$, $v = \rho$. For $\mu > 0$, $\mu' \geq 0$, the second-order operator

$$\mu\Delta\mathbf{u} + (\mu+\mu') \text{ grad } (\text{div }\mathbf{u}) =: \sum_{i,j=1}^{3} A_{ij}D_iD_j\mathbf{u}$$

is strongly parabolic since

$$\sum_{i,j=1}^{3} \langle y_i,\ A_{ij} y_j \rangle = \mu \sum_{i=1}^{3} |y_i|^2 + (\mu + \mu') \Big(\sum_{i=1}^{3} y_i^{(i)} \Big)^2 .$$

Here $y_i \in \mathbf{R}^3$ is arbitrary, and $y_i^{(i)}$ is the i-th component of y_i. Thus the uncoupled $\mathbf{u}$-equation in (6.3.4) is strongly parabolic as long as $R(\mathbf{x}, t)$ is bounded. The uncoupled ρ-equation in (6.3.4) is symmetric hyperbolic since it is of first-order and scalar.

We consider also the inviscid case $\mu = \mu' = 0$. Except for zero-order terms and forcing functions, one obtains the first-order system (2.4.2) where U, V, W, R are now (known) functions of $(\mathbf{x}, t)$. The symbol $P_1(\mathbf{x}, t, i\omega)$ equals

$$i\omega_1 \begin{pmatrix} U & 0 & 0 & \kappa/R \\ 0 & U & 0 & 0 \\ 0 & 0 & U & 0 \\ R & 0 & 0 & U \end{pmatrix} + i\omega_2 \begin{pmatrix} V & 0 & 0 & 0 \\ 0 & V & 0 & \kappa/R \\ 0 & 0 & V & 0 \\ 0 & R & 0 & V \end{pmatrix}$$

$$+ i\omega_3 \begin{pmatrix} W & 0 & 0 & 0 \\ 0 & W & 0 & 0 \\ 0 & 0 & W & \kappa/R \\ 0 & 0 & R & W \end{pmatrix},$$

and the symmetrizer is

$$H(\mathbf{x}, t, \omega) = \begin{pmatrix} 1 & 0 & 0 & 0 \\ 0 & 1 & 0 & 0 \\ 0 & 0 & 1 & 0 \\ 0 & 0 & 0 & h \end{pmatrix}, \quad h = \kappa/R^2.$$

Thus the linearized inviscid compressible equations form a strongly hyperbolic system if $\kappa = (dr/d\rho)\ (R) > 0$. Again, as in the one dimensional case, the symmetrization of the system is equivalent to the introduction of a scaled density.

6.4. Short-Time Existence for Nonlinear Systems

Local (in time) existence results for nonlinear strongly hyperbolic, parabolic, and mixed systems in s space dimensions can basically be derived in the same way as in the case of one space dimension. However, the proofs become technically more difficult because the Sobolev inequalities depend on the number s of space dimensions. We first treat nonlinear symmetric hyperbolic systems in detail. Then corresponding existence results for parabolic and mixed systems are stated.

6.4.1. Nonlinear Symmetric Hyperbolic Systems

Consider a symmetric hyperbolic problem

$$u_t = \sum_{j=1}^{s} B_j(u) D_j u, \quad B_j = B_j^*, \tag{6.4.1}$$

$$u(x,0) = f(x), \quad f \in C^\infty, \quad f(x) \equiv f(x + e_j). \tag{6.4.2}$$

As in the one dimensional case, we assume uniform bounds for the (real) coefficients and their derivatives, i.e.,

$$\sum_{j=1}^{s} |D_u^p\, B_j(u)| \le K_p, \quad p = 0, 1, 2, \ldots, \tag{6.4.3}$$

for all arguments $u \in \mathbf{R}^n$. These global bounds can be replaced by local bounds in a neighborhood of the initial data if one applies the cut-off technique described in Section 5.2.

We start with an *a priori* estimate of $\|u\|_{H^{s+2}}$. As it turns out, if this norm is estimated in some time interval, then higher derivatives can be bounded in the same interval. In the proofs we repeatedly use the Sobolev inequality (see Theorem A.3.6)

$$|v|_\infty \le c\|v\|_{H^k} \quad \text{if} \quad k > \frac{s}{2}.$$

It implies that

$$|D^\nu u|_\infty \le c\|u\|_{H^k} \quad \text{if} \quad |\nu| + [s/2] + 1 \le k.$$

Here $[s/2]$ is the largest integer $\le s/2$, and c is a numerical constant independent of $u,\ v$.

Lemma 6.4.1. *Suppose that u solves (6.4.1) in $0 \le t \le T$. Then*

$$\frac{d}{dt}\, \|u\|_{H^{s+2}}^2 \le C\{1 + \|u\|_{H^{s+2}}^{s+4}\} \quad \text{in} \quad 0 \le t \le T,$$

where C depends only on the constants $K_0, \ldots, K_{s+2}$ of (6.4.3).

Proof. Differentiation of (6.4.1) yields

$$D^\nu u_t = \sum_{j=1}^{s} B_j(u) D_j D^\nu u + \sum_{j=1}^{s} \sum_{1 \le |\alpha| \le |\nu|} c_\alpha\, D^\alpha \big(B_j(u)\big) D^{\alpha'} D_j u$$

where

$$\nu = (\nu_1, \dots, \nu_s), \quad \alpha = (\alpha_1, \dots \alpha_s), \quad \alpha' = (\alpha'_1, \dots, \alpha'_s)$$

are multi-indices with $\alpha + \alpha' = \nu$ and

$$c_\alpha = (\nu_1! \dots \nu_s!)/(\alpha_1! \dots \alpha_s! \; \alpha'_1! \dots \alpha'_s!).$$

Therefore, the time derivative

$$\frac{1}{2} \frac{d}{dt} (D^\nu u, D^\nu u) = (D^\nu u, D^\nu u_t)$$

can be estimated in terms of

$$\text{(6.4.4)} \qquad \big(D^\nu u, B_j(u) D_j D^\nu u\big), \quad j = 1, \dots, s,$$

together with

$$\text{(6.4.5)} \qquad \Big(D^\nu u, D^\alpha \big(B_j(u)\big) D^\beta u\Big), \quad 1 \le |\alpha| \le |\nu|, \; |\alpha| + |\beta| = |\nu| + 1.$$

The symmetry $B_j = B_j^*$ and integration by parts yield

$$\big(v, B_j(u) D_j v\big) = -\Big(D_j\big(B_j(u)\big)v, v\Big) - \big(B_j(u) D_j v, v\big),$$

and thus

$$\big(v, B_j(u) D_j v\big) = -\frac{1}{2}\Big(v, D_j\big(B_j(u)\big)v\Big).$$

One finds that for $|\nu| \le s + 2$,

$$\Big|\big(D^\nu u, B_j(u) D_j D^\nu u\big)\Big| \le \text{const} |D_j u|_\infty \; \|D^\nu u\|^2 \le \text{const} \; \|u\|^3_{H^{s+2}}.$$

To estimate a term (6.4.5), we apply the chain rule and write

$$D^\alpha\big(B_j(u)\big) = \sum_{r=1}^{|\alpha|} \sum_{|\sigma_1| + \dots + |\sigma_r| \le |\alpha|} C_\sigma D^{\sigma_1} u \dots D^{\sigma_r} u.$$

Here each C_σ consists of partial derivatives of the coefficient B_j w.r.t u (and C_σ acts as an r-linear map on $D^{\sigma_1} u \dots D^{\sigma_r} u$.) According to assumption (6.4.3), the C_σ are bounded. Consider first the case

$$|\alpha| + [s/2] + 1 \le s + 2.$$

Then (by Sobolev's inequality)

$$\text{(6.4.6)} \qquad |D^{\sigma_i} u|_\infty \le \text{const} \|u\|_{H^{s+2}},$$

since $|\sigma_i| + [s/2] + 1 \le s + 2$. Therefore,

$$\left|\Big(D^\nu u, D^\alpha\big(B_j(u)\big)D^\beta u\Big)\right| \le \text{const} \sum_{r=1}^{|\alpha|} \|u\|_{H^{s+2}}^r \, \|D^\nu u\| \, \|D^\beta u\|$$

$$\le \text{const} \sum_{r=1}^{|\alpha|} \|u\|_{H^{s+2}}^{r+2}.$$

Secondly, let

$$|\alpha| + [s/2] + 1 > s + 2;$$

thus

$$|\beta| + [s/2] + 1 \le s + 2.$$

(Observe that $|\alpha| + |\beta| = |\nu| + 1 \le s + 3$.) Then

$$|D^\beta u|_\infty \le \text{const} \, \|u\|_{H^{s+2}},$$

and we find that

$$\left|\Big(D^\nu u, D^\alpha\big(B_j(u)\big)D^\beta u\Big)\right| \le \text{const} \, \|u\|_{H^{s+2}}^2 \sum_\sigma \|D^{\sigma_1}u \dots D^{\sigma_r}u\|.$$

If each $|\sigma_i|$ is $\le s + 1 - [s/2]$, then (6.4.6) applies; hence

$$\|D^{\sigma_1}u \dots D^{\sigma_r}u\| \le \text{const} \, \|u\|_{H^{s+2}}^r.$$

If, say, $|\sigma_1| > s + 1 - [s/2]$, then all other $|\sigma_i|$ satisfy $|\sigma_i| < s + 1 - [s/2]$, and therefore

$$\|D^{\sigma_1}u \dots D^{\sigma_r}u\| \le \|D^{\sigma_1}u\| \, |D^{\sigma_2}u|_\infty \dots |D^{\sigma_r}u|_\infty \le \text{const} \, \|u\|_{H^{s+2}}^r.$$

Summarizing, the estimate

$$\frac{d}{dt} \|u\|_{H^{s+2}}^2 \le \text{const} \sum_{r=1}^{s+2} \|u\|_{H^{s+2}}^{r+2}$$

is shown, and the lemma is proved.

The following result is an immediate implication of the previous lemma.

Corollary 6.4.2. *There is a time $T > 0$ depending on $\|f\|_{H^{s+2}}$, but not on higher derivatives of f, with the following property: If u solves* (6.4.1), (6.4.2) *in $0 \le t \le T$, then*

$$\|u(\cdot, t)\|_{H^{s+2}} \le 2\|f\|_{H^{s+2}} \text{ in } 0 \le t \le T.$$

Proof. If $f = 0$, then $u \equiv 0$. Thus let $f \not\equiv 0$, and suppose that $y(t)$ solves

$$\frac{d}{dt} y(t) = C\big(1 + y(t)^{s/2+2}\big), \quad y(0) = \|f\|^2_{H^{s+2}},$$

where C is determined as in Lemma 6.4.1. Then (by Lemma 4.1.2)

$$\|u(\cdot, t)\|_{H^{s+2}} \le y(t) \le 2y(0) \quad \text{in} \quad 0 \le t \le T$$

for some $T > 0$.

We shall now show that all higher derivatives of the solution can be estimated in the same time interval $0 \le t \le T$. The existence of a smooth solution is assumed.

Lemma 6.4.3. *Let $T = T\big(\|f\|_{H^{s+2}}\big)$ denote the time of Corollary 6.4.2. For every $p = 0, 1, 2, \ldots$, there is a constant M_p, depending on $\|f\|_{H^{s+2+p}}$, with*

$$\|u(\cdot, t)\|_{H^{s+2+p}} \le M_p \quad \text{in} \quad 0 \le t \le T.$$

Proof. For $p = 0$ the result is shown. Assuming it is proven up to $p - 1$, we shall prove it for $p > 0$. To this end, let $|\nu| = s + 2 + p$. As in the proof of Lemma 6.4.1, we can bound

$$\frac{d}{dt} \, (D^\nu u, D^\nu u)$$

in terms of (6.4.4) and (6.4.5). Since we already have the bound

$$|D_j u|_\infty \le c_1 M_0,$$

we find that

$$\Big|(D^\nu u, B_j(u) D_j D^\nu u)\Big| \le c_2 \|D^\nu u\|^2.$$

It remains to estimate

$$\Big|\Big(D^\nu u, D^\alpha\big(B_j(u)\big) D^\beta u\Big)\Big| \le \|D^\nu u\| \; \|D^\alpha\big(B_j(u)\big) D^\beta u\| \,,$$

$$1 \le |\alpha| \le |\nu| = s + 2 + p\,, \quad |\alpha| + |\beta| = s + 3 + p.$$

By Sobolev's inequality we have that

(6.4.7) $|D^\mu u|_\infty \le c \|u\|_{H^{s+1+p}} \le c\, M_{p-1}$ for $|\mu| + [s/2] + 1 \le s + 1 + p$.

Case 1. $|\alpha| + [s/2] + 1 \leq s + 1 + p.$
Then

$$|D^{\alpha}(B_j(u))|_{\infty} \leq \text{ const } \quad \text{by (6.4.7)},$$

and $\|D^{\beta}u\| \leq \|u\|_{H^{s+2+p}}$.

Case 2. $|\alpha| + [s/2] = s + 1 + p;$ thus $|\beta| + [s/2] + 1 \leq s + 2 + p.$
Then

$$|D^{\beta}u|_{\infty} \leq c\|u\|_{H^{s+2+p}},$$

and it remains to estimate

$$|D^{\sigma_1}u \dots D^{\sigma_r}u|_{\infty} \text{ for } |\sigma_1| + \dots + |\sigma_r| \leq |\alpha|. \tag{6.4.8}$$

Case 2a. all $|\sigma_i| \leq s + p - [s/2]$.
The estimate of (6.4.8) follows by (6.4.7).

Case 2b. $|\sigma_1| > s + p - [s/2]$.
Then $\sigma_1 = \alpha$, $r = 1$, and thus

$$\|D^{\sigma_1}u\| \leq \|u\|_{H^{|\alpha|}} \leq \text{ const.}$$

Case 3. $|\alpha| + [s/2] > s + 1 + p;$ thus $|\beta| + [s/2] + 1 \leq s + 1 + p.$
In this case $|D^{\beta}u|_{\infty} \leq$ const by (6.4.7).

Case 3a. all $|\sigma_i| \leq s + p - [s/2]$.
We have bounded the term (6.4.8) in Case 2a.

Case 3b. $|\sigma_1| > s + p - [s/2]$.
We cannot have $|\sigma_2| > s + p - [s/2]$ since otherwise

$$s + 2 + p \geq |\alpha| \geq |\sigma_1| + |\sigma_2| \geq 2s + 2p - 2[s/2] + 2.$$

Thus $|D^{\sigma_i}u|_{\infty}$, $i \neq 1$, is bounded by (6.4.7), and

$$\|D^{\sigma_1}u\| \leq \|u\|_{H^{|\alpha|}} \leq \|u\|_{H^{s+2+p}}.$$

To summarize, we have shown that

$$\frac{d}{dt}(D^{\nu}u, D^{\nu}u) \leq \text{ const } \|u\|^2_{H^{s+2+p}}, \quad |\nu| = s + 2 + p.$$

If we use the notation

$$|u|^2_{H^{s+2+p}} = \sum_{|\nu|=s+2+p} (D^\nu u, D^\nu u),$$

then

$$\frac{d}{dt}|u|^2_{H^{s+2+p}} \le \text{ const } \left(M_{p-1} + |u|^2_{H^{s+2+p}}\right),$$

and the induction step can be completed.

We shall now sketch the proof for the *existence* of solutions. To this end, we consider the sequence of functions $u^k = u^k(x,t)$ defined by the iteration

$$u_t^{k+1} = \sum_{j=1}^{s} B_j(u^k) D_j u^{k+1}, \quad u^{k+1}(x,0) = f(x), \quad u^0(x,t) \equiv f(x).$$

Note that each function u^k is determined by a *linear* equation. In the same way as in Section 4.1.4, we can show the analogues of Lemma 4.1.5 and Lemma 4.1.6. The techniques to obtain the estimates are the same as used in the proofs of Lemma 6.4.1 and Lemma 6.4.3.

Lemma 6.4.4. *For a suitable time $T_1 = T_1(\|f\|_{H^{s+2}})$ it holds that*

$$\|u^k(\cdot,t)\|_{H^{s+2}} \le 2\|f\|_{H^{s+2}} \text{ in } 0 \le t \le T_1.$$

Lemma 6.4.5. *For each $j = s+3,\ s+4, \ldots$ there exists K_j with*

$$\|u^k(\cdot,t)\|_{H^j} \le K_j \text{ in } 0 \le t \le T_1,$$

where T_1 is determined in the previous lemma.

As in Section 4.1.4, we can use the uniform smoothness of the sequence u^k in $0 \le t \le T_1$ and employ Gronwall's Lemma to show the existence of a C^∞-solution in $0 \le t \le T_1$. Uniqueness of a solution follows as in the one dimensional case; see Lemma 5.1.1.

Theorem 6.4.6. *Consider the* 1*-periodic Cauchy problem* (6.4.1), (6.4.2). *The problem has a unique C^∞-solution $u = u(x,t)$, defined for $0 \le t \le T$. The time $T > 0$ depends on $\|f\|_{H^{s+2}}$, but not on higher derivatives of f.*

6.4.2. The Compressible Euler Equations

In two space dimensions the equations without forcing read

$$u_t + uu_x + vu_y + \frac{1}{\rho}p_x = 0, \qquad v_t + uv_x + vv_y + \frac{1}{\rho}p_y = 0,$$

$$\rho_t + u\rho_x + v\rho_y + \rho(u_x + v_y) = 0, \qquad p = r(\rho),$$

or in matrix form

$$\begin{pmatrix} u \\ v \\ \rho \end{pmatrix}_t + \begin{pmatrix} u & 0 & \frac{1}{\rho}r'(\rho) \\ 0 & u & 0 \\ \rho & 0 & u \end{pmatrix} \begin{pmatrix} u \\ v \\ \rho \end{pmatrix}_x + \begin{pmatrix} v & 0 & 0 \\ 0 & v & \frac{1}{\rho}r'(\rho) \\ 0 & \rho & v \end{pmatrix} \begin{pmatrix} u \\ v \\ \rho \end{pmatrix}_y = 0.$$

If $r'(\rho) = \frac{dr}{d\rho}(\rho) > 0$, then we can introduce a new density $\tilde{\rho}$ by

$$\rho = g(\tilde{\rho}), \quad \frac{dg}{d\tilde{\rho}}(\tilde{\rho}) = \frac{g(\tilde{\rho})}{\left[r'\big(g(\tilde{\rho})\big)\right]^{1/2}}.$$

This transformation leads to a symmetric hyperbolic system. In 3D we can proceed in exactly the same way.

6.4.3. More General Nonlinear Systems

The local existence result stated in Theorem 6.4.6 can be generalized without difficulties to

Symmetric hyperbolic systems

$$u_t = \sum_{j=1}^{s} B_j(x,t,u)D_j u + F(x,t,u), \quad B_j = B_j^*,$$

where B_j and F depend smoothly on all variables.

One can also treat

Strongly parabolic systems

$$u_t = \epsilon \sum_{i,j=1}^{s} D_i\big(A_{ij}(x,t)\big)D_j u + F(x,t,u, \; D_1 u, \ldots, D_s u)$$

$$\sum_{i,j=1}^{s} \langle y_i, \; A_{ij}(x,t)y_j\rangle \geq \delta \sum_{i=1}^{s} |y_i|^2, \quad \delta > 0, \; \epsilon > 0,$$

and obtain an existence interval depending on $\epsilon > 0$. If the system becomes symmetric hyperbolic for $\epsilon = 0$, then an existence interval $0 \le t \le T$, T independent of $\epsilon > 0$, can be established. For $\epsilon \to 0$ the solutions of the parabolic problem converge (along with all their derivatives) to the solution of the hyperbolic problem. Proofs of these results have been given in Chapter 5 for one space dimension, but all arguments generalize.

Similarly, we obtain local existence for

Mixed hyperbolic-parabolic systems

$$u_t = \sum_{i,j=1}^{s} D_i\big(A_{ij}(x,t)\, D_j u\big) + F_1(x,t,u,v,Du,Dv),$$

$$v_t = \sum_{j=1}^{s} B_j(x,t,u,v) D_j v + F_2(x,t,u,v,Du), \quad B_j = B_j^*.$$

Further generalizations to general parabolic systems, strongly hyperbolic systems (with smooth symmetrizers) and mixed systems can be obtained by changing the L_2-norm. We refer to Sections 6.1.3 and 6.2.2.

These results establish short-time existence for the compressible viscous or inviscid Navier-Stokes equations. In the inviscid case, the usual assumption of $dr/d\rho > 0$ for the equation of state $p = r(\rho)$ is required to ensure hyperbolicity.

6.5. A Global Existence Theorem in 2D

To begin with, we consider a parabolic system where the first-order terms are in so-called self-adjoint form:

$$(6.5.1) \quad u_t = \frac{1}{2}\left\{D_1(A(u)u) + A(u)D_1 u\right\} + \frac{1}{2}\left\{D_2(B(u)u) + B(u)D_2 u\right\} + \epsilon\Delta u,$$

$$\epsilon > 0, \quad A,\, B \in C^\infty, \quad A = A^*, \quad B = B^*.$$

(In this section ϵ is arbitrary but fixed.) As before, we assume an initial condition

$$(6.5.2) \qquad u(x,y,0) = f(x,y), \quad f \in C^\infty, \quad f \text{ } 1-\text{periodic in } x \text{ and in } y.$$

According to Section 6.4, there is a time $T > 0$ and a C^∞-solution u defined for $0 \le t < T$. We use the notation

$$J_j^2(t) = \|D_1^j u(\cdot,t)\|^2 + \|D_2^j u(\cdot,t)\|^2, \quad j = 1, 2, \ldots,$$

and start with a simple a priori estimate.

Lemma 6.5.1. *Suppose that u is a smooth solution defined for $0 \leq t < T$. Then there are constant K_1, K_2 with*

$$\|u(\cdot, t)\| \leq K_1, \quad 0 \leq t < T, \tag{6.5.3}$$

$$\int_0^T J_1^2(\tau)\, d\tau \leq K_2. \tag{6.5.4}$$

Proof. Integration by parts gives us

$$\frac{1}{2}\frac{d}{dt}\|u(\cdot, t)\|^2 = (u, u_t) = \epsilon(u, \Delta u) = -\epsilon J_1^2(t).$$

Integrating w.r.t. t, we find that

$$\|u(\cdot, t)\|^2 + 2\epsilon \int_0^t J_1^2(\tau)\, d\tau = \|f\|^2,$$

and the lemma is proved.

Now consider a parabolic system

$$u_t = A(u)u_x + B(u)u_y + \epsilon\Delta u, \quad \epsilon > 0, \tag{6.5.5}$$

where A, $B \in C^\infty$ are not necessarily symmetric. We assume that the coefficients grow at most linearly for large arguments:

$$|A(u)| + |B(u)| \leq K_3(1 + |u|), \quad u \in \mathbf{R}^n. \tag{6.5.6}$$

The following lemma contains the key result of our global existence theorem.

Lemma 6.5.2. *Suppose that u is a smooth solution of* (6.5.5), (6.5.2) *defined for $0 \leq t < T$. If we assume the bounds* (6.5.3), (6.5.4), (6.5.6), *then*

$$\sup_{0 \leq t < T} J_1^2(t), \quad \int_0^T J_2^2(\tau)\, d\tau, \quad \int_0^T |u(\cdot, \tau)|_\infty^2\, d\tau$$

are finite.

Proof. Using (6.5.5), we obtain

$$\begin{aligned}
\frac{1}{2}\frac{d}{dt}J_1^2(t) &= (u_x, u_{xt}) + (u_y, u_{yt}) \\
&= -(u_{xx}, u_t) - (u_{yy}, u_t) \\
&= -(u_{xx}, A(u)u_x) - (u_{xx}, B(u)u_y) - (u_{yy}, A(u)u_x) - (u_{yy}, B(u)u_y) \\
&\quad - \epsilon\|\Delta u\|^2
\end{aligned}$$

$$\leq \Big\{|A(u)|_\infty \|u_x\| + |B(u)|_\infty \|u_y\|\Big\}\Big\{\|u_{xx}\| + \|u_{yy}\|\Big\}$$
$$- \epsilon\|\Delta u\|^2$$
$$\leq C_1(1 + |u|_\infty)J_1 J_2 - \epsilon\|\Delta u\|^2.$$

One shows easily that

$$\|\Delta u\|^2 = \|u_{xx}\|^2 + \|u_{yy}\|^2 + 2(u_{xx}, u_{yy})$$
$$= J_2^2 + 2(u_{xy}, u_{xy}) \geq J_2^2,$$

and therefore,

$$\frac{d}{dt}J_1^2 \leq C_2(1 + |u|_\infty^2)J_1^2 - \frac{\epsilon}{2}\|\Delta u\|^2.$$

In Appendix 3 (see (A.3.7)) we prove the special Sobolev inequality

$$|u|_\infty^2 \leq C\|u\|(J_2 + \|u\|).$$

Together with assumption (6.5.3) we obtain that

$$\frac{d}{dt}J_1^2 \leq C_3(1 + J_1^2)J_1^2 - \frac{\epsilon}{4}\|\Delta u\|^2. \tag{6.5.7}$$

Let us now show the bound for $J_1(t)$. Clearly,

$$2J_1 \frac{d}{dt}J_1 = \frac{d}{dt}J_1^2 \leq C_3(1 + J_1^2)J_1^2,$$

and if we set

$$\beta(t) = \frac{1}{2}C_3(1 + J_1^2(t)),$$

then

$$\frac{d}{dt}J_1(t) \leq \beta(t)J_1(t), \quad 0 \leq t < T.$$

This yields

$$J_1(t) \leq J_1(0)\exp\{\int_0^t \beta(\tau)\,d\tau\},$$

and with assumption (6.5.4) we conclude that

$$J_1(t) \leq J_1(0)\exp\{\frac{1}{2}C_3(T + K_2)\}.$$

Thus we have proved a bound for $J_1(t)$.

If we use (6.5.7) again and observe that $J_2^2 \le \|\Delta u\|^2$, then

$$\int_0^T J_2^2(\tau)\, d\tau < \infty$$

follows. The special Sobolev inequality implies that $\int_0^T |u|_\infty^2 \, d\tau < \infty$, and the lemma is proved.

Let us assume that also the first *derivatives* of the coefficients in (6.5.5) grow at most linearly for large arguments:

$$|A_u(u)| + |B_u(u)| \le K_4(1 + |u|), \quad u \in \mathbf{R}^n. \tag{6.5.8}$$

Then we can apply the same technique to the differentiated differential equation (6.5.5). As a result, one obtains

Theorem 6.5.3. *Suppose that* $u \in C^\infty$ *is a solution of* (6.5.5), (6.5.2) *defined for* $0 \le t < T$. *We assume the bounds* (6.5.3), (6.5.4) *for* $u = u(x,t)$ *and the bounds* (6.5.6), (6.5.8) *for the coefficients of the differential equation. Then*

$$\sup_{0\le t<T} J_2(t), \quad \int_0^T J_3^2(\tau)\, d\tau, \quad \sup_{0\le t<T} |u(\cdot,t)|, \quad \int_0^T \{|D_1 u|_\infty^2 + |D_2 u|_\infty^2\}\, d\tau$$

are finite, and consequently u *can be extended as a smooth solution beyond* T. *If a priori bounds* (6.5.3), (6.5.4) *(with* $K_j = K_j(T)$*) hold for any finite time interval, then* u *exists for all time.*

Proof. Differentiate (6.5.5) w.r.t. x and w.r.t. y to obtain

$$\begin{aligned}
\frac{1}{2}\frac{d}{dt} J_2^2(t) &= (u_{xx}, u_{xxt}) + (u_{yy}, u_{yyt}) \\
&= -(u_{xxx}, u_{xt}) - (u_{yyy}, u_{yt}) \\
&= -(u_{xxx}, (A(u)u_x)_x) - (u_{xxx}, (B(u)u_y)_x) \\
&\quad - (u_{yyy}, (A(u)u_x)_y) - (u_{yyy}, (B(u)u_y)_y) \\
&\quad - \epsilon(u_{xxx}, \Delta u_x) - \epsilon(u_{yyy}, \Delta u_y) \\
&\le C_1(1 + |u|_\infty)(J_1^2 + J_2)J_3 - \epsilon J_3^2.
\end{aligned}$$

(To obtain the above inequality, we observe that

$$-(u_{xxx}, \Delta u_x) = -\|u_{xxx}\|^2 - (u_{xxy}, u_{xxy}) \le -\|u_{xxx}\|^2.)$$

The inequality for $(d/dt)J_2^2$ yields

$$\frac{d}{dt}J_2^2 \le C_2(1 + |u|_\infty^2)(J_1^4 + J_2^2) - \epsilon J_3^2,$$

and the desired bound follows from Lemma 6.5.2. This process can be continued, and by induction we obtain bounds for all derivatives. The remaining statements can be shown by the general arguments given at the beginning of Section 5.4.

Notes on Chapter 6

Petrovskii (1938) proved the existence of solutions of the Cauchy problem for strictly hyperbolic equations. The proof was simplified by Leray (1953), who constructed a symmetrizer. Kreiss (1963) generalized the construction to the case where the algebraic multiplicity of the eigenvalues of the symbol $P(x, t, iw)$ does not change. In this case the symmetrizer can be chosen as a quotient of differential operators. The smoothness of the symmetrizer is a major problem when the multiplicity of the eigenvalues changes. Besides the trivial case – when the system is symmetric and $H \equiv I$ – no general theorem is known.

Recently Clarke Hernquist (1988) was able to construct a smooth symmetrizer for a particular class of symbols with changing multiplicity. The class of corresponding differential equations was introduced by John (1978). For the construction of the symmetrizer for parabolic systems see also Mizohata (1956) and Kreiss (1963).

7 Initial-Boundary Value Problems in One Space Dimension

In applications the interesting phenomena frequently occur near the boundary, and consequently the formulation of boundary conditions plays an important rôle. In this chapter we treat problems in one space dimension with an interval as spatial domain. After discussion of the heat equation as a specific example, more general parabolic systems will be considered in Section 7.2. The energy method in its discrete and continuous form will be employed to show well-posedness under Dirichlet and Neumann boundary conditions. For more general boundary conditions, the Laplace transform method is the appropriate tool, and we present it in detail in Sections 7.4 and 7.5. If a *determinant-condition* is satisfied then the problem is *strongly well-posed in the generalized sense*. Important concepts of well-posedness for initial-boundary value problems are discussed in Section 7.3.

For hyperbolic equations the characteristics play, of course, an eminent role in determining correct boundary conditions: Values for the ingoing characteristic variables must be provided. If this is the case and if the solution is not overspecified at the boundary, then the hyperbolic problem is well-posed. Also, we will derive boundary conditions for mixed hyperbolic-parabolic systems and will apply the results to the linearized (compressible) Navier-Stokes equations.

A unified view of all results which can be obtained by the energy method is given in Section 7.8: The energy method applies if the spatial differential operators are *semibounded* on the space of functions obeying the (homogeneous) boundary conditions.

Throughout this chapter we restrict ourselves to *linear* problems, but – as in the periodic case – nonlinear equations could again be treated by an iterative process. The presence of boundary conditions does not lead to essentially new difficulties. We refer to Chapter 8 for some further remarks.

7.1. A Strip Problem for the Heat Equation

Consider the heat equation

$$u_t = u_{xx} \tag{7.1.1}$$

in the strip $0 \le x \le 1,\ t \ge 0$. As we did earlier, we prescribe an initial condition

$$u(x,0) = f(x), \quad 0 \le x \le 1. \tag{7.1.2}$$

Here f is assumed to be real. In addition, we require the boundary conditions

$$u(0,t) = u_x(1,t) = 0, \quad t \ge 0. \tag{7.1.3}$$

We want to obtain a solution $u(x,t)$ which is smooth, including at the "corners" $(x,t) = (0,0)$, $(x,t) = (1,0)$ of the strip, and need compatibility conditions to be satisfied. Before discussing this further, let us try to find a solution using a series expansion.

Solution in series form. The first step is to construct special solutions of the differential equation which satisfy the boundary conditions. These are functions in *separated variables*: Introducing

$$u(x,t) = e^{\lambda t}\hat{u}(x) \tag{7.1.4}$$

into (7.1.1) and (7.1.3), we find that $\hat{u}(x)$ must be a solution of the eigenvalue problem

$$\hat{u}_{xx} = \lambda\hat{u}, \quad \hat{u}(0) = \hat{u}_x(1) = 0. \tag{7.1.5}$$

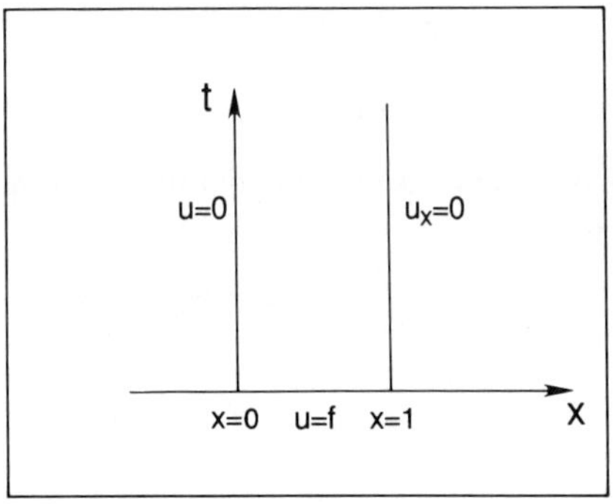

FIGURE 7.1.1. Strip problem for the heat equation.

The solutions of (7.1.5) are given by

$$\hat{u}_j(x) = a_j \sin\left(j + \frac{1}{2}\right)\pi x,$$

$$\lambda_j = -\left(j + \frac{1}{2}\right)^2 \pi^2, \quad j = 0, 1, 2 \ldots,$$

and therefore the special solutions (7.1.4) read

$$u(x,t) = a_j e^{-(j+\frac{1}{2})^2\pi^2 t} \sin\left(j + \frac{1}{2}\right)\pi x, \quad j = 0, 1, 2, \ldots.$$

Any finite sum of these functions clearly satisfies the differential equation and boundary conditions. If the given initial function $f(x)$ can be expanded in a series

$$f(x) = \sum_{j=0}^{\infty} \hat{f}_j \sin\left(j + \frac{1}{2}\right)\pi x$$

which converges sufficiently fast, then

$$u(x,t) = \sum_{j=0}^{\infty} \hat{f}_j \, e^{-(j+\frac{1}{2})^2\pi^2 t} \sin\left(j + \frac{1}{2}\right)\pi x \tag{7.1.6}$$

is a classical solution of the initial-boundary value problem (7.1.1)–(7.1.3).

Uniqueness. For simplicity we consider only real-valued functions and define the L_2-scalar product and norm by

$$(g,h) = \int_0^1 g(x)h(x)\,dx, \quad \|g\|^2 = (g,g).$$

The rule of integration by parts reads

$$(g, h_x) = -(g_x, h) + gh\Big|_0^1, \quad gh\Big|_0^1 = g(1)h(1) - g(0)h(0).$$

Suppose u and v are two solutions of the initial-boundary value problem (7.1.1)–(7.1.3). We obtain, for $w = u - v$,

$$\begin{aligned}\frac{d}{dt}(w, w) &= 2(w, w_t) = 2(w, w_{xx})\\ &= -2\|w_x\|^2 + 2ww_x\Big|_0^1\\ &= -2\|w_x\|^2 \le 0,\end{aligned}$$

i.e., $\|w(\cdot, t)\|^2 \le \|w(\cdot, 0)\|^2 = 0$. Thus $w \equiv 0$, and the solution is unique.

Compatibility conditions, generalized solutions, smoothing. Suppose the strip problem (7.1.1)–(7.1.3) has a solution $u(x, t)$ which is C^∞-smooth in $0 \le x \le 1,\ t \ge 0$, i.e., up to the boundary and the initial line. This implies that $f \in C^\infty$, and the initial data are *compatible* with the boundary conditions:

$$f(0) = u(0, 0) = 0, \quad f_x(1) = u_x(1, 0) = 0.$$

Differentiating the boundary conditions with respect to t and using that

$$u_{tt} = u_{xxt} = u_{txx} = u_{xxxx}, \quad \text{etc.},$$

we find that

$$0 = \frac{\partial^\nu}{\partial t^\nu} u(0, 0) = \frac{\partial^{2\nu}}{\partial x^{2\nu}}\, u(0, 0) = \frac{d^{2\nu}}{dx^{2\nu}}\, f(0),$$

$$0 = \frac{\partial^\nu}{\partial t^\nu} u_x(1, 0) = \frac{\partial^{2\nu+1}}{\partial x^{2\nu+1}}\, u(1, 0) = \frac{d^{2\nu+1}}{dx^{2\nu+1}}\, f(1)\ , \quad \nu = 0, 1, 2, \ldots.$$

This yields necessary conditions for f if we want a solution which is C^∞-smooth in $0 \le x \le 1,\ t \ge 0$. The easiest way to comply with these conditions is to assume that $f \in C^\infty$ is identically zero in a neighborhood of $x = 0$ and $x = 1$. Let

$$C_0^\infty(0, 1) = \Big\{ f \in C^\infty(0, 1) \Big| \text{ there is } \epsilon = \epsilon(f) > 0 \\ \text{such that } f \equiv 0 \text{ outside } \epsilon \le x \le 1 - \epsilon \Big\}.$$

One can show: If $f \in C_0^\infty(0, 1)$, then the formula (7.1.6) defines a solution $u(x, t)$ of the strip problem, which is C^∞-smooth in $0 \le x \le 1,\ t \ge 0$. An estimate as in the uniqueness argument given above shows that

$$\|u(\cdot,t)\|^2 \le \|f\|^2, \quad t \ge 0.$$

Since the set $C_0^\infty(0,1)$ is dense in L_2, we obtain a *generalized solution* for all initial functions $f \in L_2$ by the usual extension of the solution operator. One can show that the generalized solution is C^∞-smooth for $0 \le x \le 1,\ t > 0$. Indeed, the formula (7.1.6) remains valid for $f \in L_2,\ t > 0$. For all $f \in L_2$ the solution $u(x,t)$ depends even analytically on the argument (x,t) for $t > 0$. This follows from the exponential decay of the coefficients in (7.1.6).

A priori estimates of derivatives. We want to give another existence proof using a difference approximation. First let us derive the *a priori* estimates for the solution. Suppose that $u(x,t)$ solves the strip problem and is C^∞-smooth in $0 \le x \le 1,\ t \ge 0$. We have seen that

$$\frac{d}{dt}(u,u) = -2\|u_x\|^2 \le 0, \quad \text{and thus} \quad \|u(\cdot,t)\| \le \|f\|.$$

Differentiation of (7.1.1) and (7.1.3) with respect to t gives us for $v = u_t$:

$$\begin{aligned} v_t &= v_{xx}, && 0 \le x \le 1, \quad t \ge 0, \\ v(x,0) &= f_{xx}(x), && 0 \le x \le 1, \\ v(0,t) &= v_x(1,t) = 0, && t \ge 0. \end{aligned}$$

Therefore,

$$\|u_{xx}(\cdot,t)\| = \|u_t(\cdot,t)\| \le \|f_{xx}\|, \quad t \ge 0.$$

Repeating this process, we find estimates for all time derivatives and the space derivatives of even order. Sobolev inequalities imply bounds for the remaining space derivatives, and estimates for mixed derivatives follow from the differential equation.

The difference scheme. We shall now approximate the strip problem by an ordinary initial value problem. Let N be a natural number, let $h = 1/N$ denote the gridsize, and let $x_\nu = \nu h, \nu = 0,\ldots,N$, denote the gridpoints. The gridfunction $v(t)$ with components

$$v_\nu(t) = v_\nu^h(t), \quad t \ge 0, \quad \nu = 0,\ldots,N,$$

will approximate $u(x_\nu,t)$. We replace (7.1.1), (7.1.2) by

$$\begin{aligned} \frac{d}{dt}v_\nu(t) &= D_-\ D_+ v_\nu(t), && \nu = 1,\ldots,N-1, \\ v_\nu(0) &= f(x_\nu), && \nu = 1,\ldots,N-1. \end{aligned} \tag{7.1.7}$$

The boundary conditions (7.1.3) are replaced by

$$v_0(t) = 0, \quad D_+ v_{N-1}(t) = 0. \tag{7.1.8}$$

The equations (7.1.8) can be used to eliminate v_0 and v_N from (7.1.7), and one obtains an ordinary (linear) initial value problem for $v_1, \ldots, v_{N-1}$. Thus there is a unique solution $v(t) = v^h(t)$ of the above difference approximation.

Estimates for $v^h(t)$ and the limit $h \to 0$. We shall estimate all difference-differential quotients of $v = v^h$ independently of h. To this end, let us introduce some notation. If v, w are (real) gridfunctions, their discrete L_2-scalar product and norm* are defined by

$$(v, w)_h = h \sum_{\nu=1}^{N-1} v_\nu w_\nu, \quad \|v\|^2 = (v, v)_h. \tag{7.1.9}$$

In analogy to integration by parts we have

Lemma 7.1.1. *For all gridfunctions v, w it holds that*

$$(v, D_- w)_h = -(D_+ v, w)_h + v_N w_{N-1} - v_1 w_0.$$

Proof. The formula follows from

$$\begin{aligned}(v, D_- w)_h + (D_+ v, w)_h &= \sum_{\nu=1}^{N-1} \big[v_\nu(w_\nu - w_{\nu-1}) + w_\nu(v_{\nu+1} - v_\nu)\big] \\ &= \sum_{\nu=1}^{N-1} \big[-v_\nu w_{\nu-1} + v_{\nu+1} w_\nu\big] = v_N w_{N-1} - v_1 w_0.\end{aligned}$$

In the next lemma, we let $v(t) = v^h(t)$ denote the solution of the difference approximation defined above.

Lemma 7.1.2. *Assume that $f \in C_0^\infty(0, 1)$ and define $f(x) = 0$ for $x \notin (0, 1)$. For every $q = 0, 1, 2, \ldots$, there is an $h_0 > 0$ such that*

$$\left\| \frac{d^q}{dt^q} v^h(t) \right\|_h \leq \left\| (D_+ D_-)^q f^h \right\|_h, \quad h \leq h_0.$$

(Here f^h denotes the initial function restricted to the proper grid.)

*Since the expressions (7.1.9) ignore the boundary data v_0, v_N, etc., we obtain a scalar product and norm on the space of gridfunctions defined on the interior grid $x_1, \ldots, x_{N-1}$.

Proof. Using the discrete boundary conditions and Lemma 7.1.1, we find that

$$\begin{aligned}\frac{1}{2}\frac{d}{dt}(v,v)_h &= \left(v, \frac{dv}{dt}\right)_h = (v, D_-D_+v)_h \\ &= -\|D_+v\|_h^2 + v_N\, D_+v_{N-1} - v_1\, D_+v_0 \\ &= -\|D_+v\|_h^2 - h|D_+v_0|^2 \le 0,\end{aligned}$$

and therefore,

$$\|v(t)\|_h^2 \le \|v(0)\|_h^2 = \|f^h\|_h^2.$$

This proves the statement for $q = 0$. Now let $w = dv/dt$. Differentiation of (7.1.7) and (7.1.8) with respect to t yields

$$\begin{aligned}\frac{d}{dt}\, w_\nu(t) &= D_-D_+w_\nu(t), \quad \nu = 1, \ldots, N-1, \\ w_0(t) &= D_+w_{N-1}(t) = 0.\end{aligned}$$

The initial condition for w reads

$$w_\nu(0) = \frac{d}{dt}v_\nu(0) = D_-D_+v_\nu(0), \quad \nu = 1, \ldots, N-1.$$

Here $v_\nu(0) = f_\nu^h, \quad \nu = 1, \ldots, N-1$. Furthermore, using the boundary conditions to determine v_0, v_N, we obtain that

$$v_0(0) = 0, \quad v_N(0) = v_{N-1}(0) = f_{N-1}^h = 0 = f_N^h$$

if h is small enough. This shows that

$$v_\nu(0) = f_\nu^h, \quad \nu = 0, \ldots, N, \quad h \le h_0,$$

and therefore,

$$w_\nu(0) = D_-D_+f_\nu^h, \quad \nu = 1, \ldots, N-1, \quad h \le h_0.$$

Thus we obtain, as before,

$$\|w(t)\|_h = \left\|\frac{dv}{dt}(t)\right\|_h \le \|D_+D_-f^h\|_h, \quad h \le h_0.$$

This process can be continued and the lemma follows.

The discrete norms $\|(D_+D_-)^q f^h\|_h$ approach $\|d^{2q}f/dx^{2q}\|$ as $h \to 0$, and thus they are bounded independently of h. Therefore we have estimates for all time derivatives of $v^h(t)$ which are independent of h. Interpolation leads to a solution of the strip problem for $h \to 0$:

Theorem 7.1.3. *Assume that* $f \in C_0^\infty(0,1)$. *Then the initial-boundary value problem* (7.1.1)–(7.1.3) *has a unique solution*

$$u \in C^\infty(0 \le x \le 1,\ t \ge 0).$$

Proof. Note that

$$\frac{dv_\nu}{dt}(t) = D_- D_+ v_\nu(t), \quad \nu = 1, \ldots, N-1,$$

implies that

$$\frac{d^2 v_\nu}{dt^2}(t) = (D_- D_+)^2 v_\nu(t), \quad \nu = 2, \ldots, N-2.$$

In general,

$$(D_- D_+)^q v_\nu(t) = \frac{d^q v_\nu}{dt^q}(t), \quad \nu = q, \ldots, N-q.$$

If we introduce the notation

$$(u, v)_{r,s} = h \sum_{\nu=r}^{s} u_\nu v_\nu, \quad \|u\|_{r,s}^2 = (u, u)_{r,s},$$

then

$$\|(D_- D_+)^q v^h(t)\|_{q,N-q} \le \|(D_- D_+)^q f^h\|_h, \quad h \le h_0(q).$$

Thus we have bounded the divided differences with respect to x. By the results of Appendix 2 we can interpolate the gridfunctions $v_\nu^h(t)$ with respect to x and obtain C^∞-functions $w^h(x,t)$ which are uniformly (with respect to h) smooth. In the same way as in the periodic case, we can send $h \to 0$ and apply the Arzela-Ascoli theorem to obtain a C^∞-solution u of the strip problem.

Smoothness of the generalized solution for t > 0 . Let $f \in L_2$, and consider a sequence $f^k \in C_0^\infty(0,1)$ with

$$\lim_{k \to 0} \|f - f^k\| = 0.$$

By definition, the corresponding solutions u^k converge to the generalized solution u belonging to the initial function f.

For $t > 0$, we can derive estimates of the derivatives of u^k which depend only on $\|f^k\|$ but not on derivatives of f^k. As before, this implies smoothness of the limit u for $t > 0$. To show the estimates, we first recall that

$$\frac{d}{dt}\|u^k\|^2 = -2\|u_x^k\|^2 \le 0,$$

i.e.,

$$\|u^k(\cdot,t)\|^2 \le \|f^k\|^2, \quad 2\int_0^t \|u_x^k(\cdot,\tau)\|^2\,d\tau \le \|f^k\|^2. \tag{7.1.10}$$

Also,

$$\begin{aligned}\frac{d}{dt}\left\{t(u_x^k, u_x^k)\right\} &= 2t(u_{xt}^k, u_x^k) + \|u_x^k\|^2 \\ &= 2t(u_{xxx}^k, u_x^k) + \|u_x^k\|^2 \\ &= \|u_x^k\|^2 - 2t\|u_{xx}^k\|^2.\end{aligned}$$

(Note that $u_{xx}^k(0,t) = u_t^k(0,t) = 0$.) Using (7.1.10), integration from 0 to t gives us that

$$t\|u_x^k\|^2 \le \frac{1}{2}\|f^k\|^2 - \int_0^t 2\tau\|u_{xx}\|^2\,d\tau.$$

This process can be continued, and we can bound

$$t^p\|\frac{\partial^p u^k}{\partial x^p}(\cdot,t)\|^2$$

in terms of $\|f^k\|^2$. Then we can argue as in Section 3.2.6 and obtain

Theorem 7.1.4. *Let $f \in L_2$, and let u denote the generalized solution to the strip problem. The function u is a C^∞-smooth in $0 \le x \le 1,\ t > 0$; i.e., u is smooth up to the boundary for $t > 0$.*

7.2. Strip Problems for Strongly Parabolic Systems

In this section we consider parabolic systems

$$u_t = A(x,t)u_{xx} + B(x,t)u_x + C(x,t)u =: P(x,t,\partial/\partial x)u \tag{7.2.1}$$

in the strip $0 \le x \le 1,\ t \ge 0$. At time $t = 0$ we give initial data

$$u(x,0) = f(x), \quad 0 \le x \le 1. \tag{7.2.2}$$

As boundary conditions we require n linearly independent relations between the components of u and u_x at each boundary point $x = 0, x = 1$; i.e., the boundary conditions have the form

$$L_{j0}u(j,t) + L_{j1}u_x(j,t) = 0, \quad j = 0,1, \tag{7.2.3}$$

with constant $n \times n$ matrices L_{j0}, L_{j1}. The $n \times 2n$ matrix (L_{j0}, L_{j1}) has rank n for $j = 0$ and $j = 1$ since the boundary conditions are linearly independent. The matrix coefficients A, B, C in (7.2.1) are assumed to be C^∞-smooth.

Furthermore, we require that

$$A(x,t) = A^*(x,t) \geq \delta I \text{ in } 0 \leq x \leq 1, \quad 0 \leq t \leq T, \tag{7.2.4}$$

for some $\delta > 0$; i.e., the system (7.2.1) is *symmetric parabolic*. For the initial function f we assume that

$$f \in C_0^\infty(0,1).$$

All functions and matrices are taken as *real*, for simplicity. The constants c_1, c_2, etc. introduced below will depend on the time interval $0 \leq t \leq T$, where T is arbitrary but fixed.

Extensions. The reader can generalize all arguments from the real to the complex case and assume, instead of (7.2.4),

$$A(x,t) + A^*(x,t) \geq 2\delta I \text{ in } 0 \leq x \leq 1\,, \quad 0 \leq t \leq T.$$

This generalization (to a strongly parabolic system) essentially requires one to replace $2(u, Au_{xx})$ by $(Au_{xx}, u) + (u, Au_{xx})$ in the arguments given below. We refer to Section 3.1. Also, without difficulty, one can add a smooth forcing $F = F(x,t)$ in (7.2.1). Furthermore, the coefficients in the boundary conditions could depend on t. This would make the proofs technically slightly more complicated, however.

Solutions in series form. Boundary conditions of the general form (7.2.3) can be motivated as follows: If the coefficients A, B, C do not depend on time, then we can try to solve the strip problem by the same technique as in the last section and first construct special solutions in separated variables. The ansatz

$$u(x,t) = e^{\lambda t}\hat{u}(x)$$

leads to the eigenvalue problem

$$\begin{aligned} &A\hat{u}_{xx} + B\hat{u}_x + C\hat{u} = \lambda\hat{u}, \\ &L_{j0}\hat{u}(j) + L_{j1}\hat{u}_x(j) = 0, \quad j = 0, 1. \end{aligned} \tag{7.2.5}$$

For the ordinary differential equation (7.2.5), the above boundary conditions are well established. In fact, under quite general additional assumptions, one can show that the eigenvalues form a sequence λ_k with $\operatorname{Re}\lambda_k \to -\infty$, and that an arbitrary $f \in L_2$ can be expanded in terms of the functions which span the invariant eigenspaces. If this is the case, the strip problem can be solved as in the last section in series form.

7.2.1. Solution-Estimates under Various Boundary Conditions

The basic energy estimate. If the coefficients $A,\ B,\ C$ are time dependent, the above approach of series expansion does not easily apply. When can we expect a unique solution of the strip problem (7.2.1)–(7.2.3)? In order to derive sufficient requirements on the boundary conditions, let us suppose that u is a C^∞-solution and let us try to derive the basic energy estimate. Recall the rule of integration by parts:

$$
\begin{aligned}
(g, h_x) &= \int_0^1 \langle g(x), h_x(x)\rangle\, dx \\
&= -(g_x, h) + \langle g(1), h(1)\rangle - \langle g(0), h(0)\rangle.
\end{aligned}
$$

Equation (7.2.1) gives us that

$$
\begin{aligned}
\frac{1}{2}\frac{d}{dt}(u,u) &\le (u, Au_{xx}) + c_1\big\{\|u\|\ \|u_x\| + \|u\|^2\big\} \\
&= \big(u, (Au_x)_x\big) - (u, A_x u_x) + c_1\big\{\|u\|\ \|u_x\| + \|u\|^2\big\} \\
&\le -(u_x, Au_x) + \langle u, Au_x\rangle\big|_0^1 + c_2\big\{\|u\|\ \|u_x\| + \|u\|^2\big\} \\
&\le -\delta\|u_x\|^2 + \langle u, Au_x\rangle\big|_0^1 + \frac{\delta}{2}\ \|u_x\|^2 + c_3\|u\|^2.
\end{aligned}
$$

In order to obtain the basic energy estimate, we arrive at the following condition, which we call the

Requirement for an energy estimate. For all functions $w = w(x),\ w \in C^\infty$, which satisfy the boundary conditions

(7.2.6) $$L_{j0}w(j) + L_{j1}w_x(1) = 0, \quad j = 0, 1,$$

the estimate

(7.2.7) $$\langle w,\ A(\cdot, t)w_x\rangle\big|_0^1 \le \frac{\delta}{2}\ \|w_x\|^2 + c\|w\|^2, \quad 0 \le t \le T,$$

holds.* Here c may be dependent on T but not on w.

*The value $\delta/2$ of the coefficient of $\|w_x\|^2$ can be changed to any other positive value without altering the requirement. This will follow from the considerations below.

Before we discuss this requirement further, we shall treat the important special cases of Dirichlet and Neumann boundary conditions. The above requirement, which leads to the basic energy estimate, does in fact imply the existence of a unique smooth solution u of the strip problem. We shall prove this below using a difference approximation.

Dirichlet conditions. The simplest boundary conditions are the so-called Dirichlet conditions

$$u(0,t) = u(1,t) = 0.$$

If they are imposed, then the boundary term in (7.2.7) is zero.

Neumann conditions. These are of the form

$$u_x(j,t) + L_{j0}u(j,t) = 0, \quad j = 0, 1.$$

If w fulfills (7.2.6), then

$$\langle w, A(\cdot,t)w_x\rangle\big|_0^1 \leq c\{|w(0)|^2 + |w(1)|^2\},$$

and we can use the Sobolev inequality (see Appendix 3)

$$|w|_\infty^2 \leq \epsilon\|w_x\|^2 + c(\epsilon)\|w\|^2, \quad \epsilon > 0,$$

to show (7.2.7).

Mixed conditions. Clearly, we can also require a Dirichlet condition at one boundary point and a Neumann condition on the other. If $A(0,t)$, say, is diagonal, we can use a Dirichlet condition for some components of $u(0,t)$ and a Neumann condition for the others. If $A(0,t)$ is not diagonal but has a complete set of eigenvectors, then we can introduce new dependent variables $\tilde{u}$ so that the matrix $\tilde{A}(0,t)$ becomes diagonal. For $\tilde{u}$ we can impose boundary conditions of the form described.

The general case. To discuss the general case of boundary conditions (7.2.3), first note that the requirement imposed by such a condition remains the same if we multiply the conditions with a nonsingular matrix on the left. For example, if L_{j1} is nonsingular, we can assume a Neumann condition at $x = j$; and if $L_{j1} = 0$, we can assume a Dirichlet condition at $x = j$. If rank $L_{j1} = r_j$, $1 \leq r_j \leq n-1$, we can assume that

$$L_{j1} = \begin{pmatrix} L_{j1}^I \\ 0 \end{pmatrix}, \quad L_{j1}^I \text{ of size } r_j \times n, \quad \text{rank } L_{j1}^I = r_j.$$

Partitioning L_{j0} correspondingly, we "split" the boundary conditions into a Neumann and a Dirichlet part:

$$L_{j1}^I u_x(j,t) + L_{j0}^I u(j,t) = 0, \quad L_{j0}^{II} u(j,t) = 0, \quad j = 0, 1.$$

The following *result from linear algebra* will be applied at each boundary point.

Lemma 7.2.1. *Let $A \in \mathbf{R}^{n,n}$ be arbitrary and let $L_0, L_1 \in \mathbf{R}^{n,n}$ be of the form*

$$L_1 = \begin{pmatrix} L_1^I \\ 0 \end{pmatrix}, \quad L_0 = \begin{pmatrix} L_0^I \\ L_0^{II} \end{pmatrix} \quad \begin{matrix} \} \, r \\ \} \, n-r \end{matrix},$$

where rank $L_1^I = r$, rank $(L_0, L_1) = n$. The following conditions are equivalent:
(i) *There exists a constant $c > 0$ such that*

(7.2.8) $$|\langle w, Aw_x \rangle| \le c|w|^2$$

for all w, $w_x \in \mathbf{R}^n$ which satisfy

(7.2.9) $$L_1 w_x + L_0 w = 0.$$

(ii) *If $a, b \in \mathbf{R}^n$ are vectors with*

(7.2.10) $$L_1^I b = 0, \quad L_0^{II} a = 0,$$

then

(7.2.11) $$\langle a, Ab \rangle = 0.$$

Proof. (i) $\Rightarrow$ (ii): Let a, b satisfy (7.2.10), and choose $\tilde{w}_x \in \mathbf{R}^n$ with

$$L_1 \tilde{w}_x + L_0 a = 0.$$

Considering the vectors

$$w_x = \tilde{w}_x + \beta b, \quad \beta \in \mathbf{R}, \quad w = a,$$

we find that

$$\langle w, Aw_x \rangle = \langle a, A\tilde{w}_x \rangle + \beta \langle a, Ab \rangle.$$

Since (7.2.9) is fulfilled for any β, there exists – by assumption – a bound of the right-hand side which is independent of β. This implies $\langle a, Ab \rangle = 0$.

(ii) $\Rightarrow$ (i): Let w, w_x satisfy (7.2.9) and decompose*

*ker $L = \{s \,|\, Ls = 0\}$ is the kernel or nullspace of L. By $S^\perp = \{s \,|\, \langle s, v \rangle = 0$ for all $v \in S\}$ we denote the orthogonal complement of a subspace S.

$$w_x = w_x' + w_x'', \quad w_x' \in \ker L_1^I, \quad w_x'' \in \left(\ker L_1^I\right)^\perp.$$

Clearly, $L_0^{II} w = 0$, and therefore (ii) implies that

$$\langle w, Aw_x \rangle = \langle w, Aw_x'' \rangle.$$

If $s_1, \ldots, s_{n-r}$ denotes a basis of $\ker L_1^I$, then w_x'' fulfills the matrix equation

$$\begin{pmatrix} L_1^I \\ s_1^T \\ \vdots \\ s_{n-r}^T \end{pmatrix} w_x'' = -L_0\, w.$$

The system-matrix is nonsingular; thus one can solve for w_x'', and the estimate (7.2.8) follows.

Remark. Condition (ii) states *geometrically* that the kernel (= nullspace) of L_0^{II} is orthogonal to the A-image of the kernel of L_1^I. Since ker L_0^{II} has dimension r and ker L_1^I has dimension $n-r$, and since A is nonsingular in our application, we obtain that the equality

$$\text{ker } L_0^{II} = \left\{A(\text{ker } L_1^I)\right\}^\perp \tag{7.2.12}$$

is an equivalent formulation of (ii). Formally, condition (ii) is also applicable for Dirichlet and Neumann conditions characterized by

$$\ker L_1^I = \mathbf{R}^n, \quad \ker L_0^{II} = \{0\}$$

and

$$\ker L_1^I = \{0\}, \quad \ker L_0^{II} = \mathbf{R}^n,$$

respectively. If the boundary conditions are neither of Dirichlet nor of Neumann type, then (7.2.12) puts a severe restriction on A.

We apply the previous lemma to show

Lemma 7.2.2. *There is an energy estimate (see (7.2.7)) if and only if*

$$L_{j1}^I\, b_j = 0, \quad L_{j0}^{II}\, a_j = 0, \quad j = 0, 1, \qquad a_j,\, b_j \in \mathbf{R}^n, \tag{7.2.13}$$

implies that

$$\langle a_j,\, A(j,t) b_j \rangle = 0, \quad j = 0,\, 1, \quad 0 \le t \le T. \tag{7.2.14}$$

Proof. First assume that (7.2.14) holds for all vectors with (7.2.13). By the previous lemma,

$$\left|\langle w(j),\ A(j,t)w_x(j)\rangle\right| \leq c(j,t)\left|w(j)\right|^2.$$

The proof of Lemma 7.2.1 shows that $c(j,t)$ can be bounded uniformly for $0 \leq t \leq T$. An application of Sobolev's inequality

$$|w|_\infty^2 \leq \epsilon\|w_x\|^2 + c(\epsilon)\|w\|^2, \quad \epsilon > 0,$$

shows (7.2.7).

Conversely, suppose that (7.2.7) holds for all functions w with (7.2.6). Since – for each fixed t – the vectors $w(j,t),\ w_x(j,t),\ j = 0,1$, can be prescribed arbitrarily, it is necessary that an estimate

$$\left|\langle w(j,t),\ A(j,t)w_x(j,t)\rangle\right| \leq c(j,t)\left|w(j,t)\right|^2$$

holds. (The right-hand side of (7.2.7) cannot be used to bound a term $\left|w_x(j,t)\right|^2$.) Another application of Lemma 7.2.1 finishes the proof.

For later purposes we show

Lemma 7.2.3. *Suppose the matrix function A satisfies* (7.2.4) *and suppose the requirement for an energy estimate to be valid. Then the $n \times n$ matrices*

$$\begin{pmatrix} L_{j1}^{I} \\ L_{j0}^{II} \end{pmatrix}, \quad j = 0,1,$$

are nonsingular.

Proof. Otherwise there exists a vector $a_j \neq 0$ with

$$L_{j1}^{I} a_j = 0, \quad L_{j0}^{II} a_j = 0.$$

Condition (7.2.14) implies that

$$\langle a_j,\ A(j,t)a_j\rangle = 0,$$

in contradiction to (7.2.4).

A priori estimates of the solution and its derivatives. Suppose that the boundary conditions are such that (7.2.7) holds; i.e., there is an energy estimate. Furthermore, assume that u is a C^∞-solution of the strip problem. Clearly,

$$\frac{d}{dt}\ \|u\|^2 \leq c\|u\|^2, \quad 0 \leq t \leq T,$$

and thus

$$\|u(\cdot,t)\|^2 \leq K\|f\|^2, \quad 0 \leq t \leq T.$$

We want to show that we can also estimate the derivatives of u; we set $v = u_t$. Differentiation of (7.2.1), (7.2.3) yields

$$v_t = Av_{xx} + Bv_x + Cv + P_t u, \quad P_t u = A_t u_{xx} + B_t u_x + C_t u,$$

$$L_{j0} v(j,t) + L_{j1} v_x(j,t) = 0, \quad j = 0, 1.$$

Thus, except for the inhomogeneous term $P_t u$, the time derivative $v = u_t$ satisfies the same differential equation and boundary conditions as u. By (7.2.1),

$$\begin{aligned} \|u_{xx}\| &= \|A^{-1}v - A^{-1}Bu_x - A^{-1}Cu\| \\ &\le c_1\{\|v\| + \|u_x\| + \|u\|\}. \end{aligned}$$

Using the Sobolev inequality

$$\|u_x\| \le \epsilon\|u_{xx}\| + c(\epsilon)\|u\|,$$

we obtain that

$$\|u_{xx}\| + \|u_x\| \le c_2\{\|v\| + \|u\|\}, \tag{7.2.15}$$

i.e.,

$$\|P_t u\| \le c_3\{\|v\| + \|u\|\}.$$

As before,

$$\begin{aligned} \frac{d}{dt}\|v\|^2 &\le 2(v, Av_{xx}) + c_4\{\|v\|\ \|v_x\| + \|v\|^2 + \|u\|^2\} \\ &\le c_5(\|v\|^2 + \|u\|^2). \end{aligned}$$

Since we have already a bound for $\|u\|$, we obtain an estimate for $v = u_t$. By (7.2.15) the space derivatives u_x, u_{xx} are also bounded. Repeated differentiation with respect to t gives us the desired estimates. We summarize the result in

Lemma 7.2.4. *Suppose that the boundary conditions are such that the requirement for an energy estimate is fulfilled. Given any nonnegative integers p, q and any time $T > 0$, there exists a constant $K = K(p, q, T)$ such that*

$$\left\| \frac{\partial^{p+q}}{\partial t^q \partial x^p} u(\cdot, t) \right\|^2 \le K\|f\|^2_{H^{p+2q}}$$

in $0 \le t \le T$. The constant K is independent of $f \in C_0^\infty(0,1)$.

7.2.2. Existence of a Solution via Difference Approximations

The difference scheme. The notations for the gridsize $h = 1/N$, the gridpoints $x_0, \dots, x_N$, etc., are the same as in the last section. We replace the strip problem (7.2.1)–(7.2.3) by

$$\text{(7.2.16)} \qquad \begin{aligned} dv_\nu(t)/dt &= A_\nu D_- D_+ v_\nu(t) + B_\nu D_0 v_\nu(t) + C_\nu v(t) \\ &=: Q(x_\nu, t)v_\nu(t), \quad \nu = 1, \dots, N-1, \end{aligned}$$

$$\text{(7.2.17)} \qquad \begin{aligned} v_\nu(0) &= f(x_\nu), \qquad \nu = 1, \dots, N-1, \\ L_{00}v_0(t) + L_{01}D_+ v_0(t) &= 0, \\ L_{10}v_N(t) + L_{11}D_+ v_{N-1}(t) &= 0, \qquad t \ge 0 \end{aligned}$$

The discretized boundary conditions can also be written in the form

$$\text{(7.2.18)} \qquad \begin{aligned} (L_{01} - hL_{00})v_0(t) &= L_{01}v_1(t), \\ (L_{11} + hL_{10})v_N(t) &= L_{11}v_{N-1}(t). \end{aligned}$$

We assume the *requirement for an energy estimate* to be valid. Then it follows from Lemma 7.2.3 that the above equations can be solved for $v_0(t)$, $v_N(t)$ if h is sufficiently small. Thus we can eliminate v_0 and v_N from (7.2.16). Therefore the difference scheme has a unique solution $v(t) = v^h(t)$ for $0 < h \le h_0$.

The basic estimate. In the following we estimate the solution $v^h(t)$ of the difference scheme independently of h; we start with

Lemma 7.2.5. *There is a constant K and a step-size $h_0 > 0$ with*

$$\|v^h(t)\|_h^2 \le K\|f^h\|_h^2, \quad 0 \le t \le T,$$

for $0 < h \le h_0$.

Proof. a) We remind the reader of the notations

$$(v, w)_h = h\sum_{\nu=1}^{N-1} \langle v_\nu, w_\nu \rangle, \quad (v, w)_{0,N-1} = h\sum_{\nu=0}^{N-1} \langle v_\nu, w_\nu \rangle,$$

the discrete 'integration-by-parts rule'

$$\begin{aligned} (v, D_- w)_h &= -(D_+ v, w)_h + \langle v_N, w_{N-1} \rangle - \langle v_1, w_0 \rangle \\ &= -(D_+ v, w)_{0,N-1} + \langle v_N, w_{N-1} \rangle - \langle v_0, w_0 \rangle, \end{aligned}$$

and the product formula

$$AD_-w = D_-(Aw) - (D_-A)E^{-1}w.$$

b) The difference equations (7.2.16) yield

$$\frac{1}{2}\frac{d}{dt}(v,v)_h \le (v, AD_-D_+v)_h + c_1\big\{\|v\|_h\,\|D_0v\|_h + \|v\|_h^2\big\},$$

where

$$\begin{aligned}(v, A\,D_-D_+v)_h &= \big(v, D_-(AD_+v)\big)_h - \big(v,(D_-A)D_-v\big)_h\\ &= -\,(D_+v, AD_+v)_{0,N-1}\\ &\quad + \langle v_N,\ A_{N-1}D_+v_{N-1}\rangle - \langle v_0,\ A_0D_+v_0\rangle\\ &\quad - \big(v,\ (D_-A)D_-v\big)_h.\end{aligned}$$

Since the pair $v_0(t), D_+v_0(t)$ satisfies the boundary conditions at $x = 0$, we obtain, from (7.2.8),

$$\Big|\langle v_0, A_0\ D_+v_0\rangle\Big| \le c_2|v_0|^2.$$

Also,

$$\begin{aligned}\langle v_N,\ A_{N-1}D_+v_{N-1}\rangle &= \langle v_N,\ A_ND_+v_{N-1}\rangle - \langle v_N,\ (A_N - A_{N-1})D_+v_{N-1}\rangle\\ &\le c_2|v_N|^2 + c_3h|v_N|\ |D_+v_{N-1}|\\ &\le c_4\big\{|v_N|^2 + h^2|D_+v_{N-1}|^2\big\}\\ &\le c_4\big\{|v_N|^2 + h\|D_+v\|_{0,N-1}^2\big\}.\end{aligned}$$

Finally, $\|D_0v\|_h \le \|D_+v\|_h$, and

$$\begin{aligned}|(v,(D_-A)D_-v)_h| &\le c_5\|v\|_h\|D_+v\|_{0,N-1}\\ &\le c_6\|v\|_h^2 + \frac{\delta}{4}\|D_+v\|_{0,N-1}^2.\end{aligned}$$

Therefore we have the estimate

$$\begin{aligned}\frac{1}{2}\frac{d}{dt}(v,v)_h &\le \Big(-\frac{\delta}{2} + c_4h\Big)\ \|D_+v\|_{0,N-1}^2\\ &\quad + c_4\big(|v_0|^2 + |v_N|^2\big) + c_7\|v\|_h^2.\end{aligned}$$

Using the discrete Sobolev inequality (see Appendix 3, formula (A.3.15))

$$\max_{1\le\nu\le N-1}|v_\nu|^2 \le \epsilon\|D_+v\|_h^2 + c(\epsilon)\|v\|_h^2,$$

we obtain

$$|v_0|^2 = |v_1 - hD_+v_0|^2 \le 2|v_1|^2 + 2h^2|D_+v_0|^2$$
$$\le (2\epsilon + 2h)\|D_+v\|^2_{0,N-1} + c_8\|v\|^2_h.$$

A similar estimate holds for $|v_N|^2$. Choosing $\epsilon = \delta/(16c_4)$, we find that

$$\frac{1}{2}\frac{d}{dt}\|v\|^2_h \le \left(-\frac{\delta}{4} + ch\right)\|D_+v\|^2_{0,N-1} + c\|v\|^2_h.$$

This proves the basic estimate for all sufficiently small h.

Estimates of derivatives, existence, smoothing. In a similar way as in the last section, by considering the difference equations satisfied by $w = dv/dt$, we also can estimate derivatives.

Lemma 7.2.6. *Suppose that $f \in C_0^\infty(0,1)$ and define $f(x) = 0$ for $x \notin (0,1)$. As before, assume the requirement for an energy estimate to be valid. Given any nonnegative integer q and any $T > 0$, there is a constant K and a step-size $h_0 > 0$ with*

$$\left\|\frac{d^q}{dt^q}v^h(t)\right\|^2_h \le K\left\{\|(D_-D_+)^q f^h\|^2_h + \|f^h\|^2_h\right\}, \quad 0 \le t \le T,$$

for $0 < h \le h_0$.

Proof. The function $w = dv/dt$ satisfies the difference equations

$$dw_\nu/dt = Q(x_\nu, t)w_\nu(t) + Q_t(x_\nu, t)v_\nu(t), \quad \nu = 1, \ldots, N-1,$$

and the same boundary conditions as v. The initial data are

$$w_\nu(0) = dv_\nu(0)/dt = Q(x_\nu, 0)v_\nu(0), \quad \nu = 1, \ldots, N-1.$$

Here $v_\nu(0) = f(x_\nu), \nu = 1, \ldots, N-1$, by (7.2.17). Using (7.2.18) one obtains, for sufficiently small h,

$$v_0(0) = v_1(0) = 0 = f(x_0),$$

and thus

$$w_\nu(0) = Q(x_\nu, 0)f(x_\nu), \quad \nu = 1, \ldots, N-1.$$

We can proceed in exactly the same way as for the *a priori* estimate of u_t and show that

$$\|D_-D_+v\|_h \le c_1\left\{\|w\|_h + \|D_+v\|_h + \|v\|_h\right\},$$

$$\|D_-D_+v\|_h + \|D_+v\|_h \le c_2\left\{\|w\|_h + \|v\|_h\right\}.$$

Thus the extra term $Q_t(x_\nu, t)v_\nu(t)$ poses no problem, and we obtain the desired estimate for $w = dv/dt$. The lemma follows by repeated differentiation with respect to t.

Clearly, we can use the time derivatives to estimate the divided differences

$$\|(D_- D_+)^q v^h\|_{q,N-q}$$

independently of h. Therefore, as in the last section, we can use interpolation to obtain existence of a solution for $h \to 0$. We summarize the main result in

Theorem 7.2.7. *Suppose that $f \in C_0^\infty(0,1)$, and assume that the boundary conditions are such that (7.2.7) is valid for all functions with (7.2.6); i.e., we have the basic energy estimate. Then the strip problem (7.2.1)–(7.2.3) has a unique solution u. The solution is C^∞-smooth in $0 \le x \le 1, t \ge 0$. The solution and its derivatives can be estimated in terms of f and its derivatives.*

For all initial data $f \in L_2$ there is a unique generalized solution. The smoothing properties of parabolic equations remain valid for the initial-boundary value problem. The proof proceeds in the same way as in the last section by showing an estimate of

$$t^p \|\partial^p u/\partial x^p\|^2$$

in terms of $\|f\|^2$. The generalized solution is a C^∞-function in $0 \le x \le 1$, $t > 0$.

7.3. Discussion of Concepts of Well-Posedness

In this section we give a somewhat informal discussion of different concepts of well-posedness for initial-boundary value problems. We do not restrict ourselves to the case of one space dimension and consider a system of differential equations

$$u_t = Pu + F(x,t), \quad x \in \Omega, \quad t \ge 0, \tag{7.3.1}$$

with initial data

$$u(x,0) = f(x), \quad x \in \Omega, \tag{7.3.2}$$

and boundary conditions

$$Lu = g(x,t), \quad x \in \partial\Omega, \quad t \ge 0. \tag{7.3.3}$$

Here P is a linear spatial differential operator whose coefficients may depend smoothly on x and t, and L is a linear operator combining values of u and its derivatives at the boundary. In general, L may also involve time derivatives of u. The boundary

$$\Gamma = \partial\Omega$$

of the domain $\Omega \subset \mathbf{R}^s$ is assumed to be smooth. We consider F, f, and g as the data of the problem, the operators P and L and the domain Ω being fixed.

Roughly speaking, the initial-boundary value problem (7.3.1)–(7.3.3) is called *well-posed* if for all smooth compatible data F, f, and g there is a unique smooth solution u, and in every finite time interval $0 \le t \le T$ the solution can be estimated in terms of the data. Clearly, to arrive at a definition, we must specify the norms which enter the estimates.* In the following we shall use the notation

$$(u, v) = \int_\Omega \langle u(x), v(x)\rangle \, dx, \quad \|u\|^2 = (u, u),$$

$$(u, v)_\Gamma = \int_\Gamma \langle u(x), v(x)\rangle \, dx, \quad \|u\|_\Gamma^2 = (u, u)_\Gamma.$$

For example, in case of the strip problem in one space dimension we have

$$\Omega = [0, 1], \quad \Gamma = \{0, 1\},$$

$$(u, v)_\Gamma = \langle u(0), v(0)\rangle + \langle u(1), v(1)\rangle, \quad \|u\|_\Gamma^2 = |u(0)|^2 + |u(1)|^2.$$

The Cauchy problem. Let us recall the definition of well-posedness for the spatially periodic Cauchy problem or the pure Cauchy problem. In those cases well-posedness required an estimate of the form

$$\|u(\cdot, t)\|^2 \le K_T \left\{ \|f\|^2 + \int_0^t \|F(\cdot, \tau)\|^2 \, d\tau \right\} \tag{7.3.4}$$

for $0 \le t \le T$. Using Duhamel's principle, one can replace such an estimate by estimates for the case $F \equiv 0$. We have used this for constant-coefficient operators P in Chapter 2. Also, one could develop a theory of well-posed problems where the estimate (7.3.4) is weakened in the following way: First, without much restriction, one can assume that $f(x) \equiv 0$ because the transformation

*It is not essential to make the smoothness assumptions for F, f, g, and u more precise. The reader can always replace "smooth" by C^∞. Once estimates are derived for this case, more general – less smooth – data can be treated by approximation as long as the norms for the data are defined. One obtains a generalized solution.

$$v(x,t) = u(x,t) - e^{-t} f(x)$$

leads to zero initial data. Then, instead of (7.3.4), one could require an estimate

$$\int_0^t \|u(\cdot,\tau)\|^2\, d\tau \le K_T \int_0^t \|F(\cdot,\tau)\|^2\, d\tau. \tag{7.3.5}$$

Such a concept of well-posedness would be satisfactory in the sense that "no derivatives are lost". If $F \in C_0^\infty$, then all derivaties of u vanish on the initial line $t = 0$, and – by differentiating the differential equation (7.3.1) – one obtains equations like (7.3.1) for the derivatives of u. Only the lower order terms will have changed. Thus, as we have seen, in many cases one can derive estimates like (7.3.5) for the derivatives of u in terms of the derivatives of F.

Clearly, (7.3.5) is – at least formally – a weaker requirement than (7.3.4). However, for the pure Cauchy problem or the periodic Cauchy problem not much seems to be gained by such a generalization of well-posedness. We do not know of any problem for which (7.3.5) holds but (7.3.4) is not valid. (If the coefficients of P are allowed to become singular for $t = 0$ it seems likely that (7.3.5) can indeed become a truly weaker requirement.)

Strongly well-posed problems. Consider now the initial-boundary value problem (7.3.1)–(7.3.3). Generalizing (7.3.4) we give

Definition 1. The initial-boundary value problem (7.3.1)–(7.3.3) is called *strongly well-posed* if for all smooth compatible data F, f, g, there is a unique smooth solution u, and for every finite time interval $0 \le t \le T$ there is a constant K_T such that

$$\begin{aligned} &\|u(\cdot,t)\|^2 + \int_0^t \|u(\cdot,\tau)\|_\Gamma^2\, d\tau \\ &\quad \le K_T \left\{ \|f\|^2 + \int_0^t \|F(\cdot,\tau)\|^2\, d\tau + \int_0^t \|g(\cdot,\tau)\|_\Gamma^2\, d\tau \right\} \end{aligned} \tag{7.3.6}$$

in $0 \le t \le T$. The constant K_T may not depend on F, f, or g.

The above estimate holds for parabolic equations with Neumann boundary conditions in any number of space dimensions (see Section 8.1) and for hyperbolic equations in one space dimension. One can give examples to show that there are real difficulties in estimating the boundary term

$$\int_0^t \|u(\cdot,\tau)\|_\Gamma^2\, d\tau$$

in more than one space dimension for hyperbolic problems. The failure to estimate this term is not due to the techniques employed.

Homogeneous boundary conditions. We can always transform to homogeneous boundary conditions by constructing a function $\psi = \psi(x, t)$ with

$$L\psi = g.$$

Then $v = u - \psi$ satisfies the homogeneous condition $Lv = 0$. This motivates

Definition 2. Consider the initial-boundary value problem (7.3.1)–(7.3.3) with $g = 0$. We call the problem *well-posed* if for all smooth compatible data F and f there is a unique smooth solution u, and we have, instead of (7.3.6),

$$\|u(\cdot, t)\|^2 \le K_T \Big\{ \|f\|^2 + \int_0^t \|F(\cdot, \tau)\|^2 \, d\tau \Big\} \tag{7.3.7}$$

in $0 \le t \le T$.

There are technical difficulties in working with this concept of well-posedness. To illustrate these, assume we have obtained the basic estimate (7.3.7), for example by integration by parts. We would like to show similar estimates of the derivatives of the solution in terms of the derivatives of the data. To this end, we differentiate the given equations (7.3.1) and the boundary conditions (7.3.3) with respect to t and in the tangential directions to obtain equations of a similar type for the derivatives. In this process, however, inhomogeneous boundary terms will appear, in general, if the boundary operator L has variable coefficients. These can be subtracted out by other functions ψ, as before. However, derivatives of ψ appear as inhomogeneous terms in the differential equation, and in the resulting estimate we "loose" derivatives; i.e., we need higher derivatives of the data to bound lower derivatives of the solution. This is intolerable if one wants to go over to nonlinear problems, for example. To summarize, the above concept of well-posedness (with $g = 0$) leads to technical difficulties if the differentiation of the boundary conditions introduces inhomogeneous terms.

An example. To illustrate the foregoing, consider the following simple example

$$u_t = u_{xx} + F(x, t) \ \text{ in } \ 0 \le x \le 1, \quad t \ge 0,$$

$$u(x, 0) = f(x), \quad 0 \le x \le 1,$$

$$u(0, t) = g_0(t), \quad u_x(1, t) = g_1(t), \quad t \ge 0.$$

The data are assumed to be compatible. We try to derive the basic estimate and proceed as in Section 7.1:

$$\frac{1}{2}\frac{d}{dt}\|u(\cdot,t)\|^2 = (u,u_t) = (u,u_{xx}) + (u,F)$$

$$= -(u_x,u_x) + \langle u,u_x\rangle\Big|_0^1 + (u,F).$$

The boundary term is

$$\langle u,u_x\rangle\Big|_0^1 = u(1,t)u_x(1,t) - u(0,t)u_x(0,t)$$

$$= u(1,t)g_1(t) - g_0(t)u_x(0,t).$$

Using Sobolev inequalities, the term

$$u(1,t)g_1(t)$$

originating from an inhomogeneous Neumann condition can be estimated properly:

$$\big|u(1,t)g_1(t)\big| \le \frac{1}{2}\big|u(1,t)\big|^2 + \frac{1}{2}\big|g_1(t)\big|^2$$

$$\le \epsilon\|u_x\|^2 + c(\epsilon)\|u\|^2 + \frac{1}{2}\big|g_1(t)\big|^2.$$

However, the term

$$g_0(t)u_x(0,t)$$

originating from an inhomogeneous Dirichlet condition cannot be treated. We can only derive energy estimates if we first transform to *homogeneous* Dirichlet conditions. To this end, choose a function $\phi \in C^\infty$ with

$$\phi(x) = \begin{cases} 1 & \text{for } x \le 1/4, \\ 0 & \text{for } x \ge 1/2\ . \end{cases}$$

Then

$$v(x,t) = u(x,t) - g_0(t)\phi(x)$$

satisfies homogeneous Dirichlet conditions at $x = 0$. Therefore v can be estimated in terms of the data. (The first derivative of g_0 is needed but this is not important.) We can proceed as in Section 7.2 to estimate derivatives of v. No problems occur here since the coefficient which defines the homogeneous condition,

$$v(0,t) = 0,$$

is constant. Thus the relation remains homogeneous if we differentiate it to derive a condition for v_t.

Another example. Consider a parabolic strip problem

$$u_t = A(x,t)u_{xx} + F(x,t), \quad A + A^* \geq 2\delta I > 0, \quad 0 \leq x \leq 1, \quad t \geq 0,$$

$$u(x,0) = f(x), \quad 0 \leq x \leq 1,$$

$$L_{j0}(t)u(j,t) + L_{j1}(t)u_x(j,t) = g_j(t), \quad t \geq 0, \quad j = 0,1.$$

Here $L_{j0}(t),\ L_{j1}(t),\ j = 0,1$, are $n \times n$ matrices, and rank $\big(L_{j0}(t),\ L_{j1}(t)\big) = n$.

Neumann boundary conditions. If $L_{j1}(t), j = 0,1$, are nonsingular for all $t \geq 0$, there are no difficulties in deriving an estimate of the form (7.3.6). We refer to Section 8.1.2 where we treat this situation in more than one space dimension. The above strip problem is strongly well-posed if the boundary conditions allow us to express $u_x(j,t)$ in terms of $u(j,t),\ g_j(t)$.

Dirichlet conditions. Another extreme case is $L_{j1}(t) \equiv 0, j = 0,1$. The boundary conditions read

$$u(j,t) = L_{j0}(t)^{-1}\ g_j(t).$$

After transforming to homogeneous conditions as indicated above, we obtain the basic estimate and estimates of derivatives.

Mixed conditions. If the boundary conditions contain a Dirichlet and a Neumann part, we must transform the Dirichlet part to become homogeneous with a constant coefficient matrix. Then, if we can derive the basic estimate, we can also estimate derivatives.

Strongly well-posed problems in the generalized sense. Instead of transforming to homogeneous boundary conditions, it is also of interest to study the case $f \equiv 0$. (This will become more apparent below, when we will use the Laplace transform in time; defining $u(x,t) = 0$ for $t < 0$, we obtain a continuous function u only if $u(x,0) = f(x) \equiv 0$.)

Definition 3. Consider the problem (7.3.1)–(7.3.3) with $f \equiv 0$. We call the problem *strongly well-posed in the generalized sense* if for all smooth compatible data F and g there is a unique smooth solution u and we have, instead of (7.3.6),

$$\int_0^t \|u(\cdot,\tau)\|^2\,d\tau + \int_0^t \|u(\cdot,\tau)\|_\Gamma^2\,d\tau \le K_T\Big\{\int_0^t \|F(\cdot,\tau)\|^2\,d\tau + \int_0^t \|g(\cdot,\tau)\|_\Gamma^2\,d\tau\Big\}$$

in $0 \le t \le T$.

The theory of strongly well-posed problems in the generalized sense is well developed for parabolic and hyperbolic systems. In case of constant coefficients, there are necessary and sufficient conditions of an algebraic nature. For variable-coefficient problems the estimates are valid if all relevant frozen-coefficient equations are strongly well-posed in the generalized sense.

As we shall show, initial-boundary value problems for strictly hyperbolic and symmetric hyperbolic systems which are strongly well-posed in the generalized sense are also strongly well-posed in the sense of Definition 1. For parabolic systems such a result does not hold.

As mentioned before, the assumption $f \equiv 0$ does not pose technical problems if one wants to estimate derivatives. We need only to assume that F vanishes identically in a neighborhood of $t = 0$; then all derivatives of u are zero at $t = 0$. The set of all smooth F vanishing identically near $t = 0$ is still dense in L_2.

Weakly well-posed problems. Finally one can weaken the required estimate further by assuming that both the initial function f and the boundary data g vanish. We give

Definition 4. The problem (7.3.1)–(7.3.3) with $f \equiv 0$, $g \equiv 0$ is called *weakly well-posed* if for all $F \in C_0^\infty$ there is a unique smooth solution u and, instead of (7.3.6), we have

$$\int_0^t \|u(\cdot,\tau)\|^2\,d\tau \le K_T \int_0^t \|F(\cdot,\tau)\|^2\,d\tau$$

in $0 \le t \le T$.

At present, there is no general theory available for weakly well-posed problems. We anticipate that any such theory would be extemely complicated.

7.4. Half-Space Problems and the Laplace Transform

One aim of this section is to indicate the limitations of the energy method in a discussion of well-posedness. We shall employ the Laplace transform in time to solve parabolic initial-boundary value problems and shall show that these problems are strongly well-posed in the generalized sense if and only if a certain determinant condition is fulfilled. Using this technique, one can *decide* the question of well-posedness in the generalized sense. In contrast to this, if the boundary conditions are neither of Dirichlet nor of Neumann type, the energy method applies only in exceptional cases.

7.4.1. Parabolic Half-Space Problems*; Main Result

Consider an equation

$$u_t = Au_{xx} + F(x,t), \quad A + A^* \geq 2\delta I, \quad \delta > 0, \tag{7.4.1}$$

in the domain $0 \leq x < \infty$, $t \geq 0$, under n linearly independent boundary conditions at $x = 0$,

$$L_0 u(0,t) + L_1 u_x(0,t) = g(t), \quad t \geq 0,$$

and an initial condition

$$u(x,0) = f(x), \quad 0 \leq x < \infty. \tag{7.4.2}$$

Assumptions. The $n \times n$ matrices A, L_0, L_1 are assumed to be constant; results for variable coefficients are stated at the end of this section, without proof however. Throughout, the assumption of strong parabolicity

$$A + A^* \geq 2\delta I, \quad \delta > 0,$$

is essential. Concerning the boundary conditions: If rank $L_1 = r$, then we can assume, without restriction, that the matrices are of the form

$$L_0 = \begin{pmatrix} L_0^I \\ L_0^{II} \end{pmatrix}, \quad L_1 = \begin{pmatrix} L_1^I \\ 0 \end{pmatrix},$$

where L_0^I, L_1^I have size $r \times n$, and the boundary conditions take the form

$$L_0^I u(0,t) + L_1^I u_x(0,t) = g^I(t), \quad L_0^{II} u(0,t) = g^{II}(t). \tag{7.4.3}$$

(The cases of a Dirichlet condition, $r = 0$, and a Neumann condition, $r = n$, are formally included in our discussion. However, the reader may focus attention on the mixed case where $1 \leq r \leq n-1$; only in the latter case do the Laplace transform method and the energy method differ significantly.)

As previously, it is not a severe restriction to put strong smoothness assumptions on the data, namely

$$F \in C_0^\infty(0 < x, t < \infty), \quad f \in C_0^\infty(0 < x < \infty), \quad g \in C_0^\infty(0 < t < \infty).$$

*Since we work in one space dimension, the half-space is actually only the half-line $0 \leq x < \infty$ in this section.

Once solution-estimates are derived under these assumptions, more general data can be treated by approximation.

Solution concept. We seek a solution

$$u \in C^\infty(0 \le x, t < \infty)$$

for which all derivatives are L_2-functions of $0 \le x < \infty$ at each time t:

$$\|\frac{\partial^{p+q}}{\partial t^p \partial x^q} u(\cdot, t)\| < \infty, \quad t \ge 0.$$

This requirement can be thought of as a boundary condition at $x = +\infty$.

The main result. Before giving further motivations, let us formulate the main result of this section; it will be proved below via Laplace transformation. Henceforth, we assume that the matrix A has n *distinct* eigenvalues μ_k. We transform A to diagonal form

$$\begin{aligned} &\Phi^{-1} A \Phi = \text{diag}\,(\mu_1, \ldots, \mu_n), \quad &&\text{Re}\, \mu_k \ge \delta > 0, \\ &A\Phi_k = \mu_k \Phi_k, &&\Phi = (\Phi_1 | \ldots | \Phi_n), \end{aligned} \tag{7.4.4a}$$

introduce the notation

$$\Lambda = \text{diag}\,(\frac{1}{\sqrt{\mu_1}}, \ldots, \frac{1}{\sqrt{\mu_n}}), \quad \text{Re}\, \frac{1}{\sqrt{\mu_k}} > 0, \tag{7.4.4b}$$

and define the square matrix

$$C_0 = \begin{pmatrix} L_1^I \Phi \Lambda \\ L_0^{II} \Phi \end{pmatrix}. \tag{7.4.4c}$$

Theorem 7.4.1. *If* $\det C_0 \neq 0$, *then the parabolic initial-boundary value problem* (7.4.1)–(7.4.3) *has a unique solution. For* $f \equiv 0$, *the problem is strongly well-posed in the generalized sense. The same result is valid (with exactly the same determinant condition) for an equation with lower-order terms,*

$$u_t = Au_{xx} + Bu_x + Cu + F(x, t).$$

7.4.2. The Energy Method

Assume that the problem (7.4.1)–(7.4.3) has a solution u and let $g^{II} \equiv 0$. We try to obtain a solution-estimate by the energy method; thus we consider the time derivative of the energy:

$$\begin{aligned}\frac{d}{dt}\|u(\cdot,t)\|^2 &= (u,u_t)+(u_t,u)\\ &= (u,Au_{xx})+(Au_{xx},u)+(u,F)+(F,u)\\ &= -\Big((A+A^*)u_x,u_x\Big)+\Big\{\langle u,Au_x\rangle+\langle Au_x,u\rangle\Big\}\Big|_{x=0}^{x=\infty}\\ &\quad + (u,F)+(F,u)\\ &= -\Big((A+A^*)u_x,u_x\Big) - 2\,\mathrm{Re}\,\langle u(0,t),Au_x(0,t)\rangle+(u,F)+(F,u)\\ &\le -2\delta\|u_x\|^2+\|u\|^2+\|F\|^2-2\,\mathrm{Re}\,\langle u(0,t),Au_x(0,t)\rangle.\end{aligned}$$

Since the value $|u_x(0,t)|^2$ cannot be bounded in terms of the L_2-norms

$$\|u(\cdot,t)\|^2+\|u_x(\cdot,t)\|^2,$$

we must bring the boundary conditions into play. The proof of the following result, which slightly generalizes Lemma 7.2.1, is left to the reader.

Lemma 7.4.2. *Let $A,\ L_0,\ L_1\in\mathbf{C}^{n,n}$,*

$$L_0=\begin{pmatrix}L_0^I\\ L_0^{II}\end{pmatrix},\quad L_1=\begin{pmatrix}L_1^I\\ 0\end{pmatrix},$$

where L_0^I, L_1^I have size $r\times n$, $\operatorname{rank}L_1^I=r$, $\operatorname{rank}(L_0,L_1)=n$. The following conditions are equivalent:

(i) *There exists $c>0$ such that*

$$|\langle w,Aw_x\rangle|\le c\Big\{|w|^2+|g^I|^2\Big\}$$

for all $w,w_x\in\mathbf{C}^n$, $g^I\in\mathbf{C}^r$ with

$$L_0w+L_1w_x=\begin{pmatrix}g^I\\ 0\end{pmatrix}.$$

(ii) *If $a,b\in\mathbf{C}^n$ are vectors with*

$$L_1^Ib=0,\quad L_0^{II}a=0,$$

then $\langle a,Ab\rangle=0$.

If the matrices A, L_0, L_1 meet the requirements of the previous lemma, then

$$\begin{aligned}\frac{d}{dt}\|u\|^2 \leq & -2\delta\|u_x\|^2 + \|u\|^2 + \|F\|^2 \\ & + 2c\Big\{|u(0,t)|^2 + |g^I(t)|^2\Big\} \\ \leq & -\delta\|u_x\|^2 - |u(0,t)|^2 \\ & + c_1\|u\|^2 + \|F\|^2 + 2c|g^I(t)|^2,\end{aligned}$$

and we obtain an estimate of the solution in terms of the data. More precisely,

$$\begin{aligned}&\|u(\cdot,t)\|^2 + \int_0^t |u(0,\tau)|^2\, d\tau \\ &\leq K_T\Big\{\|f\|^2 + \int_0^t \|F(\cdot,\tau)\|^2\, d\tau + \int_0^t |g^I(\tau)|^2\, d\tau\Big\}\end{aligned}$$

in $0 \leq t \leq T$. One can also show the *existence* of a smooth solution, and therefore the problem is strongly well-posed in the sense of Definition 1, Section 7.3. If the conditions of Lemma 7.4.2 are not met, then the energy method does not apply, and the question of well-posedness may be investigated by other means only.

Example. To illustrate the differences, we apply Theorem 7.4.1 and the energy method to the following simple example:

$$u_t = \begin{pmatrix} 1 & 0 \\ 0 & \frac{1}{4} \end{pmatrix} u_{xx} + F(x,t), \qquad u(x,t) = \begin{pmatrix} v(x,t) \\ w(x,t) \end{pmatrix},$$

$$v_x(0,t) + w_x(0,t) = g^I(t), \qquad v(0,t) + \alpha w(0,t) = 0,$$

$$t \geq 0, \qquad u(x,0) = f(x), \qquad 0 \leq x < \infty.$$

Here

$$A = \begin{pmatrix} 1 & 0 \\ 0 & \frac{1}{4} \end{pmatrix}, \qquad \Phi = I, \qquad A = \begin{pmatrix} 1 & 0 \\ 0 & 2 \end{pmatrix},$$

$$L_0^I = (0,0), \qquad L_1^I = (1,1), \qquad L_0^{II} = (1,\alpha),$$

$$(L_0, L_1) = \begin{pmatrix} 0 & 0 & 1 & 1 \\ 1 & \alpha & 0 & 0 \end{pmatrix}.$$

The matrix C_0 reads

$$C_0 = \begin{pmatrix} 1 & 2 \\ 1 & \alpha \end{pmatrix},$$

with $\det C_0 = \alpha - 2$; thus Theorem 7.4.1 applies whenever $\alpha \neq 2$.

When does the energy method apply? We check condition (ii) of Lemma 7.4.2 and find

$$a = c_1 \begin{pmatrix} \alpha \\ -1 \end{pmatrix}, \quad b = c_2 \begin{pmatrix} 1 \\ -1 \end{pmatrix}, \quad \langle a, Ab \rangle = c_3(\bar{\alpha} + \frac{1}{4}).$$

Thus the energy method applies only for $\alpha = -\frac{1}{4}$.

This example is typical for parabolic problems under mixed boundary conditions: The energy method applies only in exceptional cases whereas the Laplace transform method applies in most cases, and yields well-posedness in the generalized sense. What can be said if $\det C_0 = 0$, i.e., if Theorem 7.4.1 does not apply? We will show in Section 7.5 that the problem indeed becomes ill-posed in the generalized sense; again, this result remains valid if lower-order terms $Bu_x + Cu$ are added to the differential equation.

7.4.3. An Eigenvalue Problem

Suppose that $F \equiv 0$, $g \equiv 0$ in (7.4.1)–(7.4.3) and substitute

$$u(x,t) = e^{st} f(x), \quad s \in \mathbf{C}, \quad f \in C^\infty,$$

into the equations. One obtains a solution* if and only if

$$(7.4.5) \qquad \begin{aligned} sf(x) &= Af_{xx}(x), & 0 \le x < \infty, \\ L_0 f(0) + L_1 f_x(0) &= 0, & f \in L_2. \end{aligned}$$

(Note that *all* derivatives of $u(\cdot, t)$ are in L_2 if $f \in L_2$; this follows from $sf = Af_{xx}$.) Considering $s \in \mathbf{C}$ as a parameter, we have obtained an eigenvalue problem and give

Definition 1. A number $s \in \mathbf{C}$ is called an eigenvalue of (7.4.5) if – for the number s in question – there is a non-trivial solution $f = f(x)$, $f \in C^\infty \cap L_2$.

Using elementary means of linear algebra, we shall discuss the constant-coefficient problem (7.4.5) below. First let us indicate the relevance of eigenvalues for the question of well- or ill-posedness.

Lemma 7.4.3. *If there is a sequence s_ν of eigenvalues with* Re $s_\nu \to \infty$, *then the problem* (7.4.1)–(7.4.3) *is ill-posed in any sense.*

Proof. If such a sequence s_ν exists, then there is no bound on the exponential growth rate (in time) of genuine solutions.

*Recall that the requirement $u(\cdot, t) \in L_2$ is part of our solution-concept; consequently we exclude all solutions f of $sf = Af_{xx}$ which grow exponentially as $x \to \infty$.

Now we solve the eigenvalue problem (7.4.5) explicitly; the reader is reminded of the notations (7.4.4). For any $s \in \mathbf{C}$, $\operatorname{Re} s > 0$, the general solution of

$$sf = Af_{xx}, \quad f \in L_2(0 \le x < \infty),$$

is given by*

$$f(x) = \sum_{k=1}^{n} \sigma_k \exp(-\sqrt{s/\mu_k}\, x)\, \Phi_k, \quad \sigma \in \mathbf{C}^n.$$

Clearly,

$$f(0) = \sum_{k=1}^{n} \sigma_k \Phi_k = \Phi\sigma,$$

$$f_x(0) = -\sqrt{s}\sum_{k=1}^{n} \sigma_k \frac{1}{\sqrt{\mu_k}} \Phi_k = -\sqrt{s}\Phi\Lambda\sigma.$$

If we introduce these expressions into the (homogeneous) boundary conditions (7.4.3), we obtain a linear system for the vector $\sigma \in \mathbf{C}^n$. After division of the first r equations by $-\sqrt{s}$, the system takes the form

$$\begin{pmatrix} -\frac{1}{\sqrt{s}} L_0^I \Phi + L_1^I \Phi\Lambda \\ L_0^{II}\Phi \end{pmatrix} \sigma = 0. \tag{7.4.6}$$

Thus we have shown

Lemma 7.4.4. *A number $s \in \mathbf{C}$, $\operatorname{Re} s > 0$, is an eigenvalue of* (7.4.5) *if and only if the determinant of the matrix* (7.4.6) *vanishes.*

These results motivate Theorem 7.4.1 (but they do not prove it): If $\operatorname{Re} s \to \infty$, then the matrices (7.4.6) converge to C_0; consequently, if $\det C_0 \neq 0$, there are no eigenvalues with arbitrarily large real parts. If one had a converse of Lemma 7.4.3, then well-posedness would follow. Unfortunately, a complete proof of Theorem 7.4.1 is more elaborate. A simple conclusion can be drawn, however, from the results shown:

Lemma 7.4.5. *Let* $\det C_0 = 0$, *and assume that* $L_0^I = 0$, *i.e., that the Neumann part of the boundary conditions has no lower-order terms. Then the problem* (7.4.1)–(7.4.3) *is ill-posed.*

Proof. All numbers $s \in \mathbf{C}$, $\operatorname{Re} s > 0$, are eigenvalues. The result follows from Lemma 7.4.3.

*For any complex number c with $\operatorname{Re} c > 0$ let $\sqrt{c}$ denote the root with $\operatorname{Re}\sqrt{c} > 0$.

7.4.4. The Laplace Transform; Elementary Properties

In this section we introduce the Laplace transform $\hat{u} = \hat{u}(s)$ of a function $u = u(t)$ and prove some elementary results, which will be needed below. Essentially all the results follow quite easily from corresponding properties of the Fourier transform. To avoid confusion, we use here the following notation for the Fourier transform of $v = v(t)$:

$$(\mathcal{F}v)(\xi) = \frac{1}{\sqrt{2\pi}} \int_{-\infty}^{\infty} e^{-i\xi t} v(t)\, dt, \quad \xi \in \mathbf{R}.$$

Laplace transform vs. Fourier transform. Suppose $u = u(t)$, $0 \le t < \infty$, is a continuous function with values in $\mathbf{C}^n$, which satisfies an estimate

$$|u(t)| \le Ce^{\alpha t}, \quad t \ge 0,$$

for some real constants C, α. The analytic function*

$$\hat{u}(s) = \int_0^{\infty} e^{-st} u(t)\, dt, \quad s \in \mathbf{C}, \quad \operatorname{Re} s > \alpha,$$

is the *Laplace transform* of u. The Laplace and Fourier transform are closely related; to explain the relation, we set

$$u_0(t) = \begin{cases} u(t) & t > 0, \\ \frac{1}{2}u(0) & t = 0, \\ 0 & t < 0, \end{cases}$$

and note that each function

$$v_\eta(t) = e^{-\eta t} u_0(t), \quad \eta > \alpha,$$

has the Fourier transform

$$(\mathcal{F}v_\eta)(\xi) = \frac{1}{\sqrt{2\pi}} \int_{-\infty}^{\infty} e^{-i\xi t} e^{-\eta t} u_0(t)\, dt, \quad \xi \in \mathbf{R}.$$

Therefore we have the relation

$$\hat{u}(\eta + i\xi) = \sqrt{2\pi}(\mathcal{F}v_\eta)(\xi), \quad \eta > \alpha, \quad \xi \in \mathbf{R}.$$

The Fourier inversion formula gives us

$$\begin{aligned} e^{-\eta t} u_0(t) &= \frac{1}{\sqrt{2\pi}} \int_{-\infty}^{\infty} e^{i\xi t} (\mathcal{F}v_\eta)(\xi)\, d\xi \\ &= \frac{1}{2\pi} \int_{-\infty}^{\infty} e^{i\xi t} \hat{u}(\eta + i\xi)\, d\xi, \end{aligned}$$

*The integral is defined componentwise.

and thus we have shown the following *inversion formula* for the Laplace transform:

$$u_0(t) = \frac{1}{2\pi}\int_{-\infty}^{\infty} e^{(\eta+i\xi)t}\,\hat{u}(\eta+i\xi)\,d\xi, \quad \eta > \alpha, \quad t \in \mathbf{R}. \tag{7.4.7}$$

Applications of Parseval's relation. For a function $v = v(t)$ and its Fourier transform $\mathcal{F}v = (\mathcal{F}v)(\xi)$ it holds that

$$\int_{-\infty}^{\infty} |v(t)|^2\,dt = \int_{-\infty}^{\infty} |(\mathcal{F}v)(\xi)|^2\,d\xi.$$

In terms of the Laplace transform, the formula reads

$$\int_0^{\infty} e^{-2\eta t}|u(t)|^2\,dt = \frac{1}{2\pi}\int_{-\infty}^{\infty} |\hat{u}(\eta+i\xi)|^2\,d\xi, \quad \eta > \alpha. \tag{7.4.8}$$

As a consequence one obtains – for each finite time T – an estimate of u in terms of its Laplace transform:

$$\int_0^{T} |u(t)|^2\,dt \le \frac{e^{2\eta T}}{2\pi}\int_{-\infty}^{\infty} |\hat{u}(\eta+i\xi)|^2\,d\xi \quad \text{if}\,\eta > \alpha \ge 0. \tag{7.4.9}$$

In our applications below we consider continous functions

$$u = u(x,t), \quad 0 \le x < \infty, \quad t \ge 0.$$

Suppose that u satisfies an estimate

$$|u(x,t)| \le Ce^{\alpha t}, \quad 0 \le x < \infty, \quad t \ge 0;$$

then

$$\hat{u}(x,s) = \int_0^{\infty} e^{-st}u(x,t)\,dt, \quad 0 \le x < \infty, \quad \mathrm{Re}\,s > \alpha,$$

denotes the Laplace transform in time. We want to show

Lemma 7.4.6. *Under the above assumptions let $u(\cdot,t) \in L_2$ for each $t \ge 0$, and assume we have a bound*

$$\|u(\cdot,t)\|^2 \le K^2 e^{2\alpha t}, \quad t \ge 0.$$

Then

$$\int_0^{\infty} e^{-2\eta t}\|u(\cdot,t)\|^2 dt = \frac{1}{2\pi}\int_{-\infty}^{\infty} \|\hat{u}(\cdot,\eta+i\xi)\|^2 d\xi, \quad \eta > \alpha, \tag{7.4.10a}$$

and consequently,

$$\int_0^T \|u(\cdot,t)\|^2 dt \le \frac{e^{2\eta T}}{2\pi} \int_{-\infty}^{\infty} \|\hat{u}(\cdot,\eta+i\xi)\|^2 d\xi \quad \text{if } \eta > \alpha \ge 0. \tag{7.4.10b}$$

(*Under our assumptions the integrals exist and are finite.*)

Proof. The result follows if one integrates the relation

$$\int_0^{\infty} e^{-2\eta t} |u(x,t)|^2\, dt = \frac{1}{2\pi} \int_{-\infty}^{\infty} |\hat{u}(x,\eta+i\xi)|^2\, d\xi, \quad \eta > \alpha,$$

with respect to x and interchanges the order of integration. Integrability of the function

$$e^{-2\eta t} |u(x,t)|^2, \quad x \ge 0, \quad t \ge 0,$$

and Fubini's theorem justify the interchange.

Laplace transform of time derivatives. Suppose that $u = u(t),\ 0 \le t < \infty$, is continuously differentiable and

$$|\frac{du}{dt}(t)| = |u_t(t)| \le Ce^{\alpha t}, \quad t \ge 0.$$

Then integration by parts gives us that

$$\hat{u}_t(s) = \int_0^{\infty} e^{-st} u_t(t)\, dt = -u(0) + s\hat{u}(s) \quad \text{for } \operatorname{Re} s > \alpha.$$

Future does not affect past. We will need a result which states – roughly speaking – that we may alter data $g(t)$ for $t > T$ without changing the solution for $t \le T$. For illustration, we consider an ordinary initial value problem

$$\frac{du}{dt}(t) = Lu(t) + g(t), \quad t \ge 0, \quad u(0) = 0,$$

where $L \in \mathbf{C}^{n,n},\ L + L^* \le 2\beta I$, and $g = g(t)$ is continuous with

$$|g(t)| \le Ce^{\alpha t}, \quad t \ge 0.$$

Laplace transformation gives us

$$s\hat{u}(s) = L\hat{u}(s) + \hat{g}(s),$$

and thus

$$\hat{u}(s) = (sI - L)^{-1} \hat{g}(s) \quad \text{for } \operatorname{Re} s > \max\{\alpha, \beta\}.$$

Using the inversion formula (7.4.7), we obtain an explicit expression for the solution $u(t)$. Since $\hat{g}(s)$ depends on *all* values $g(t)$, one might be tempted to conclude from this formula that each value $u(t_0)$ depends on *all* values $g(t)$, too. However, as is well-known for the above initial value problem, if we alter the data $g(t)$ for $t > T$, the solution $u(t)$ remains unchanged for $0 \le t \le T$. A slight generalization of this result is contained in

Lemma 7.4.7. *Suppose that* $u = u(t)$, $g = g(t)$, $0 \le t < \infty$, *are continuous functions bounded in norm by some exponential* $ce^{\alpha t}$, *and suppose that their Laplace transforms satisfy an estimate*

$$|\hat{u}(s)|^2 \le c_1 |\hat{g}(s)|^2 \quad \text{if } \operatorname{Re} s > \alpha_1,$$

for some constants c_1, $\alpha_1 \ge \alpha$. *If* $g(t) \equiv 0$ *for* $0 \le t \le T$, *then* $u(t) \equiv 0$ *for* $0 \le t \le T$.

Proof. We fix a time $T_1 < T$ and use (7.4.9) for $\eta > \alpha_1$:

$$\begin{aligned}
\int_0^{T_1} |u(t)|^2\, dt &\le \frac{1}{2\pi} e^{2\eta T_1} \int_{-\infty}^{\infty} |\hat{u}(\eta + i\xi)|^2\, d\xi \\
&\le \frac{c_1}{2\pi} e^{2\eta T_1} \int_{-\infty}^{\infty} |\hat{g}(\eta + i\xi)|^2\, d\xi \\
&= c_1 e^{2\eta T_1} \int_T^{\infty} e^{-2\eta t} |g(t)|^2\, dt \quad \text{(by (7.4.8))} \\
&= c_1 e^{2\eta T_1} \int_0^{\infty} e^{-2\eta (T+t)} |g(T+t)|^2\, dt \\
&\le c_1 e^{-2\eta (T - T_1)} \int_0^{\infty} e^{-2\eta t} c^2 e^{2\alpha (T+t)}\, dt \\
&= c_1 c^2 e^{-2\eta (T-T_1)} e^{2\alpha T} \frac{1}{2\eta - 2\alpha}.
\end{aligned}$$

For $\eta \to \infty$ we obtain

$$\int_0^{T_1} |u(t)|^2\, dt = 0.$$

Since $T_1 < T$ is arbitrary and u is continuous, the result follows.

Estimates of the Laplace transform. Let $g \in C_0^\infty(0 < t < \infty)$. Then integration by parts gives us

$$\hat{g}(s) = \int_0^\infty e^{-st} g(t)\, dt = -\frac{1}{s}\int_0^\infty e^{-st} g'(t)\, dt,$$

$$|\hat{g}(s)| \le K_1/|s|, \quad \operatorname{Re} s > 0.$$

This process can be repeated. Therefore, for any $p = 1, 2, \ldots$, there is a constant $K_p = K_p(g)$ independent of s with

$$|\hat{g}(s)| \le K_p/|s|^p, \quad \operatorname{Re} s > 0.$$

In other words, as $|s| \to \infty$, $\operatorname{Re} s > 0$, the function $|\hat{g}(s)|$ decays faster than any power $|s|^{-p}$.

7.4.5. Solution via Laplace Transform

Consider the parabolic initial-boundary value problem (7.4.1)–(7.4.3) under the assumptions stated in Section 7.4.1. In addition, we assume first that $f \equiv 0$ and $L_0^I = 0$; thus the Neumann part of the boundary condition at $x = 0$ has no lower-order terms. If the matrix C_0 introduced in (7.4.4c) is singular then, according to Lemma 7.4.5, the problem is ill-posed, and consequently we assume that $\det C_0 \neq 0$. We shall construct a solution via Laplace transformation and derive solution-estimates which imply strong well-posedness in the generalized sense.

Estimate of the Laplace transform. To begin with, assume that there is a solution $u = u(x, t)$ which is "well-behaved" in the sense that all the following operations are permitted. The solution formula derived below can subsequently be used to justify this assumption. Laplace transformation gives us

$$\begin{aligned} s\hat{u}(x, s) &= A\hat{u}_{xx}(x, t) + \hat{F}(x, s), \\ L_1^I \hat{u}_x(0, s) = \hat{g}^I(s), &\quad L_0^{II}\hat{u}(0, s) = \hat{g}^{II}(s), \\ \|\hat{u}(\cdot, s)\| < \infty &\quad \text{for} \quad \operatorname{Re} s > 0. \end{aligned}$$

We prove the following estimate of $\hat{u}$ in terms of $\hat{F}$ and $\hat{g}$:

Lemma 7.4.8. *There is a constant c independent of the data $F = F(x, t)$ and $g = g(t)$ with*

$$\begin{aligned} &|\hat{u}(0, s)|^2 + \frac{1}{|s|}|\hat{u}_x(0, s)|^2 + |s|^{1/2}\|\hat{u}(\cdot, s)\|^2 + \frac{1}{|s|^{1/2}}\|\hat{u}_x(\cdot, s)\|^2 \\ &\le c\left\{\frac{1}{|s|^{3/2}}\|\hat{F}(\cdot, s)\|^2 + \frac{1}{|s|}|\hat{g}^I(s)|^2 + |\hat{g}^{II}(s)|^2\right\} \end{aligned}$$

for all $s \in \mathbf{C}$, $\operatorname{Re} s > 0$. *(Here and in the following, the powers of* $|s|$ *have significance for large* $|s|$*, not for* $s \approx 0$.)

Proof. a) For brevity we set $\rho = \sqrt{s}$, $|\arg \rho| < \frac{\pi}{4}$. Introducing the vector

$$w(x,s) = \begin{pmatrix} \hat{u}(x,s) \\ \frac{1}{\rho}\hat{u}_x(x,s) \end{pmatrix},$$

we can write the second-order differential equation for $\hat{u}(\cdot, s)$ as a first-order system:

$$w_x(x,s) = \rho \begin{pmatrix} 0 & I \\ A^{-1} & 0 \end{pmatrix} w(x,s) - \frac{1}{\rho} \begin{pmatrix} 0 \\ A^{-1}\hat{F}(x,s) \end{pmatrix}.$$

The boundary conditions become

$$\begin{pmatrix} 0 & L_1^I \\ L_0^{II} & 0 \end{pmatrix} w(0,s) = \begin{pmatrix} \frac{1}{\rho}\hat{g}^I(s) \\ \hat{g}^{II}(s) \end{pmatrix}, \qquad \|w(\cdot, s)\| < \infty.$$

The system-matrix

$$M_0 = \begin{pmatrix} 0 & I \\ A^{-1} & 0 \end{pmatrix}$$

can be diagonalized; using the notation (7.4.4), we set

$$S_0 = \begin{pmatrix} \Phi\Lambda^{-1} & 0 \\ 0 & \Phi \end{pmatrix} \begin{pmatrix} -I & I \\ I & I \end{pmatrix} = \begin{pmatrix} -\Phi\Lambda^{-1} & \Phi\Lambda^{-1} \\ \Phi & \Phi \end{pmatrix}$$

and obtain (note that $\Phi^{-1}A\Phi = \Lambda^{-2}$)

$$S_0^{-1} M_0 S_0 = \begin{pmatrix} -\Lambda & 0 \\ 0 & \Lambda \end{pmatrix}.$$

In the new variable

$$\tilde{w}(x,s) = \begin{pmatrix} \tilde{w}^{(0)}(x,s) \\ \tilde{w}^{(1)}(x,s) \end{pmatrix} = S_0^{-1} w(x,s)$$

the system is diagonal:

(7.4.11)
$$\tilde{w}_x(x,s) = \rho \begin{pmatrix} -\Lambda & 0 \\ 0 & \Lambda \end{pmatrix} \tilde{w}(x,s) - \frac{1}{\rho}\tilde{F}(x,s),$$

$$\tilde{F}(x,s) = S_0^{-1} \begin{pmatrix} 0 \\ A^{-1}\hat{F}(x,s) \end{pmatrix}.$$

The boundary condition at $x = 0$ reads

$$B_0\tilde{w}^{(0)}(0,s) + B_1\tilde{w}^{(1)}(0,s) = \begin{pmatrix} \frac{1}{\rho}\hat{g}^I(s) \\ \hat{g}^{II}(s) \end{pmatrix} =: \tilde{g}(s) ,$$

where

$$(B_0, B_1) = \begin{pmatrix} 0 & L_1^I \\ L_0^{II} & 0 \end{pmatrix} S_0 .$$

Thus the matrices B_0, B_1 have size $n \times n$, and

$$B_0 = -\begin{pmatrix} 0 \\ L_0^{II} \end{pmatrix} \Phi\Lambda^{-1} + \begin{pmatrix} L_1^I \\ 0 \end{pmatrix} \Phi = \begin{pmatrix} L_1^I\Phi \\ -L_0^{II}\Phi\Lambda^{-1} \end{pmatrix}$$

is nonsingular since $\det C_0 \neq 0$ by assumption. Therefore we obtain that

(7.4.12) $$\tilde{w}^{(0)}(0,s) = B_0^{-1}\tilde{g}(s) - B_0^{-1}B_1\tilde{w}^{(1)}(0,s).$$

b) With a number $d > 1$, to be specified below, we set

$$D = \begin{pmatrix} I & 0 \\ 0 & -dI \end{pmatrix}.$$

The differential equation (7.4.11) yields for each fixed s, $\operatorname{Re} s > 0$,

$$\begin{aligned}(\tilde{w}, D\tilde{w}_x) + (D\tilde{w}_x, \tilde{w}) &= 2\operatorname{Re}(\tilde{w}, D\tilde{w}_x) \\ &= -2\operatorname{Re}\left(\tilde{w}, \rho\begin{pmatrix} \Lambda & 0 \\ 0 & d\Lambda \end{pmatrix}\tilde{w}\right) - 2\operatorname{Re}\left(\tilde{w}, \frac{1}{\rho}D\tilde{F}\right).\end{aligned}$$

Since all diagonal entries λ_k of Λ have positive real part and since $|\arg\rho| < \frac{\pi}{4}$, there is δ_1 independent of ρ with

$$\operatorname{Re}(\rho\lambda_k) \geq \delta_1|\rho|, \quad \delta_1 > 0,$$

and therefore

$$2\operatorname{Re}(\tilde{w}, D\tilde{w}_x) \leq -2\delta_1|\rho|\|\tilde{w}\|^2 + \frac{2d}{|\rho|}\|\tilde{w}\|\|\tilde{F}\|.$$

On the other hand, using integration by parts and the boundary condition (7.4.12), we find that

$$\begin{aligned}(\tilde{w}, D\tilde{w}_x) + (D\tilde{w}_x, \tilde{w}) &= \langle\tilde{w}, D\tilde{w}\rangle\Big|_{x=0}^{x=\infty} \\ &= -|\tilde{w}^{(0)}(0,s)|^2 + d|\tilde{w}^{(1)}(0,s)|^2 \\ &= |\tilde{w}(0,s)|^2 + (d-1)|\tilde{w}^{(1)}(0,s)|^2 - 2|\tilde{w}^{(0)}(0,s)|^2 \\ &\geq |\tilde{w}(0,s)|^2 + (d-1-c_1)|\tilde{w}^{(1)}(0,s)|^2 - c_1|\tilde{g}(s)|^2.\end{aligned}$$

Here the constant c_1 depends only on the matrices B_0, B_1 in (7.4.12). If we choose $d \geq 1 + c_1$ and combine the two estimates, we obtain that

$$|\tilde{w}(0,s)|^2 + 2\delta_1|\rho|\|\tilde{w}\|^2 \leq c_1|\tilde{g}(s)|^2 + \frac{2d}{|\rho|}\|\tilde{w}\|\|\tilde{F}\|$$
$$\leq c_1|\tilde{g}(s)|^2 + \delta_1|\rho|\|\tilde{w}\|^2 + \frac{d^2}{\delta_1|\rho|^3}\|\tilde{F}\|^2.$$

Thus there is a constant c_2 independent of the data and of $s \in \mathbf{C}$, $\operatorname{Re} s > 0$, with

$$|\tilde{w}(0,s)|^2 + |\rho|\|\tilde{w}(\cdot,s)\|^2 \leq c_2\Big\{|\tilde{g}(s)|^2 + \frac{1}{|\rho|^3}\|\tilde{F}(\cdot,s)\|^2\Big\}.$$

In terms of the original variables $\hat{u}$, $\hat{g}^I$, $\hat{g}^{II}$, $\hat{F}$, the desired estimate follows, and the lemma is proven.

Estimate of the solution. The previous lemma gives us an accurate bound of the Laplace transform $\hat{u}$ of the solution u in terms of the Laplace transforms $\hat{F}$, $\hat{g}^I$, $\hat{g}^{II}$ of the data F, g^I, g^{II}. We can now apply the results of Section 7.4.4 – which were themselves simple applications of Parseval's relation – to estimate u in terms of F and g^I, g^{II}. One obtains

Lemma 7.4.9. *First assume that $g^{II} \equiv 0$. For each time $T > 0$ there is a constant K_T, independent of the data F, g^I, with*

$$\int_0^T \Big\{|u(0,t)|^2 + |u_x(0,t)|^2 + \|u(\cdot,t)\|^2 + \|u_x(\cdot,t)\|^2\Big\}\,dt$$
$$\leq K_T \int_0^T \Big\{|g^I(t)|^2 + \|F(\cdot,t)\|^2\Big\}\,dt.$$

If g^{II} is not necessarily zero then

$$\int_0^T \Big\{|u(0,t)|^2 + \|u(\cdot,t)\|^2\Big\}\,dt \leq K_T \int_0^T \Big\{|g(t)|^2 + \|F(\cdot,t)\|^2\Big\}\,dt.$$

Proof. Assume that $g^{II} \equiv 0$ and consider, for example, the integral of $|u_x(0,t)|^2$. We fix $\eta \geq 1$ and obtain, from (7.4.9),

$$\int_0^T |u_x(0,t)|^2\,dt \leq c_1 \int_{-\infty}^{\infty} |\hat{u}_x(0,\eta + i\xi)|^2\,d\xi, \quad c_1 = \frac{e^{2\eta T}}{2\pi}.$$

According to Lemma 7.4.7, the integral on the right-hand side can be bounded by

$$c_2 \int_{-\infty}^{\infty} \left\{ \|\hat{F}(\cdot, \eta + i\xi)\|^2 \; + \; |\hat{g}^I(\eta + i\xi)|^2 \right\} d\xi.$$

Here

$$\frac{1}{2\pi} \int_{-\infty}^{\infty} \|\hat{F}(\cdot, \eta + i\xi)\|^2 d\xi \leq \int_0^{\infty} \|F(\cdot, t)\|^2 dt$$

by (7.4.10a). The integral of $|\hat{g}^I|^2$ is estimated similarly using (7.4.8). In this way one obtains a bound by

$$c_3 \int_0^{\infty} \left\{ \|F(\cdot, t)\|^2 \; + \; |g^I(t)|^2 \right\} dt.$$

Now we change the data in such a way that they vanish for $t \geq T + \epsilon, \; \epsilon > 0$. Reasoning as in the proof of Lemma 7.4.7, we note that the solution remains unaffected for $t \leq T$; since $\epsilon > 0$ is arbitrary, the result follows.

Existence of a solution. Thus far we have not shown the *existence* of a "well-behaved" solution for which the above computations are justified. On the Laplace-transform side, existence can be shown quite easily. All we need is the following simple existence result for scalar equations on the interval $0 \leq x < \infty$.

Lemma 7.4.10. *Consider a scalar equation*

$$\frac{dv}{dx}(x) = \lambda v(x) + q(x), \quad 0 \leq x < \infty,$$

where $\mathrm{Re}\,\lambda > 0, \; q \in C^{\infty}, \; \|q\| < \infty$. *The problem has a unique solution* $v = v(x)$ *with* $\|v\| < \infty$.

Proof. All solutions of the differential equation have the form

$$v(x) = e^{\lambda x}\{v_0 + \int_0^x e^{-\lambda y} q(y)\, dy\}.$$

If we set

$$v_0 = -\int_0^{\infty} e^{-\lambda y} q(y)\, dy,$$

then $v(x) = \int_x^{\infty} e^{\lambda(x-y)} q(y)\, dy$, and therefore,

$$|v(x)|^2 = |\int_x^{\infty} e^{\lambda(x-y)} q(y)\, dy|^2.$$

Integration in x yields

$$\|v\|^2 \leq \int_0^\infty \int_x^\infty e^{2\,\mathrm{Re}\,\lambda(x-y)} |q(y)|^2\, dy\, dx$$
$$= \int_0^\infty \left\{ \int_0^y e^{2\,\mathrm{Re}\,\lambda(x-y)}\, dx \right\} |q(y)|^2\, dy \leq \frac{1}{2\,\mathrm{Re}\,\lambda} \|q\|^2.$$

For an initial value $v(0) = v_0$ different from the one given above, the solution shows exponential growth, hence is not in L_2.

Now consider the diagonal system (7.4.11) with boundary condition (7.4.12) at $x = 0$; here $s \in \mathbf{C}$, $\mathrm{Re}\, s > 0$, is fixed. Using the previous lemma, we find that the side condition $\|\tilde{w}(\cdot, s)\| < \infty$ determines a unique solution $\tilde{w}(x, s)$. Then, reversing the transformations used in the proof of Lemma 7.4.7, we obtain $\hat{u}(x, s)$. The inverse Laplace transform (7.4.7) gives us

$$u(x, t) = \frac{1}{2\pi i} \int_{\mathrm{Re}\, s = \eta} e^{st} \hat{u}(x, s)\, ds, \qquad \eta > 0.$$

By assumption, $g \in C_0^\infty(0 < t < \infty)$. Hence $\hat{g}(s)$ is an analytic function of s, $\mathrm{Re}\, s > 0$, and the same is true for $\hat{u}(x, s)$. Also, $\hat{u}(x, s)$ is a smooth function of x, and $|\hat{u}(x, s)|$ decays rapidly for $|s| \to \infty$. Therefore,

$$u_t - Au_{xx} - F = \frac{1}{2\pi i} \int_{\mathrm{Re}\, s = \eta} e^{st} \left(s\hat{u} - A\hat{u}_{xx} - \hat{F} \right) ds = 0, \qquad \eta > 0.$$

This shows that u is a solution of the differential equation. Clearly, u satisfies the boundary condition. It remains to show that $u(x, 0) = 0$. By Residue Calculus we can choose any positive value for η without affecting the result $u = u(x, t)$ of the inverse Laplace transform. Therefore,

$$u(x, 0) = \lim_{\eta \to \infty} \frac{1}{2\pi i} \int_{\mathrm{Re}\, s = \eta} \hat{u}(x, s)\, ds = 0.$$

This completes the proof of Theorem 7.4.1 for $f \equiv 0$ under the assumption that there are no lower-order terms, i.e.,

$$L_0^I = 0, \quad B = C = 0.$$

Lower-order terms. The case of nonzero lower-order terms can be treated in much the same way. For the vector

$$w(x, s) = \begin{pmatrix} \hat{u}(x, s) \\ \frac{1}{\rho} \hat{u}_x(x, s) \end{pmatrix}, \qquad \rho = \sqrt{s},$$

one obtains a system

$$w_x = \rho M(\rho) w - \frac{1}{\rho} \begin{pmatrix} 0 \\ A^{-1}\hat{F} \end{pmatrix}$$

where

$$M(\rho) = M_0 + O(\frac{1}{|\rho|}) \quad \text{for} \quad |\rho| \quad \text{large},$$

and M_0 is the matrix introduced in the proof of Lemma 7.4.8. Since the eigenvalues of M_0 are all different, one can diagonalize $M(\rho)$ by a transformation

$$S(\rho) = S_0 + O(\frac{1}{|\rho|}).$$

The boundary conditions for

$$\tilde{w}(x,s) = S(\rho)^{-1} w(x,s)$$

take the form

$$B_0(\rho)\tilde{w}^{(0)}(0,s) + B_1(\rho)\tilde{w}^{(1)}(0,s) = \tilde{g}(s)$$

with

$$B_0(\rho) = B_0 + O(\frac{1}{|\rho|}).$$

Thus, if $|\rho|$ is sufficiently large, the estimates for $\tilde{w}$ follow as before; the crucial condition

$$\det B_0 \neq 0 \quad (\Leftrightarrow \det C_0 \neq 0)$$

remains unchanged. In this way, we obtain well-posedness in the generalized sense for problems with lower-order terms. Thus far we have restricted ourselves to homogeneous initial data $f \equiv 0$, however.

Inhomogeneous initial data. If $f \not\equiv 0$, we can introduce the function

$$v(x,t) = u(x,t) - e^{-t} f(x)$$

which satisfies homogeneous initial conditions at $t = 0$. Unfortunately the derivative f_{xx} enters the forcing term of the differential equation for v, and therefore, if we apply Lemma 7.4.8 in a straightforward way, we need

$$\|f\|_{H^2}^2$$

to estimate the solution u.

It is possible, however, to derive an estimate in terms of $\|f\|^2$ only. For simplicity we restrict ourselves to an equation

$$u_t = Au_{xx}, \quad u(x,0) = f(x), \tag{7.4.13}$$

and assume boundary conditions (7.4.3) with $g \equiv 0,\ L_0^I = 0$. (Again, lower-order terms could be treated by a perturbation argument.) Laplace transformation of (7.4.13) yields

$$s\hat{u}(x,s) = A\hat{u}_{xx}(x,s) + f(x);$$

thus the function $f(x)$ enters the discussion in the same way as the function $\hat{F}(x,s)$ did previously. If $\det C_0 \neq 0$ we find, as in Lemma 7.4.7,

$$\begin{aligned} |\hat{u}(0,s)|^2 + \frac{1}{|s|}|\hat{u}_x(0,s)|^2 + |s|^{1/2}\|\hat{u}(\cdot,s)\|^2 + \frac{1}{|s|^{1/2}}\|\hat{u}_x(\cdot,s)\|^2 \\ \leq \frac{c}{|s|^{3/2}}\|f\|^2. \end{aligned}$$

Using (7.4.9) for fixed $\eta > 0,\ T > 0$, we obtain an estimate of the solution on the boundary $x = 0$,

$$\begin{aligned} \int_0^T |u(0,t)|^2\,dt &\leq c_1 \int_{-\infty}^{\infty} |\hat{u}(0,\eta + i\xi)|^2\,d\xi \\ &\leq c_2\|f\|^2 \int_{-\infty}^{\infty} \frac{d\xi}{|\eta + i\xi|^{3/2}} = c_3\|f\|^2. \end{aligned}$$

The time integral of $\|u(\cdot,t)\|^2$ can be treated in the same way using (7.4.10b). To summarize, we have shown

Lemma 7.4.11. *If* $\det C_0 \neq 0,\ F \equiv 0,\ g \equiv 0$, *then the solution* $u = u(x,t)$ *of the initial-boundary value problem satisfies*

$$\int_0^T \|u(\cdot,t)\|^2 dt + \int_0^T |u(0,t)|^2\,dt \leq K_T\|f\|^2.$$

7.4.6. Extensions

We want to sketch some extensions of the previous results without giving the proofs in detail.

Strip problems. The results can be extended to strongly parabolic systems in a strip

$$0 \leq x \leq 1, \quad t \geq 0$$

with boundary conditions of the form (7.4.3) on either side of the strip, i.e., at $x = 0$ and $x = 1$:

$$eL^I_{j0}u(j,t) + L^I_{j1}u_x(j,t) = g^I_j(t), \quad L^{II}_{j0}u(j,t) = g^{II}_j(t), \quad j = 0, 1.$$

To check for well-posedness, one considers now the two matrices

$$C_j = \begin{pmatrix} L^I_{j1}\Phi\Lambda \\ L^{II}_{j0}\Phi \end{pmatrix}, \quad j = 0, 1,$$

instead of the single matrix (7.4.4c); one can show that the strip problem is well-posed in the generalized sense if

$$\det C_0 \neq 0 \quad \text{and} \quad \det C_1 \neq 0.$$

To sketch a proof of this result, we note that the general solution of

$$s\hat{u}(x,s) = A\hat{u}_{xx}(x,s), \quad 0 \le x \le 1,$$

is given by

$$\hat{u}(x,s) = \hat{u}_0(x,s) + \hat{u}_1(x,s),$$

$$\hat{u}_0(x,s) = \sum_{k=1}^{n} \sigma_{0k} \exp(-\sqrt{s/\mu_k}\; x)\Phi_k, \qquad \sigma_0 \in \mathbf{C}^n,$$

$$\hat{u}_1(x,s) = \sum_{k=1}^{n} \sigma_{1k} \exp(-\sqrt{s/\mu_k}\; (1-x))\Phi_k, \quad \sigma_1 \in \mathbf{C}^n.$$

The vectors σ_0, σ_1 have to be determined by the boundary conditions. If we substitute the above function $\hat{u}(x,s)$ into the boundary condition at $x = 0$, we note that σ_1 contributes a term which is exponentially small for large Re s. Similarly, the contribution of σ_0 is exponentially small at $x = 1$. For this reason, if Re s is sufficiently large, the vectors $\sigma_0 = \sigma_0(s)$, $\sigma_1 = \sigma_1(s)$ are uniquely determined.

To derive the proper solution-estimate, we can basically proceed as in the proof of Lemma 7.4.7. Only the scaling-matrix $D = D(x)$ must be chosen more carefully:

$$D(x) = \begin{pmatrix} d^{(0)}(x)I & 0 \\ 0 & -d^{(1)}(x)I \end{pmatrix}.$$

Here $d^{(0)}$, $d^{(1)}$ are positive smooth functions for which the ratios

$$\frac{d^{(1)}(0)}{d^{(0)}(0)}, \quad \frac{d^{(0)}(1)}{d^{(1)}(1)}$$

are sufficiently large (depending on the boundary conditions).

Variable coefficients. The previous results can be generalized to parabolic systems with variable coefficients $A = A(x,t)$, etc. Also, the coefficients of the boundary conditions are allowed to depend smoothly on time,

$$L_{jk}^{I,II} = L_{jk}^{I,II}(t), \quad j,k = 0,1.$$

Assuming that $A(x,t)$ can be diagonalized at $x = 0$, $x = 1$, we obtain matrices $C_j(t)$, $j = 0,1$. If the determinant condition

$$\det C_j(t) \neq 0 \quad \text{for all } t \geq 0, \quad j = 0,1,$$

is fulfilled, then the initial-boundary value problem is well-posed in the generalized sense. (See also the remarks at the end of Chapter 8.)

7.5. Mildly Ill-Posed Half-Space Problems

Consider a parabolic problem

$$u_t = Au_{xx} + Bu_x + Cu, \quad A + A^* \geq 2\delta I, \quad \delta > 0, \tag{7.5.1}$$

in $0 \leq x < \infty$, $t \geq 0$, with initial condition

$$u(x,0) = 0, \quad 0 \leq x < \infty, \tag{7.5.2}$$

and boundary conditions at $x = 0$,

$$L_0^I u(0,t) + L_1^I u_x(0,t) = g^I(t), \quad L_0^{II} u(0,t) = g^{II}(t), \quad t \geq 0. \tag{7.5.3}$$

All matrices are constant (in general, complex), A, B, C have size $n \times n$, L_0^I, L_1^I have size $r \times n$, and L_0^{II} has size $(n-r) \times n$. We make the reasonable assumption that L_1^I and L_0^{II} have full rank in order to have n linearly independent boundary conditions. Furthermore, we assume for simplicity that the eigenvalues

$$\mu_1, \ldots, \mu_n, \quad \operatorname{Re} \mu_k \geq \delta,$$

of A are distinct. First we want to use the ideas of the previous section to derive a formal solution formula. Second, we will show that the problem is not strongly well-posed in the generalized sense (see Definition 3 of Section 7.3) unless

$$\det C_0 \neq 0, \quad C_0 = \begin{pmatrix} L_1^I \Phi \Lambda \\ L_0^{II} \Phi \end{pmatrix}.$$

Here

$$\Phi^{-1} A \Phi = \operatorname{diag}(\mu_1, \ldots, \mu_n), \quad \Lambda = \operatorname{diag}\left(\frac{1}{\sqrt{\mu_1}}, \ldots, \frac{1}{\sqrt{\mu_n}}\right), \quad \operatorname{Re} \frac{1}{\sqrt{\mu_k}} > 0.$$

Hence there are two different ways in which a problem (7.5.1)–(7.5.3) can become ill-posed: First, there can be a sequence of eigenvalues s_ν with $\operatorname{Re} s_\nu \to \infty$; see Lemma 7.4.3. Second, no such sequence exists, but $\det C_0 = 0$. In the second case the ill-posedness is of a milder nature: One can still estimate the solution if one uses derivatives of the data. In contrast to the weakly ill-posed Cauchy problems of Chapter 2, a perturbation by lower-order terms still does not lead to arbitrarily fast exponential growth.

7.5.1. Formal Solution

Laplace transformation of (7.5.1)–(7.5.3) gives us

$$s\hat{u}(x,s) = A\hat{u}_{xx}(x,s) + B\hat{u}_x(x,s) + C\hat{u}(x,s), \tag{7.5.4}$$

$$L_0^I\hat{u}(0,s) + L_1^I\hat{u}_x(0,s) = \hat{g}^I(s), \quad L_0^{II}\hat{u}(0,s) = \hat{g}^{II}(s), \tag{7.5.5}$$

$$\|\hat{u}(\cdot,s)\| < \infty. \tag{7.5.6}$$

For each fixed s with sufficiently large real part, we want to solve this constant-coefficient problem. The general solution of the differential equation (7.5.4) with side-conditon (7.5.6) reads

$$\hat{u}(x,s) = \sum_{k=1}^{n} \sigma_k \exp(\kappa_k(s)x)\phi_k(s), \quad \sigma = \sigma(s) \in \mathbf{C}^n. \tag{7.5.7}$$

Here $\kappa_k = \kappa_k(s)$ are the roots of

$$\det(\kappa^2 A + \kappa B + C - sI) = 0 \quad \text{with} \quad \operatorname{Re}\kappa_k < 0$$

and $\phi_k = \phi_k(s) \in \mathbf{C}^n$ are corresponding eigenvectors,

$$(\kappa_k^2 A + \kappa_k B + C - sI)\phi_k = 0, \quad \phi_k \neq 0.$$

(It is assumed that $\operatorname{Re} s > 0$ is so large that the $2n$ roots κ of the above characteristic equation are distinct and exactly n of them have negative real parts.) Perturbation arguments show that

$$\kappa_k(s) = -\sqrt{\frac{s}{\mu_k}} + O(1), \quad k = 1, \ldots, n,$$

and one can normalize the eigenvectors $\phi_k(s)$ so that

$$\phi_k(s) = \Phi_k + O(\frac{1}{|\rho|}), \quad \rho = \sqrt{s},$$

where $A\Phi_k = \mu_k\Phi_k$, $\Phi_k \neq 0$.

If we introduce the notations

$$\Phi(s) = \Big(\phi_1(s)\Big| \dots \Big|\phi_n(s)\Big),$$

$$\operatorname{diag}(\kappa_k(s)) = \operatorname{diag}\Big(\kappa_1(s), \dots, \kappa_n(s)\Big),$$

then we can write for short

$$\hat{u}(0, s) = \Phi(s)\sigma, \quad \hat{u}_x(0, s) = \Phi(s) \operatorname{diag}(\kappa_k(s))\sigma,$$

and the boundary conditions (7.5.5) take the form

$$\begin{pmatrix} L_0^I \Phi(s) + L_1^I \Phi(s) \operatorname{diag}(\kappa_k(s)) \\ L_0^{II} \Phi(s) \end{pmatrix} \sigma = \begin{pmatrix} \hat{g}^I(s) \\ \hat{g}^{II}(s) \end{pmatrix}. \tag{7.5.8}$$

There are two possibilities:

(i) There is a sequence $s_\nu \in \mathbf{C}$ with Re $s_\nu \to \infty$ for which the above system-matrix is singular. Then we can argue as in Section 7.4.3 and show that there is no bound on the exponential growth rate (in time) of genuine solutions. Therefore the initial-boundary value problem is ill-posed in any sense.

(ii) If Re s is sufficiently large then the matrix in (7.5.8) is nonsingular. In this case let

$$\sigma = \sigma(s), \quad s = \eta + i\xi, \quad \eta \text{ sufficiently large},$$

denote the solution of (7.5.8). The function $\hat{u}(x, s)$ defined in (7.5.7) solves (7.5.4)–(7.5.6); inverting the Laplace transform, we have formally solved the half-space problem.

There are no difficulties in extending the formal solution process to inhomogeneous differential equations and inhomogeneous initial conditions. If all data are sufficiently "well-behaved" then – by the same arguments as before – the formal solution is a genuine C^∞-solution. Nevertheless, as we will prove in the next section, the problem is ill-posed in the generalized sense if $\det C_0 = 0$.

7.5.2. Necessity of the Determinant Condition

Using the previous notations, we show

Theorem 7.5.1. *The half-space problem* (7.5.1)–(7.5.3) *is not well-posed in the sense of Definition* 3, *Section* 7.3, *if* $\det C_0 = 0$.

Proof. a) We divide the I-part of system (7.5.8) by $-\rho = -\sqrt{s}$ and obtain

$$\Big(C_0 + C_1(\rho)\Big)\sigma = \begin{pmatrix} -\rho^{-1}\hat{g}^I(s) \\ \hat{g}^{II}(s) \end{pmatrix}, \quad |C_1(\rho)| = O\Big(\frac{1}{|\rho|}\Big). \tag{7.5.9}$$

By assumption, C_0 is singular; furthermore, taking $\operatorname{Re} s$ sufficiently large, we may assume that

$$\det\Big(C_0 + C_1(\rho)\Big) \neq 0.$$

(Otherwise we are in case (i) mentioned in the previous section.) From the rank-assumption for L_1^I, L_0^{II}, it follows that there exists a vector

$$q = \begin{pmatrix} 0 \\ q^{II} \end{pmatrix} \notin \text{range}\, C_0.$$

b) Let us outline the rest of the proof. We will construct a sequence of scalar functions

$$\gamma^{(\nu)}(\xi), \quad \xi \in \mathbf{R}, \quad \nu = 1, 2, \ldots,$$

and define data by

$$\hat{g}^{(\nu)}(\eta + i\xi) = \gamma^{(\nu)}(\xi)\begin{pmatrix} 0 \\ q^{II} \end{pmatrix}, \quad \eta \text{ large, fixed.}$$

For these data we solve (7.5.9) and obtain vectors $\sigma^{(\nu)}(\eta + i\xi)$ with

$$|\sigma^{(\nu)}(\eta + i\xi)| \geq \delta_1|\rho||\gamma^{(\nu)}(\xi)|, \quad \delta_1 > 0 \text{ fixed.} \tag{7.5.10}$$

According to (7.5.7), we obtain solutions $\hat{u}^{(\nu)}(x, \eta + i\xi)$ of (7.5.4)–(7.5.6), and the inverse Laplace transform gives us functions $u^{(\nu)}(x,t)$. We will show that

$$\int_0^1 |u^{(\nu)}(0,t)|^2\, dt \to \infty \quad \text{as} \quad \nu \to \infty,$$

whereas the data satisfy

$$\int_0^1 |g^{(\nu)}(t)|^2\, dt \leq \text{const.}$$

c) The lower bound (7.5.10) is a consequence of the following perturbation result from linear algebra.

Lemma 7.5.2. *Suppose that C_0 is a singular $n \times n$ matrix and $q \notin \text{range}\, C_0$. Let*

$$\det\Big(C_0 + \epsilon C_1(\epsilon)\Big) \neq 0, \quad 0 < \epsilon \leq \epsilon_0, \quad |C_1(\epsilon)| = O(1),$$

and solve

$$\Big(C_0 + \epsilon C_1(\epsilon)\Big) y(\epsilon) = q, \quad 0 < \epsilon \le \epsilon_0.$$

Then

$$|y(\epsilon)| \ge \frac{\delta_0}{\epsilon}, \quad \delta_0 > 0. \tag{7.5.11}$$

Proof. There is a nonsingular matrix P such that

$$PC_0 = \begin{pmatrix} D_0 \\ 0 \end{pmatrix}$$

where D_0 has full rank, say rank $D_0 = m$. There is a component j with

$$(Pq)_j \ne 0, \quad m+1 \le j \le n,$$

since otherwise $Pq \in \text{range}\,(PC_0)$. If $\delta_0 > 0$ with (7.5.11) did not exist then

$$\epsilon |y(\epsilon)| \to 0$$

for some sequence $\epsilon \to 0$, and the j-th equation of

$$\Big(PC_0 + \epsilon PC_1(\epsilon)\Big) y(\epsilon) = Pq$$

would yield a contradiction.

d) To continue the proof of Theorem 7.5.1, we define a sequence of functions $\gamma^{(\nu)}$ by

$$\gamma^{(\nu)}(\xi) = \begin{cases} \sqrt{\nu} & \text{for } \nu \le \xi \le \nu + \frac{1}{\nu}, \\ 0 & \text{otherwise}, \end{cases} \qquad \nu = 1, 2, \ldots.$$

Then the L_2-norm of $\gamma^{(\nu)}$ equals 1, and therefore

$$\int_{-\infty}^{\infty} |\hat{g}^{(\nu)}(\eta + i\xi)|^2 \, d\xi \le c_1.$$

According to (7.4.9), the inverse Laplace transforms $g^{(\nu)}(t)$ have bounded L_2-norms over $0 \le t \le 1$, say. Now invert the Laplace transform of data

$$\delta^{(\nu)}(\eta + i\xi) = \nu \gamma^{(\nu)}(\xi)$$

and obtain

$$
\begin{aligned}
v^{(\nu)}(t) &= \frac{1}{2\pi}\int_{-\infty}^{\infty} e^{(\eta+i\xi)t}\delta^{(\nu)}(\eta+i\xi)\,d\xi \\
&= \frac{\nu^{3/2}}{2\pi}e^{\eta t}\int_{\nu}^{\nu+1/\nu} e^{i\xi t}d\xi \\
&= \frac{\nu^{3/2}}{2\pi}\,e^{\eta t}\,\frac{1}{it}\left(e^{i(\nu+1/\nu)t} - e^{i\nu t}\right) \\
&= \frac{\nu^{3/2}}{\pi}\,e^{\eta t}\,\frac{1}{t}\,e^{i(\nu+1/2\nu)t}\,\sin\frac{t}{2\nu} \\
&\approx \frac{\nu^{1/2}}{2\pi}\,e^{\eta t}\,e^{i(\nu+1/2\nu)t}, \quad 0 \le t \le 1.
\end{aligned}
$$

Thus one finds

$$\int_0^1 |v^{(\nu)}(t)|^2 dt \to \infty \quad \text{as} \quad \nu \to \infty.$$

If one observes the lower bound (7.5.10) and notes that the supports of $\sigma^{(\nu)}$ are confined to small intervals for large ν, then one obtains a similar result for

$$\int_0^1 |u^{(\nu)}(0,t)|^2\,dt.$$

Also, for each fixed ν we can apply arbitrarily small changes to obey compatibility conditions at $t = 0$, and the theorem is proved.

7.6. Initial-Boundary Value Problems for Hyperbolic Equations

7.6.1. The Method of Characteristics

In this section we consider hyperbolic systems

(7.6.1) $$u_t = B(x,t)u_x + C(x,t)u + F(x,t)$$

in the strip $0 \le x \le 1,\ t \ge 0$ with initial data

(7.6.2) $$u(x,0) = f(x), \quad 0 \le x \le 1.$$

At each boundary point $x = 0,\ x = 1$ we prescribe (inhomogeneous) linear relations for u; i.e., we consider boundary conditions of the general form

(7.6.3) $$L_0 u(0,t) = g_0(t), \quad L_1 u(1,t) = g_1(t), \quad t > 0.$$

Specific assumptions about the matrices L_0, L_1 will be derived below. The coefficients and data functions are assumed to be C^∞-smooth with respect to all variables.

Hyperbolicity of the system (7.6.1) requires the eigenvalues $\lambda_j(x,t)$ of $B(x,t)$ to be real and assumes the existence of a smooth transformation $S = S(x,t)$ such that

$$S^{-1}BS = \Lambda = \text{diag}\,(\lambda_1, \dots, \lambda_n), \quad \lambda_j = \lambda_j(x,t).$$

In the interior $0 < x < 1$ the eigenvalues may change sign. However, we assume that

$$\lambda_j(0,t) \text{ and } \lambda_j(1,t), \quad j = 1, \dots, n, \tag{7.6.4}$$

have a constant sign as a function of time; i.e., each function (7.6.4) is either > 0 for all t, $= 0$ for all t, or < 0 for all t.

If we introduce new variables (so-called *characteristic variables*)

$$\tilde{u}(x,t) = S^{-1}(x,t)u(x,t),$$

then the system (7.6.1) transforms to an equation where B is replaced by Λ. To simplify notation, we assume that the given system is written in characteristic variables already, thus $B = \Lambda$ in (7.6.1).

The case of n scalar equations. To discuss the system, we use the method of characteristics and start with the case $C = F = 0$. The differential system separates into n scalar problems

$$u_{jt} = \lambda_j(x,t)u_{jx}, \quad j = 1, \dots, n.$$

Thus $u_j(x,t)$ is constant along the characteristics $(x(t),t)$ defined by

$$\frac{dx}{dt} = -\lambda_j(x,t). \tag{7.6.5}$$

The case of constant λ_j. Assume first that

$$\lambda_j(x,t) = \lambda_j = \text{ const.}, \quad j = 1, \dots, n.$$

Then the characteristics are straight lines, and thus

$$u_j(x,t) = f_j(x + \lambda_j t), \quad 0 \le x + \lambda_j t \le 1.$$

Let u_+, u_0, u_- consist of the variables u_j corresponding to indices j with $\lambda_j > 0, \lambda_j = 0, \lambda_j < 0$, respectively. Clearly, u_0 is determined by the initial data

$$u_0(x,t) = f_0(x), \quad 0 \le x \le 1, \quad t \ge 0,$$

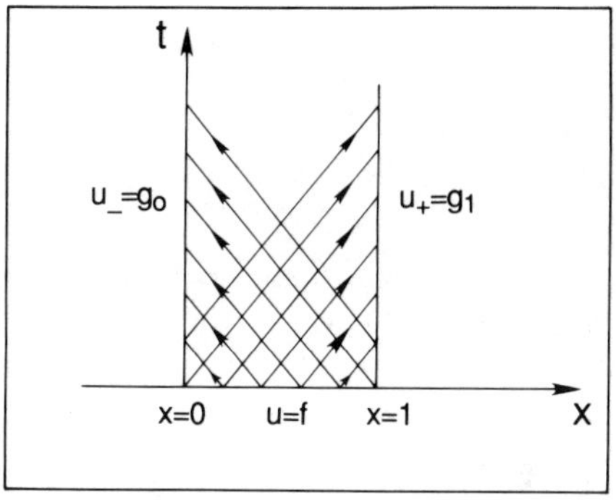

FIGURE 7.6.1. Strip with characteristics.

but we need boundary conditions to determine u_+ and u_-. Acceptable boundary conditions are

$$u_-(0,t) = g_0(t), \quad u_+(1,t) = g_1(t), \quad t > 0; \tag{7.6.6}$$

i.e., we prescribe the ingoing characteristic variables at each boundary.

If the initial function $f(x)$ and the boundary data $g_0(t)$, $g_1(t)$ are not compatible, then the solution will have discontinuities along the characteristics which start at the corners $(x,t) = (0,0)$, $(x,t) = (1,0)$. Thus, in contrast to the parabolic case, there is no smoothing during time evolution here. To avoid difficulties connected with nonsmoothness, we will assume that the data f, g_0, g_1 vanish near the corners. As usual, once solution estimates are derived for this case, the more general situation can be treated by a limiting process.

The boundary conditions (7.6.6) can immediately be generalized to

$$\begin{aligned} u_-(0,t) &= S_0(t)u_+(0,t) + g_0(t), \\ u_+(1,t) &= S_1(t)u_-(1,t) + g_1(t). \end{aligned} \tag{7.6.7}$$

One says that the ingoing characteristic variables are described in terms of the outgoing ones. Boundary conditions for the characteristic variables u_0 belonging to speeds $\lambda_j = 0$ are neither necessary nor allowed.

The case $\boldsymbol{\lambda_j = \lambda_j(x,t)}$. For simplicity we assume that all eigenvalues (7.6.4) are different from zero on the boundary; one says that *the boundary is not characteristic*. If an eigenvalue $\lambda_j(x,t)$ changes sign as a function of x, then possibly the variable u_j belongs to a positive characteristic speed at $x = 0$ and to a negative speed at $x = 1$. Nevertheless, we use the notation u_+ and u_- to assemble the variables u_j with $\lambda_j > 0$ and $\lambda_j < 0$, respectively, at each boundary point. As in the case of constant λ_j, we obtain a unique solution $u(x,t)$ if the boundary conditions have the form (7.6.7). Also, as follows from Section 3.3.2, we can add a diagonal term

$$C_0 u, \quad C_0 = \operatorname{diag}(c_1, \ldots, c_n), \quad c_j = c_j(x,t),$$

and a forcing function $F(x,t)$. The assumption for the boundary to be noncharacteristic is not essential. By solving scalar equations, one obtains

Theorem 7.6.1. *Assume that f, F, g_0, and g_1 vanish in a neighborhood of the corners $(x,t) = (0,0)$, $(x,t) = (1,0)$. The system*

$$u_t = \Lambda(x,t)u_x + C_0(x,t)u + F(x,t)$$

with initial and boundary conditions (7.6.2), (7.6.7) *has a unique smooth solution. The solution vanishes near the corners.*

7.6.2. Solution Estimates

If the coefficient $C = C(x,t)$ is not diagonal we want to use the iteration

$$\begin{aligned} u_t^{k+1} &= \Lambda u_x^{k+1} + Cu^k + F, \\ u^{k+1}(x,0) &= f(x), \quad u^0 \equiv 0, \\ u_-^{k+1}(0,t) &= S_0(t)u_+^{k+1}(0,t) + g_0(t), \\ u_+^{k+1}(1,t) &= S_1(t)u_-^{k+1}(1,t) + g_1(t) \end{aligned} \tag{7.6.8}$$

to show the existence of a solution. To start with, we show the following basic estimate:

Lemma 7.6.2. *Assume that the boundary is not characteristic. For every finite time interval $0 \le t \le T$ there is a constant K_T independent of f, F, g_0, g_1 with the following property: If u solves*

$$u_t = \Lambda u_x + Cu + F \text{ in } 0 \le x \le 1, \quad 0 \le t \le T \tag{7.6.9}$$

and satisfies (7.6.2), (7.6.7) *then*

$$\begin{aligned} &\int_0^t \{|u(0,\tau)|^2 + |u(1,\tau)|^2\}\,d\tau + \|u(\cdot,t)\|^2 \\ &\le K_T\|f\|^2 + K_T \int_0^t \{|g_0(\tau)|^2 + |g_1(\tau)|^2 + \|F(\cdot,\tau)\|^2\}\,d\tau \end{aligned} \tag{7.6.10}$$

for $0 \le t \le T$.

Proof. 1) To begin with, let us scale the variables u_+, u_- such that the matrices $S_0(t)$, $S_1(t)$ become "small". Let

$$D = \operatorname{diag}(d_1, \ldots, d_n), \quad d_j = d_j(x,t) > 0,$$

and introduce new variables $v = Du$. Equation (7.6.9) transforms to

$$v_t = \Lambda v_x + \{D_t D^{-1} - \Lambda D_x D^{-1} + DCD^{-1}\}v + DF,$$

and thus only lower-order terms change. The boundary conditions (7.6.7) become

$$\begin{aligned}
v_-(0,t) &= \tilde{S}_0(t)v_+(0,t) + D_-(0,t)g_0(t),\\
v_+(1,t) &= \tilde{S}_1(t)v_-(1,t) + D_+(1,t)g_1(t),\\
\tilde{S}_0(t) &= D_-(0,t)\ S_0(t)D_+^{-1}(0,t),\\
\tilde{S}_1(t) &= D_+(1,t)\ S_1(t)D_-^{-1}(1,t),
\end{aligned}$$

where $D_-(0,t)$ is a diagonal submatrix of $D(0,t)$, etc. Clearly, the matrices $\tilde{S}_j(t)$ become as small as we please by choosing $D(x,t)$ appropriately. To simplify notation, we assume $|S_0| + |S_1|$ to be sufficiently small from the beginning.
2) From (7.6.9) we obtain that

$$\begin{aligned}
\frac{d}{dt}\|u(\cdot,t)\|^2 &= (u, u_t) + (u_t, u)\\
&\leq (u, \Lambda u_x) + (\Lambda u_x, u) + c_1\{\|u\|^2 + \|F\|^2\}.
\end{aligned}$$

Integration by parts gives us

$$\begin{aligned}
(u, \Lambda u_x) &= (\Lambda u, u_x)\\
&= -(\Lambda u_x, u) - (\Lambda_x u, u) + \langle u, \Lambda u\rangle\big|_0^1,
\end{aligned}$$

and thus

$$\frac{d}{dt}\ \|u\|^2 \leq \langle u, \Lambda u\rangle\big|_0^1 + c_2\{\|u\|^2 + \|F\|^2\}.$$

3) Consider, for example, the boundary point $x = 1$ and note that

$$\Lambda_-(1,t) \leq -\gamma I, \quad 0 \leq t \leq T, \quad \gamma > 0.$$

Using the boundary condition at $x = 1$, we find that

$$\begin{aligned}\langle u(1,t),\ \Lambda(1,t)u(1,t)\rangle &= \langle u_-,\ \Lambda_- u_-\rangle + \langle u_+, \Lambda_+ u_+\rangle \\ &\le -\gamma|u_-|^2 + |\Lambda_+|\,|u_+|^2 \\ &\le -\frac{\gamma}{2}\{|u_-|^2 + |u_+|^2\} + c_3|g_1|^2.\end{aligned}$$

Here we have assumed S_1 to be so small that

$$2|S_1|^2\{|\Lambda_+| \ + \frac{\gamma}{2}\} \le \frac{\gamma}{2}.$$

A similar consideration applies at $x = 0$, and we obtain that

$$\frac{d}{dt}\ \|u\|^2 + \frac{\gamma}{2}\{|u_-|^2 + |u_+|^2\} \ \le\ c_2\big\{\|u\|^2 + \|F\|^2\big\} + c_3\big\{|g_0|^2 + |g_1|^2\big\}.$$

The lemma follows by integration.

We show next how to estimate derivatives of the solution of (7.6.9), (7.6.2), (7.6.7) if the solution is smooth; we let $v = u_x$. Differentiation of (7.6.9) with respect to x gives us

$$v_t = \Lambda v_x + (\Lambda_x + C)v + C_x u + F_x.$$

To obtain boundary conditions for v, we first differentiate (7.6.7) with respect to t,

$$u_{-t}(0,t) = S_0(t)u_{+t}(0,t) + S_{0t}(t)u_+(0,t) + g_{0t}(t),$$

and similarly at $x = 1$. Then we use (7.6.9) to replace u_t by u_x:

$$u_{-x}(0,t) = \Lambda_-^{-1}(0,t)S_0(t)\Lambda_+(0,t)u_{+x}(0,t) + \tilde{g}_0(t)$$

where

$$|\tilde{g}_0(t)| \le c_1\big\{|u(0,t)| + |g_{0t}(t)| + |F(0,t)|\big\}.$$

Here

$$|F(0,t)|^2 \le \frac{1}{2}\big\{\|F_x(\cdot,t)\|^2 + \|F(\cdot,t)\|^2\big\}.$$

Thus we have equations for $v = u_x$ which are of the same type as the equations for u. By Lemma 7.6.2 we can estimate

$$\int_0^t \big\{|u_x(0,\tau)|^2 + |u_x(1,\tau)|^2\big\}\,d\tau + \|u_x(\cdot,t)\|^2$$

in terms of g_0, g_1, f, F and their first derivatives.

Differentiation of (7.6.9) with respect to t gives us, for $w = u_t$,

$$w_t = \Lambda w_x + Cw + \tilde{F},$$

$$\tilde{F} = \Lambda_t u_x + C_t u + F_t.$$

Therefore we can also estimate $w = u_t$ in terms of $g_0,\ g_1,\ f,\ F$ and their first derivatives. This process can be continued, and we obtain

Lemma 7.6.3. *Assume that the initial-boundary value problem* (7.6.9), (7.6.2), (7.6.7) *has a smooth solution and that the boundary is not characteristic. The derivatives of the solution satisfy estimates of type* (7.6.10).

7.6.3. Existence of a Smooth Solution

Assume that the compatibility conditions of Theorem 7.6.1 are satisfied and that the boundary is not characteristic. We consider the iteration (7.6.8). By Theorem 7.6.1 each function of the sequence

$$u^k = u^k(x,t), \quad k = 0, 1, 2, \ldots,$$

is C^∞-smooth. Also,

$$\frac{d}{dt}\|u^{k+1}(\cdot,t)\|^2 \le \text{const}\left\{\|u^{k+1}(\cdot,t)\|^2 + \|u^k(\cdot,t)\|^2 + 1\right\}$$

and corresponding estimates for x-derivatives hold. Thus, by Gronwall's Lemma 3.1.1 and Picard's Lemma 3.3.4, the sequence u^k is uniformly smooth in any finite time interval. Furthermore,

$$\frac{d}{dt}\|u^{k+1} - u^k\|^2 \le \text{const}\left\{\|u^{k+1} - u^k\|^2 + \|u^k - u^{k-1}\|^2\right\}.$$

As before, Gronwall's Lemma and Picard's Lemma imply convergence of the sequence u^k to a C^∞-limit u. The limit u solves the initial-boundary value problem and satisfies the estimates of Lemma 7.6.3. We summarize:

Theorem 7.6.4. *Assume that the boundary is not characteristic and that the data $g_0,\ g_1,\ f,\ F$ are compatible at $t = 0$. The hyperbolic initial-boundary value problem* (7.6.9), (7.6.2), (7.6.7) *has a unique solution. The solution is a C^∞-function, which satisfies the estimate* (7.6.10). *Similar estimates hold for all derivatives.*

If the initial and boundary data do not satisfy compatibility conditions, then we can approximate the data and go to the limit. In this way we can introduce generalized solutions. As the uncoupled case shows, these will – in general –

have discontinuities travelling along the characteristics which start in the corners of the strip.

7.6.4. The Strip Problem vs. the Half-Space and Pure Cauchy Problems

Suppose that the coefficients $B = \Lambda$, C and the data F, f are defined for $-\infty < x < \infty$, $t \geq 0$. The finite speed of propagation for hyperbolic systems makes it possible to split the strip problem

$$\begin{aligned} u_t &= \Lambda(x,t)u_x + C(x,t)u + F(x,t), && 0 \leq x \leq 1, \quad t \geq 0, \\ u(x,0) &= f(x), && 0 \leq x \leq 1, \\ u_-(0,t) &= S_0(t)u_+(0,t) + g_0(t), && t \geq 0, \\ u_+(1,t) &= S_1(t)u_-(1,t) + g_1(t), && t \geq 0, \end{aligned}$$

into two half-space problems and a pure Cauchy problem. To this end, let $\phi_1 \in C^\infty$ denote a monotone function with

$$\phi_1(x) = \begin{cases} 1 & \text{for } x \leq 1/6, \\ 0 & \text{for } x \geq 1/3, \end{cases}$$

and define

$$\phi_2(x) = \phi_1(1-x)\ , \quad \phi_3(x) = 1 - \phi_1(x) - \phi_2(x)\ .$$

First we consider the following three strip problems (here the index $j = 1, 2, 3$ numbers the problem, not a component):

$$\begin{aligned} u_{jt} &= \Lambda u_{jx} + Cu_j + \phi_j F_j, && 0 \leq x \leq 1, \quad t \geq 0, \\ u_j(x,0) &= \phi_j(x) f(x), && 0 \leq x \leq 1, \\ u_{j-}(0,t) &= S_0(t)u_{j+}(0,t) + \phi_j(0)g_0(t), && t \geq 0, \\ u_{j+}(1,t) &= S_1(t)u_{j-}(1,t) + \phi_j(1)g_1(t), && t \geq 0. \end{aligned}$$

Clearly,

$$u = u_1 + u_2 + u_3$$

solves the given system. The finite speed of propagation implies that there is a time interval $0 \leq t \leq T_1$, $T_1 > 0$, where $u_1(x,t) \equiv 0$ for $5/6 \leq x \leq 1$. Therefore, we can consider u_1 as the solution of the right half-space problem

$$\begin{aligned} u_{1t} &= \Lambda u_{1x} + Cu_1 + \phi_1 F, && x \geq 0, \quad t \geq 0, \\ u_1(x,0) &= \phi_1(x) f(x), && x \geq 0, \\ u_{1-}(0,t) &= S_0(t) u_{1+}(0,t) + g_0(t), && t \geq 0, \end{aligned}$$

at least for $0 \leq t \leq T_1$. Similarly, u_2 and u_3 can be considered locally as solutions of a left half-space and a pure Cauchy problem, respectively. These considerations are valid in some finite time interval $0 \leq t \leq T_0$; at time T_0 one has obtained a new initial function and can restart. Thus, in principle, we can solve the strip problem by solving half-space and pure Cauchy problems. The strip problem is well-posed if the corresponding half-space and pure Cauchy problems are well-posed.

As in Sections 7.4, 7.5, one can employ the Laplace transform in time to discuss boundary conditions which are much more general than the conditions (7.6.7). We return to this technique in Chapter 8 to treat problems in more than one space dimension.

7.6.5. Equations Not in Characteristic Variables

In many applications the differential equation (7.6.1) is not given in characteristic variables, i.e., the matrix B is not diagonal. To test for well-posedness under boundary conditions, one can try to derive energy estimates directly without diagonalizing the system. If $B = B^*$ then

$$\begin{aligned} \frac{d}{dt} \|u(\cdot,t)\|^2 &= (u, u_t) + (u_t, u) \\ &\leq (u, Bu_x) + (Bu_x, u) + c_1 \{ \|u\|^2 + \|F\|^2 \}, \end{aligned}$$

where

$$(u, Bu_x) = (Bu, u_x) = -(Bu_x, u) - (B_x u, u) + \langle Bu, u \rangle \big|_0^1.$$

Thus the boundary conditions have to provide bounds for

$$\langle Bu, u \rangle \big|_0^1 = \langle B(1,t)u(1,t)\, ,\, u(1,t) \rangle \; - \; \langle B(0,t)u(0,t)\, ,\, u(0,t) \rangle.$$

Consider, for example, the boundary point $x = 1$ and assume that $B(1,t)$ has r positive and $n - r$ negative eigenvalues; hence there are r ingoing and $n - r$ outgoing characteristics. This suggests specifying r independent boundary conditions at $x = 1$, for example,

$$L_1 u(1,t) = g_1(t), \quad L_1 \text{ of size } \; r \times n, \text{ rank } L_1 = r.$$

Using the boundary conditions, one can eliminate r variables $u_j(1,t)$ in the quadratic term

$$\langle B(1,t)\, u(1,t), u(1,t)\rangle.$$

Similar considerations apply at $x = 0$. If the remaining quadratic terms have the correct signs, then an energy estimate can be derived. In this way one has avoided the transformation to characteristic variables.

A difficulty of this approach is that it provides only *sufficient* conditions for well-posedness. It might well happen that quadratic terms with the wrong sign appear, the problem being well-posed nevertheless. For the case where we first transform to characteristic variables, this is reflected in our additional assumption that $|S_0(t)| + |S_1(t)|$ is sufficiently small; compare the proof of the basic estimate of Lemma 7.6.2.

In applications hyperbolic differential equations often appear as second-order equations. By introduction of auxiliary variables, these can be written as first-order systems, however. We refer to Sections 2.1.4 and 3.3.3 for two simple examples.

7.7. Boundary Conditions for Hyperbolic-Parabolic Problems

In this section we treat mixed systems in the strip $0 \le x \le 1, t \ge 0$, under initial and boundary conditions. For the uncoupled systems we assume a form as described in Section 7.2 (parabolic case) and Section 7.6 (hyperbolic case). Then we allow certain coupling terms in the differential equation and in the boundary conditions. The resulting systems are shown to be well-posed. We give an application to the linearized compressible Navier-Stokes equations.

7.7.1. The Basic Estimate for Mixed Systems

Consider a parabolic system

$$u_t = A(x,t)u_{xx}, \qquad A + A^* \ge 2\delta I, \qquad \delta > 0,$$

in the strip $0 \le x \le 1,\ t \ge 0$, with boundary conditions

$$L^I_{j1}(t)u_x(j,t) + L^I_{j0}(t)u(j,t) = h_j(t),$$

$$L^{II}_{j0}(t)u(j,t) = 0, \qquad j = 0, 1.$$

For each fixed t we assume the conditions formulated in Theorem 7.2.7 to be fulfilled (see also Lemma 7.2.1). We allow the matrices L^I_{j1}, etc. to depend smoothly on t but assume that the rank r_j of L^I_{j1} is constant.

Consider further a hyperbolic system

$$v_t = \Lambda(x,t)v_x, \quad \Lambda \text{ real diagonal},$$

in the same domain with boundary conditions

$$v_-(0,t) = S_0(t)v_+(0,t) + g_0(t),$$

$$v_+(1,t) = S_1(t)v_-(1,t) + g_1(t);$$

i.e., the ingoing characteristic variables are expressed at each boundary point in terms of the outgoing ones. (See Section 7.6.1 for notations.) We want to discuss the coupled system

(7.7.1) $$u_t = Au_{xx} + B_{11}u_x + B_{12}v_x + C_{11}u + C_{12}v + F,$$

(7.7.2) $$v_t = \Lambda v_x + B_{21}u_x + C_{21}u + C_{22}v + G$$

with boundary conditions

(7.7.3) $$L_{j1}^{I}(t)u_x(j,t) + L_{j0}^{I}(t)u(j,t) + M_j(t)v(j,t) = h_j(t), \quad j = 0, 1,$$

(7.7.4) $$L_{j0}^{II}(t)u(j,t) = 0, \qquad j = 0, 1,$$

(7.7.5) $$v_-(0,t) = S_0(t)v_+(0,t) + R_0(t)u(0,t) + g_0(t),$$

(7.7.6) $$v_+(1,t) = S_1(t)v_-(1,t) + R_1(t)u(1,t) + g_1(t),$$

and initial conditions

(7.7.7) $$u(x,0) = f(x), \quad v(x,0) = \phi(x), \quad 0 \le x \le 1.$$

The coefficients $B_{11} = B_{11}(x,t)$, etc. and all inhomogeneous terms are assumed to be C^∞-smooth. We first assume the existence of a solution and show

Lemma 7.7.1. *Suppose that the boundary is not characteristic for the hyperbolic system; i.e., $\Lambda(0,t)$ and $\Lambda(1,t)$ are nonsingular. In any finite time interval $0 \le t \le T$ we can estimate*

$$\|u(\cdot,t)\|^2 + \|v(\cdot,t)\|^2 + \int_0^t \sum_{j=0,1} \left(|u(j,\tau)|^2 + |v(j,\tau)|^2\right) d\tau$$

in terms of

$$\|f\|^2 + \|\phi\|^2 + \int_0^t \left\{\|F(\cdot,\tau)\|^2 + \|G(\cdot,\tau)\|^2 + \sum_{j=0,1} \left(|g_j(\tau)|^2 + |h_j(\tau)|^2\right)\right\} d\tau.$$

Proof. a) First note that one can easily extend Lemma 7.2.1 to inhomogeneous equations as follows: If condition (ii) of the lemma is met and

$$L_1^I w_x + L_0^I w = h, \quad L_0^{II} w = 0,$$

then

$$|\langle w, Aw_x \rangle| \le c\{|w|^2 + |h|^2\}$$

for some constant c.

b) As explained in the proof of Lemma 7.6.2, we can assume without loss of generality that $|S_0| + |S_1|$ is sufficiently small. We let

$$\Lambda_+(0,t) \ge \gamma I, \quad \Lambda_-(1,t) \le -\gamma I, \quad 0 \le t \le T, \quad \gamma > 0.$$

c) First consider the u-equation (7.7.1),

$$\begin{aligned}\frac{d}{dt}\, \|u(\cdot,t)\|^2 &= (u, u_t) + (u_t, u) \\ &\le (u, Au_{xx}) + (Au_{xx}, u) \\ &\quad + (u, B_{12}v_x) + (B_{12}v_x, u) \\ &\quad + c_1\{\|u\|\ \|u_x\| + \|u\|^2 + \|v\|^2 + \|F\|^2\}.\end{aligned}$$

Integration by parts gives us that

$$\begin{aligned}&(u, Au_{xx}) + (Au_{xx}, u) \\ &\quad = -\big(u_x, (A + A^*)u_x\big) - \big(u, (A + A^*)_x u_x\big) + \langle Au_x, u\rangle\big|_0^1 + \langle u, Au_x\rangle\big|_0^1 \\ &\quad \le -2\delta\|u_x\|^2 + c_2\Big\{\|u\|\ \|u_x\| + \sum_{j=0,1}\big(|u(j,t)|^2 + |v(j,t)|^2 + |h_j(t)|^2\big)\Big\}\end{aligned}$$

and that

$$|(u, B_{12}v_x)| \le c_3\Big\{\|u_x\|\ \|v\| + \|u\|\ \|v\| + \sum_{j=0,1}\big(|u(j,t)|^2 + |v(j,t)|^2\big)\Big\}.$$

Using the Sobolev inequality

$$|u|_\infty^2 \le \epsilon\|u_x\|^2 + c(\epsilon)\|u\|^2, \tag{7.7.8}$$

we obtain that

$$\frac{d}{dt}\, \|u\|^2 \le -\delta\|u_x\|^2 + c_4\Big\{\|u\|^2 + \|v\|^2 + \|F\|^2 + \sum_{j=0,1}\big(|v(j,t)|^2 + |h_j(t)|^2\big)\Big\}.$$

d) Second, the v-equation (7.7.2) yields

$$\frac{d}{dt}\ \|v(\cdot,t)\|^2 = (v, v_t) + (v_t, v)$$
$$\leq (v, \Lambda v_x) + (\Lambda v_x, v) + c_5\big\{\|v\|\ \|u_x\| + \|u\|^2 + \|v\|^2 + \|G\|^2\big\}.$$

Integration by parts and the boundary conditions imply

$$(v, \Lambda v_x) + (\Lambda v_x, v) = -\,(\Lambda_x v, v) + \langle v, \Lambda v\rangle\Big|_0^1$$
$$\leq -\frac{\gamma}{2}\sum_{j=0,1} |v(j,t)|^2$$
$$+\, c_6\Big\{\|v\|^2 + \sum_{j=0,1}\big(|u(j,t)|^2 + |g_j(t)|^2\big)\Big\}.$$

e) We take $\alpha > 0$ sufficiently large, and the above estimates give us that

$$\frac{d}{dt}\ \|u\|^2 + \alpha\ \frac{d}{dt}\ \|v\|^2 \leq -\frac{\delta}{2}\|u_x\|^2 - \alpha\ \frac{\gamma}{4}\sum_{j=0,1} |v(j,t)|^2$$
$$+\, c(\alpha)\Big\{\|u\|^2 + \|v\|^2 + \|F\|^2 + \|G\|^2$$
$$+\sum_{j=0,1}\big(|h_j(t)|^2 + |g_j(t)|^2\big)\Big\}.$$

Another application of (7.7.8) and integration with respect to t finishes off the proof of the lemma.

7.7.2. Estimates of Derivatives and Existence of a Solution

If we differentiate the u-equation (7.7.1) with respect to t we obtain

$$u_{tt} = Au_{txx} + A_t u_{xx} + \ldots.$$

Here u_{xx} can be expressed by u_t, etc. using (7.7.1) again. Thus u_{xx} and u_x can be treated like zero-order terms as in the proof of Lemma 7.2.4. The v-equation (7.7.2) is differentiated with respect to t and with respect to x. One obtains a system

$$\begin{pmatrix} v_t \\ v_x \end{pmatrix}_t = \begin{bmatrix} \Lambda & 0 \\ 0 & \Lambda \end{bmatrix} \begin{bmatrix} v_t \\ v_x \end{bmatrix}_x + \ldots.$$

Boundary conditions for u_t and v_t follow by differentiation of the given boundary conditions with respect to t; then boundary conditions for v_x are obtained if we replace v_t by v_x using the differential equation $v_t = \Lambda v_x + \ldots$. Thus we have a system for (u_t, v_t, v_x) which has the same structure as the given system for (u, v). Consequently, estimates for u_t, u_x, u_{xx}, v_t, v_x are derived. The process can be continued, and all derivatives can be estimated.

If we assume that all data

$$f(x), \phi(x), F(x,t), G(x,t), h_j(t), g_j(t), \quad j = 0, 1,$$

vanish in a neighborhood of the corners of the strip, then existence of a smooth solution can be shown by using the iteration

$$u_t^{k+1} = Au_{xx}^{k+1} + B_{11}u_x^{k+1} + B_{12}v_x^k + C_{11}u^{k+1} + C_{12}v^k + F,$$

$$v_t^{k+1} = \Lambda v_x^{k+1} + B_{21}u_x^k + C_{21}u^k + C_{22}v^{k+1} + G.$$

The boundary conditions for u^{k+1} are the same as those for u except that v is replaced by v^k in (7.7.3). Similarly, u is replaced by u^k in (7.7.5), (7.7.6).

This proves

Theorem 7.7.2. *Consider the hyperbolic-parabolic strip problem* (7.7.1)–(7.7.7). *Assume that the boundary is not characteristic for the hyperbolic part; i.e.,* $\Lambda(0,t)$ *and* $\Lambda(1,t)$ *are nonsingular. If the compatibility conditions are satisfied, then the problem has a unique smooth solution* $\big(u(x,t),\ v(x,t)\big)$. *We can estimate the solution and its derivatives in terms of the data and their derivatives.*

7.7.3. The Case of a Characteristic Boundary

We now allow one or both of the boundaries $x = 0, x = 1$ to be characteristic for the v-equation

$$v_t = \Lambda v_x + \ldots.$$

The boundary conditions (7.7.5), (7.7.6) for v remain unchanged, hence none of the characteristic variables v_0 appear in the boundary conditions for v. For the variable u we restrict ourselves to boundary conditions of Dirichlet type

$$u(j,t) = 0, \quad j = 0, 1.$$

(If the boundary $x = 0$, say, is not characteristic, we can allow a boundary condition of the more general form (7.7.3), (7.7.4) at $x = 0$.)

To show the basic estimate, we proceed as in the proof of Lemma 7.7.1. From integration by parts we obtain the boundary terms

$$\langle u, Au_x\rangle\Big|_0^1, \quad \langle u, B_{12}v\rangle\Big|_0^1, \quad \langle v, \Lambda v\rangle\Big|_0^1.$$

The first two of these terms are zero, the third term is treated as before:

$$\langle v, \Lambda v \rangle|_0^1 \le -\frac{\gamma}{2} \sum_{j=0,1} |v_\pm(j,t)|^2 + c \sum_{j=0,1} |g_j(t)|^2.$$

Thus we obtain an estimate of

$$\|u(\cdot,t)\|^2 + \|v(\cdot,t)\|^2 + \int_0^t \sum_{j=0,1} |v_\pm(j,\tau)|^2 \, d\tau$$

in terms of the data. Bounds for derivatives can be derived as before.

To summarize:

Theorem 7.7.3. *The result formulated in Theorem 7.7.2 remains valid if one or both of the boundaries $x = 0, x = 1$ are characteristic provided that the boundary condition for u at a characteristic boundary is of Dirichlet type.*

With some restrictions, we found that coupled hyperbolic-parabolic systems are well-posed if the uncoupled systems have this property. It is possible to relax the assumptions of Theorem 7.7.3 further; however, there can be difficulties if the hyperbolic part is characteristic at the boundary and the parabolic part has a non-Dirichlet boundary condition. (The difficulties arise from the boundary term $\langle u, B_{12}v \rangle$. Actually, in one space dimension, the difficulties can be overcome since the characteristic variables satisfy ordinary differential equations along the boundary. This technique does not generalize to more than one space dimension, however.) In any concrete case, one can try to use the above tools to derive an energy estimate and estimates for derivatives. Also, the method of Laplace transform is applicable for discussing much more general boundary conditions.

7.7.4. Application to the Linearized Navier-Stokes Equations in 1D

Upon neglecting zero-order terms and forcing functions, the equations for u', ρ' derived in Section 3.5 read

$$\begin{pmatrix} u' \\ \rho' \end{pmatrix}_t = -\begin{pmatrix} U & \kappa/R \\ R & U \end{pmatrix} \begin{pmatrix} u' \\ \rho' \end{pmatrix}_x + \mu_1 \begin{pmatrix} 1 & 0 \\ 0 & 0 \end{pmatrix} \begin{pmatrix} u' \\ \rho' \end{pmatrix}_{xx}, \qquad \mu_1 = 2\mu + \mu'.$$

We assume that $\mu_1 > 0$. Clearly, the uncoupled equations

$$u'_t = -Uu'_x + \mu_1 u'_{xx},$$

$$\rho'_t = -U\rho'_x$$

are parabolic and hyperbolic, respectively. The base flow $U = U(x,t)$ is assumed to be known in the strip

$$0 \le x \le 1, \quad t \ge 0.$$

Consider the boundary $x = 0$, for definiteness. We distinguish among three cases:

$$\begin{aligned} &1) \quad \text{inflow at} \quad x = 0 \quad : U(0,t) > 0; \\ &2) \quad \text{outflow at} \quad x = 0 : U(0,t) < 0; \\ &3) \quad \text{a wall} \quad \text{at} \quad x = 0 \quad : U(0,t) = 0. \end{aligned}$$

Our results do not cover a switch between inflow and outflow in time, since we assumed in Section 7.6 that the eigenvalues $\lambda_\nu(0,t)$ have constant sign. The general theory allows boundary conditions of the form given below.

Case 1. $U(0,t) > 0$. Here ρ' is an ingoing characteristic variable. We need two boundary conditions:

$$\rho'(0,t) = r_0(t)u'(0,t) + g_0(t)$$

and

(7.7.9) $$u'(0,t) = 0 \qquad \text{or}$$

(7.7.10) $$u'_x(0,t) + \ell_0(t)u'(0,t) = m_0(t)\rho'(0,t) + h_0(t).$$

Case 2. $U(0,t) < 0$. In this case ρ' is an outgoing characteristic variable and need not be specified. For $u'(0,t)$ we can take condition (7.7.9) or (7.7.10).

Case 3. $U(0,t) = 0$. The variable ρ' need not be specified. For u' we can take (7.7.9).

Similar considerations apply at $x = 1$. By Theorem 7.7.3 the resulting strip problem for the coupled (u', ρ')-system is well-posed. Instead of the homogeneous condition (7.7.9), one could also prescribe a smooth function

$$u'(0,t) = u'_0(t).$$

The inhomogeneous case can be reduced to the homogeneous one by subtracting a suitable function from $u'(x,t)$.

7.8. Semibounded Operators

In the previous sections we have considered differential equations

$$u_t = P(x,t,\partial/\partial x)u + F(x,t)$$

in the strip $0 \le x \le 1, t \ge 0$, together with an initial condition

$$u(x,0) = f(x), \quad 0 \le x \le 1,$$

and boundary conditions. Restricting ourselves to the homogeneous time independent case, we write the boundary conditions in the form

$$L_j u(j,t) = 0, \quad j = 0, 1.$$

Here we assume that L_j is a linear operator which combines values of u and its spatial derivatives evaluated at $(x,t) = (j,t), j = 0, 1$. In most cases our proofs of an energy estimate followed from an inequality of the type

$$\begin{aligned}\frac{d}{dt}\,\|u(\cdot,t)\|^2 &= (u, u_t) + (u_t, u)\\ &= (u, Pu) + (Pu, u) + (u, F) + (F, u)\\ &\le 2\alpha\|u\|^2 + 2\|u\|\ \|F\|.\end{aligned}$$

Thus the critical question is whether we can show an estimate

$$(u, Pu) + (Pu, u) \le 2\alpha\|u\|^2$$

if u satisfies the boundary conditions.

We shall now formalize the procedure to some extent. For every fixed t the differential expression

$$P(x, t, \partial/\partial x)$$

defines an operator P^t if we specify its domain of definition D. Let us define

$$D = \{w \in C^\infty : L_0 w = 0,\ L_1 w = 0\}; \tag{7.8.1}$$

i.e., D consists of all C^∞-functions which satisfy the boundary conditions. The following definition formalizes our requirement:

Definition 1. We call the operators P^t, $t \ge 0$, semibounded on D if there exists a constant α such that

$$\text{Re}\ (w, P^t w) \le \alpha\ \|w\|^2$$

for all $t \ge 0$ and all $w \in D$.

Clearly, if the operators P^t are semibounded on D and if the strip problem has a solution u with $u(\cdot,t) \in D$ for each t, then our considerations show

$$\frac{d}{dt}\,\|u\|^2 \le 2\alpha\|u\|^2 + \|u\|^2 + \|F\|^2.$$

Thus uniqueness and the basic energy estimate follow.

One might be tempted to believe that the *existence* of a solution u with

$$u(\cdot, t) \in D$$

can also be shown once the operators $P^t, t \geq 0$, are semibounded on D. However, such a result cannot be valid without further assumptions: If we change the domain D to $D_1 \subset D$ by requiring additional boundary conditions, then the corresponding operators P_1^t are still semibounded; any solution u with

$$u(\cdot, t) \in D_1$$

would satisfy the additional boundary conditions. Obviously, this cannot be expected. In other words, a general existence result of the above type cannot be valid since one might have overspecified the solution at the boundary. This motivates the following:

Definition 2. Suppose that P^t, $t \geq 0$, is semibounded on a domain D of the form (7.8.1) and that D is determined by q linearly independent boundary conditions. Suppose further that q is minimal; i.e., if we specify a domain D_1 by less than q boundary conditions, then the corresponding operators P_1^t are not semibounded. In this case, one calls P^t *maximal semibounded* on D.

(This concept is motivated by the next theorem. It is *not* sufficient to request that P^t lose semiboundedness on all domains D_1 strictly larger than D. For example, suppose that integration by parts leads to a boundary term $u(u_x + u_{xx})$. If one requests the two boundary conditions $u_x = u_{xx} = 0$, then one cannot drop any condition. The *correct* condition is $u = 0$, however.)

In the examples which we discussed previously the operators P^t were indeed maximal semibounded; there was no overspecification and – under suitable compatibility conditions – we could show the existence of a solution. One can prove the following general result:

Theorem 7.8.1. *Consider a system of the form*

$$u_t = \sum_{\nu=0}^{p} A_\nu(x,t) \frac{\partial^\nu}{\partial x^\nu} u, \quad 0 \leq x \leq 1, \quad t \geq 0,$$

with boundary conditions

$$\sum_{\nu=0}^{r} B_{j\nu} \frac{\partial^\nu}{\partial x^\nu} u(j,t) = 0, \quad j = 0, 1, \quad t \geq 0.$$

Suppose that $A_p(x,t)$ is nonsingular for all x, t and that all coefficients are C^∞-smooth. If the associated operators P^t, $t \geq 0$, are maximal semibounded on the *corresponding domain D, then the initial-boundary value problem is well-posed. Especially, for initial data $f \in C_0^\infty(0,1)$, there is a unique solution*

$$u \in C^\infty(0 \le x \le 1, t \ge 0)$$

with $u(\cdot, t) \in D$ for $t \ge 0$.

Let us apply the concept of maximal semiboundedness to some simple examples.

Example 1. Consider the equation $u_t = u_x$ under boundary conditions

$$\sum_{\nu=0}^{p} a_{j\nu} \frac{\partial^\nu}{\partial x^\nu} u(j,t) = 0, \quad j = 0, 1.$$

Here

$$(w, w_x) + (w_x, w) = |w|^2 \Big|_0^1 = |w(1)|^2 - |w(0)|^2.$$

An estimate by $2\alpha\|w\|^2$ is possible only if the boundary conditions imply $w(1) = 0$. At $x = 0$ no boundary condition is allowed if the operator $P = \partial/\partial x$ is to be *maximal* semibounded.

Example 2. Consider a system

$$u_t = \Lambda u_x, \quad \Lambda \text{ real diagonal},$$

where Λ is constant for simplicity. Here

$$\begin{aligned}(w, \Lambda w_x) + (\Lambda w_x, w) &= \langle w,\ \Lambda w\rangle\Big|_0^1 \\ &= \langle w_+(1),\ \Lambda_+ w_+(1)\rangle + \langle w_-(1),\ \Lambda_- w_-(1)\rangle \\ &\quad - \langle w_+(0),\ \Lambda_+ w_+(0)\rangle - \langle w_-(0),\ \Lambda_- w_-(0)\rangle.\end{aligned}$$

In order to obtain the desired estimate, we need boundary conditions of the form

$$w_+(1) = S_1 w_-(1), \quad w_-(0) = S_0 w_+(0),$$

where $|S_0| + |S_1|$ must be sufficiently small. If the boundary conditions have the above form, but $|S_0| + |S_1|$ is not small, then the operator $P = \Lambda\partial/\partial x$ is not semibounded on the corresponding domain. Nevertheless, one can derive an energy estimate as we have shown in Lemma 7.6.2. Thus semiboundedness is sufficient but not necessary for an energy estimate.

Example 3. Consider the linearized Korteweg-de Vries equation

$$u_t = u_x + \delta u_{xxx}, \quad \delta > 0,$$

and assume the functions are real. Here

$$2(w, w_x + \delta w_{xxx}) = \left\{w^2 + 2\delta w w_{xx} - \delta(w_x)^2\right\}\Big|_0^1.$$

If we want an estimate by $2\alpha\|w\|^2$ then the boundary conditions must imply

$$w^2(1) + 2\delta w(1)w_{xx}(1) - \delta w_x^2(1) \leq 0, \tag{7.8.2}$$

$$w^2(0) + 2\delta w(0)w_{xx}(0) - \delta w_x^2(0) \geq 0. \tag{7.8.3}$$

At $x = 1$ we need one boundary condition, for example, $w(1) = 0$. Another possibility is

$$w_{xx}(1) = aw(1) + bw_x(1). \tag{7.8.4}$$

For the left-hand side of (7.8.2) one obtains the quadratic form

$$(1 + 2\delta a)w^2(1) + 2\delta bw(1)w_x(1) - \delta w_x^2(1)$$

$$= \big(w(1), w_x(1)\big)\begin{pmatrix} 1 + 2\delta a & \delta b \\ \delta b & -\delta \end{pmatrix}\begin{pmatrix} w(1) \\ w_x(1) \end{pmatrix}.$$

Condition (7.8.2) is satisfied if and only if both eigenvalues of the above 2×2 matrix are ≤ 0. This yields a restriction for the coefficients a, b in the boundary conditions (7.8.4). At $x = 0$ we need two boundary conditions. One possibility is

$$w(0) = w_x(0) = 0.$$

We can also require

$$w_x(0) = cw(0), \quad w_{xx}(0) = dw(0),$$

and (7.8.3) becomes equivalent to $1 + 2\delta d - \delta c^2 \geq 0$. If we impose any of these boundary conditions to the KdV equation, then the strip problems are well-posed according to Theorem 7.8.1.

Notes on Chapter 7

The Laplace transformation and expansions into eigenfunctions have been used for a long time in applied mathematics. See, for example, Carrier and Pearson (1976). Our aim was mainly to show that the energy method is rather restrictive with respect to the admissable boundary conditions, though it is very powerful when it applies.

The proof of Theorem 7.8.1 is conceptually rather simple. We obtain immediately the basic energy estimate. Then, in the same way as for parabolic equations, we can estimate the time derivatives. This provides bounds for the x-derivatives. Corresponding estimates for suitable difference approximations (Kreiss (1960)) can be obtained in a similar way, and the existence of a solution follows as in the text.

8 Initial-Boundary Value Problems in Several Space Dimensions

In Chapters 3 to 6 we have assumed spatial periodicity, and Chapter 7 treated the case of one space variable $0 \le x \le 1$ or $0 \le x < \infty$. Here we want to combine the two situations and consider a space variable

$$x = (x_1, \ldots, x_s), \quad 0 \le x_1 \le 1 \text{ or } 0 \le x_1 < \infty,$$

under periodicity assumptions in $x_2, \ldots, x_s$. These special x-domains are fairly representative in the sense that problems posed in other domains (with smooth boundary) can be broken up into subproblems which can be transformed to the special cases.

8.1. Linear Strongly Parabolic Systems

8.1.1. General Assumptions

We consider second-order systems

$$(8.1.1) \qquad u_t = \sum_{i,j=1}^{s} A_{ij} D_i D_j u + \sum_{i=1}^{s} B_i D_i u + Cu + F, \quad D_i = \partial/\partial x_i,$$

in the strip.

At time $t = 0$ we give initial data

$$u(x, 0) = f(x), \tag{8.1.3}$$

and at the surfaces

$$x_1 = 0, \quad x_1 = 1$$

we prescribe linear (inhomogeneous) boundary conditions combining u and $D_1 u$. Introducing the notation

$$y = (x_2, \ldots, x_s),$$

we can write the boundary conditions in the general form

$$L_{j0} u(j, y, t) + L_{j1} D_1 u(j, y, t) = g_j(y, t), \quad j = 0, 1, \tag{8.1.4}$$

where $L_{j0} = L_{j0}(y, t), L_{j1} = L_{j1}(y, t), j = 0, 1$, are $n \times n$ matrices. The coefficients $A_{ij} = A_{ij}(x, t)$, etc., the forcing $F = F(x, t)$, the initial function $f = f(x)$, the inhomogeneous boundary terms $g_j = g_j(y, t)$, and the coefficients $L_{j\nu} = L_{j\nu}(y, t)$ of the boundary conditions are all assumed to be C^∞-smooth, 1-periodic with respect to $x_2, \ldots, x_s$, and real, for simplicity. The differential equation (8.1.1) is assumed to be strongly parabolic;* i.e., there exists $\delta = \delta_T > 0$ such that

$$\sum_{i,j=1}^{s} \langle \xi_i, \ A_{ij}(x, t) \xi_j \rangle \geq \delta \sum_{i=1}^{s} |\xi_i|^2 \tag{8.1.5}$$

for all vectors $\xi_1, \ldots, \xi_s \in \mathbf{R}^n$ and all (x, t) in the strip (8.1.2) with $0 \leq t \leq T$.

The $n \times 2n$ matrices

$$\big(L_{j0}(y, t), \ L_{j1}(y, t)\big)$$

are assumed to have rank n for all arguments, but further assumptions about the boundary conditions are needed for well-posedness. We try to determine a solution $u = u(x, t)$ with values in $\mathbf{R}^n$ which is C^∞-smooth and 1-periodic in each variable $x_2, \ldots, x_s$. As before, if we want u to be smooth in the strip (8.1.2) including the boundaries of the domain at $t = 0$, we need compatibility conditions to be satisfied.

The data

$$f(x), F(x, t), g_j(y, t), j = 0, 1 \ \big(\text{where } x = (x_1, y)\big),$$

*The definition given here coincides with our terminology in Section 6.1 since the expressions $A_{ij} \ D_i D_j u$ and $D_i \ A_{ij} \ D_j u$ differ only by a lower-order term.

fulfill *sufficient compatibility conditions* if they vanish identically in some neighborhood of all points

$$(x_1 = 0,\ y,\ t = 0) \quad \text{and} \quad (x_1 = 1,\ y,\ t = 0).$$

Notation. The inner product and norm in this section are given by

$$(u, v) = \int_0^1 \cdots \int_0^1 \langle u(x), v(x) \rangle \, dx_1 \ldots dx_s,$$

$$\|u\| = (u, u)^{1/2}.$$

For boundary integrals we use the notation

$$\begin{aligned} (u, v)_{\Gamma_0} &= \int_0^1 \cdots \int_0^1 \langle u(0, x_2, \ldots, x_s),\ v(0, x_2, \ldots, x_s) \rangle \, dx_2 \ldots dx_s \\ &= \int_Y \langle u(0, y),\ v(0, y) \rangle \, dy, \end{aligned}$$

where $Y = [0, 1]^{s-1}$. Furthermore,

$$(u, v)_{\Gamma_1} = \int_Y \langle u(1, y),\ v(1, y) \rangle \, dy,$$

$$\|u\|_{\Gamma_j}^2 = (u, u)_{\Gamma_j}, \quad j = 0, 1,$$

$$(u, v)_\Gamma \Big|_0^1 = (u, v)_{\Gamma_1} - (u, v)_{\Gamma_0},$$

$$\|u\|_\Gamma^2 = \|u\|_{\Gamma_0}^2 + \|u\|_{\Gamma_1}^2.$$

Integration by parts. Let $u = u(x)$, $v = v(x)$ be smooth functions of x, 1-periodic in $x_2, \ldots, x_s$. Then

(8.1.6) $$(u, D_1 v) = -(D_1 u, v) + (u, v)_\Gamma \Big|_0^1,$$

(8.1.7) $$(u, D_i v) = -(D_i u, v), \quad i = 2, \ldots, s.$$

8.1.2. Neumann Boundary Conditions

For illustration, we will concentrate on so-called Neumann boundary conditions, which we write in the form

(8.1.8) $$D_1 u(j, y, t) = S_j(y, t) u(j, y, t) + g_j(y, t), \quad j = 0, 1.$$

We first show the basic energy estimate.

Lemma 8.1.1. *Suppose that the smooth function u solves the parabolic problem* (8.1.1) *with initial condition* (8.1.3) *and boundary conditions* (8.1.8). *There is a constant K_T independent of f, F, g_0, g_1 with*

$$\|u(\cdot,t)\|^2 + \int_0^t \|u(\cdot,\tau)\|_\Gamma^2\,d\tau \le K_T\|f\|^2 + K_T\int_0^t \{\|F(\cdot,\tau)\|^2 + \|g(\cdot,\tau)\|_\Gamma^2\}\,d\tau$$

in $0 \le t \le T$. *Here*

$$\|g(\cdot,\tau)\|_\Gamma^2 = \sum_{j=0,1} \|g_j(\cdot,\tau)\|_{\Gamma_j}^2.$$

Proof. From the differential equation (8.1.1), integration by parts (8.1.6), (8.1.7), and the parabolicity assumption (8.1.5) we obtain that

$$\frac{d}{dt}\|u(\cdot,t)\|^2 = 2(u,u_t)$$

$$\le -2\delta\sum_{i=1}^{s}\|D_iu\|^2 + c_1\Big\{\|u\|\sum_{i=1}^{s}\|D_iu\| + \|u\|^2 + \|F\|^2\Big\} + (u, A_{11}D_1u)_\Gamma\Big|_0^1$$

$$\le -\delta\sum_{i=1}^{s}\|D_iu\|^2 + c_2\{\|u\|^2 + \|F\|^2\} + (u, A_{11}D_1u)_\Gamma\Big|_0^1.$$

We can use the boundary conditions (8.1.8) to replace D_1u in the above boundary term by u, g_0, g_1. The one-dimensional Sobolev inequality

$$|w(0,y)|^2 + |w(1,y)|^2 \le \epsilon\int_0^1 |D_1w(x_1,y)|^2\,dx_1 + c(\epsilon)\int_0^1 |w(x_1,y)|^2\,dx_1$$

can be integrated with respect to y to yield

$$\|w\|_{\Gamma_0}^2 + \|w\|_{\Gamma_1}^2 \le \epsilon\|D_1w\|^2 + c(\epsilon)\|w\|^2. \tag{8.1.9}$$

Therefore we obtain that

$$\frac{d}{dt}\|u\|^2 \le -\frac{\delta}{2}\sum_{i=1}^{s}\|D_iu\|^2 + c_3\{\|u\|^2 + \|F\|^2 + \|g\|_\Gamma^2\}.$$

Another application of (8.1.9) and integration with respect to t proves the lemma.

We show next how to estimate derivatives; we write $D_t = \partial/\partial t$ for time differentiation. We apply one of the operators

$$D_t,\ D_2, \ldots, D_s$$

to (8.1.1) and obtain

$$(8.1.10) \qquad (D_k u)_t = \sum_{i,j=1}^{s} A_{ij}\ D_i D_j (D_k u) + \text{ terms of order } \leq 2,$$

$k = t, k = 2, \ldots, s$. (Since we do not get boundary conditions for $D_1 u$, we do not apply the operator D_1 to (8.1.1).) Introducing the vector

$$\underline{u} = (D_t u,\ D_2 u, \ldots, D_s u) \in \mathbf{R}^{ns},$$

we can write the above system (8.1.10) as

$$(8.1.11) \qquad \underline{u}_t = \sum_{i,j=1}^{s} \underline{A}_{ij}\ D_i\ D_j\ \underline{u} + \ldots,$$

where $\underline{A}_{ij}$ is the block-diagonal matrix

$$\underline{A}_{ij} = \begin{bmatrix} A_{ij} & & & 0 \\ & A_{ij} & & \\ & & \ddots & \\ 0 & & & A_{ij} \end{bmatrix} \in \mathbf{R}^{ns \times ns}.$$

The terms not written out in (8.1.10) contain

$$D_1^2 u \quad \text{and} \quad D_1 u.$$

Noting that A_{11} is nonsingular (see (8.1.5)) and using the given differential equation (8.1.1), we can express $D_1^2 u$ by

$$u_t,\ D_1 D_2 u, \ \text{ etc.}$$

Therefore, $D_1^2 u$ and $D_1 u$ can be estimated in terms of F, $\underline{u}$ and the first derivatives of $\underline{u}$.

A key observation is that we can easily derive boundary conditions for $\underline{u}$. An application of one of the operators $D_t,\ D_2, \ldots, D_s$ to (8.1.8) yields

$$D_1 D_k u = S_j D_k u + D_k g_j + (D_k S_j) u, \quad k = t, \quad k = 2, \ldots s,$$

at each boundary $x_1 = j = 0, x_1 = j = 1$. With the same notation as before,

$$(8.1.12) \qquad D_1 \underline{u}(j, y, t) = \underline{S}_j \underline{u}(j, y, t) + g_j^{(1)}(y, t).$$

Thus we have boundary conditions for $\underline{u}$ of the same structure as those for u. It is important to note that the inhomogeneous data

$$g_j^{(1)}(y, t), \quad j = 0, 1,$$

can be estimated in terms of the first derivatives of the given boundary data $g_j(y,t)$ and the u-values on the boundary. The latter have already been estimated in Lemma 8.1.1. Thus we have a parabolic system (8.1.11) for $\underline{u}$ under Neumann boundary conditions (8.1.12). Lemma 8.1.1 applied to this problem gives us estimates for $\underline{u}$, i.e., for

$$u_t, \; D_2 u, \ldots, D_s u.$$

To bound $D_1 u$, we take the inner product of (8.1.1) with u and integrate by parts. This yields

$$\begin{aligned}(u, u_t) &= (u, \; A_{11} D_1^2 u) + \ldots \\ &= -(D_1 u, \; A_{11} D_1 u) + \ldots \\ &\leq -\delta \|D_1 u\|^2 + \ldots .\end{aligned}$$

Thus we obtain bounds for $\|D_1 u\|^2$ also. This process can be continued, and inductively one obtains

Lemma 8.1.2. *Suppose that the smooth function u solves the initial-boundary value problem* (8.1.1), (8.1.3), (8.1.8). *In every finite time interval we can estimate u and its derivatives in terms of the data f, F, g_0, g_1 and their derivatives.*

Assume that compatibility conditions for the data f, F, g_0, g_1 are satisfied. To show the *existence* of a smooth solution u, we can again use a difference scheme. Proceeding as in the continuous case, we can prove estimates for the solution $v = v^h$ of the difference scheme independently of the step-size $h > 0$. There are no essential differences from the case of one space dimension treated in Section 7.2. The periodic variables $x_2, \ldots, x_s$ are discretized as in Section 3.2. One obtains

Theorem 8.1.3. *Suppose that the data f, F, g_0, g_1 are compatible. The parabolic problem* (8.1.1), (8.1.3) *under Neumann boundary conditions* (8.1.8) *has a unique smooth solution. One can estimate the solution and its derivatives in terms of the data and their derivatives.*

8.1.3. Other Boundary Conditions

If the matrix $L_{j1} = L_{j1}(y,t)$ in (8.1.4) is singular but of constant rank r_j, we can write the boundary conditions as

$$L_{j0}^I u(j,y,t) + L_{j1}^I D_1 u(j,y,t) = g_j^I(y,t), \quad L_{j0}^{II} u(j,t) = 0, \quad j = 0, 1,$$

after we transform the Dirichlet part to become homogeneous. To derive an energy estimate, we can proceed as in one space dimension, see Section 7.2. If either $L_{j1} = 0$ (the pure Dirichlet case) or if at each boundary point algebraic conditions as in Lemma 7.2.1 are met, then the problem is well-posed. Also, more general boundary conditions can be treated via localization, Laplace transform in time, and Fourier expansion in $x_2, \ldots, x_s$. We refer to Section 8.4 where we will illustrate this technique for hyperbolic problems.

8.1.4. General Domains

We want to indicate how our results can be extended to more general domains. Consider the example

$$u_t = u_{xx} + u_{yy} \quad \text{for} \quad r = x^2 + y^2 \leq 1, \tag{8.1.13}$$

with a boundary condition

$$Bu = 0 \quad \text{at} \quad r = 1. \tag{8.1.14}$$

As usual, $u(\cdot, 0) = f$ is given at $t = 0$.

Let $\varphi = \varphi_1(r)$ denote a monotone C^∞-function with

$$\varphi_1(r) = \begin{cases} 1 & \text{for } r \geq 1 - \delta, \\ 0 & \text{for } r \leq 1 - 2\delta, \end{cases} \qquad 0 < \delta \ll 1.$$

If we introduce the notations $\varphi_2 = 1 - \varphi_1$, $u_j = \varphi_j u$, $j = 1, 2$, then $u = u_1 + u_2$, and we can write (8.1.13) as a system for u_1, u_2 :

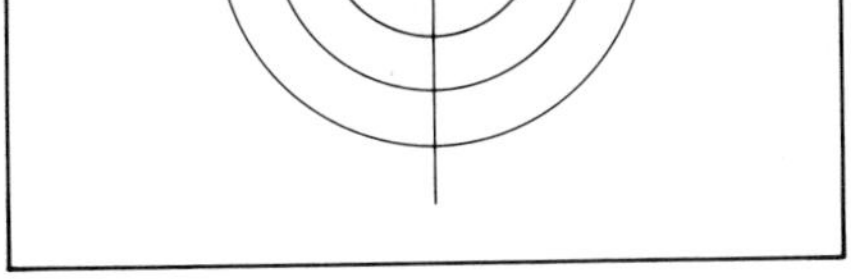

FIGURE 8.1.1. Circle and annulus.

$$u_{jt} = u_{jxx} + u_{jyy} - 2\varphi_{jx}\big(u_{1x} + u_{2x}\big) - 2\varphi_{jy}\big(u_{1y} + u_{2y}\big)$$
$$- \big(\varphi_{jxx} + \varphi_{jyy}\big)\big(u_1 + u_2\big), \quad j = 1, 2.$$

To motivate the following, we first neglect lower-order terms and write v_j instead of u_j. Then we obtain the two separate equations

$$v_{jt} = v_{jxx} + v_{jyy}, \quad j = 1, 2.$$

Since $u_1 \equiv 0$ for $r \le 1 - 2\delta$, we consider the equation for v_1 in the annulus $1 - 2\delta \le r \le 1$ under boundary conditions

$$Bv_1 = 0 \ \text{ at } r = 1, \quad v_1 = 0 \ \text{ at } r = 1 - 2\delta.$$

This problem is of the type which we have treated, because we can map the annulus onto a periodic strip by the introduction of polar coordinates. Similarly, $u_2 \equiv 0$ for $r \ge 1 - \delta$, which motivates the consideration of a pure Cauchy problem for v_2 with "boundary condition" $\|v_2(\cdot, t)\| < \infty$.

To treat the full problem, we define the iteration

$$u_{jt}^{n+1} = u_{jxx}^{n+1} + u_{jyy}^{n+1} - 2\varphi_{jx}\big(u_{1x}^n + u_{2x}^n\big) - 2\varphi_{jy}\big(u_{1y}^n + u_{2y}^n\big)$$
$$- \big(\varphi_{jxx} + \varphi_{jyy}\big)\,\big(u_1^n + u_2^n\big), \quad j = 1, 2, \quad n = 0, 1, \ldots.$$

Here the functions u_1^{n+1} are determined in the annulus subject to the boundary conditions of v_1; the pure Cauchy problem is solved to determine u_2^{n+1}. The initial conditions are $u_j^{n+1}(\cdot, 0) = \varphi_j f$. In a similar way as for the iterations discussed before (see, for example, Section 4.1.4), one can use Gronwall's Lemma 3.1.1 and Picard's Lemma 3.3.4 to show uniform smoothness and convergence as $n \to \infty$. The function $u = u_1 + u_2$, $u_j = \lim u_j^n$, solves the given problem.

A similar procedure can be used to treat problems in rather general domains with smooth boundaries.

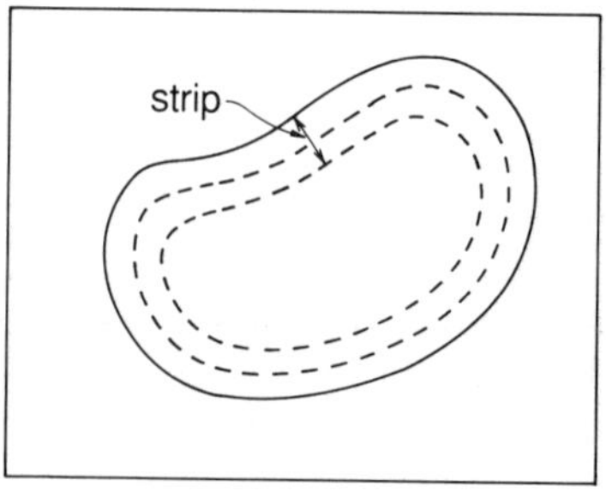

FIGURE 8.1.2. Domain and boundary strip.

8.2. Symmetric Hyperbolic Systems in Several Space Dimensions

In this section we consider first-order systems

$$u_t = \sum_{i=1}^{s} B_i(x,t)D_i u + C(x,t)u + F(x,t) \tag{8.2.1}$$

in the strip

$$0 \le x_1 \le 1, \quad -\infty < x_2, \ldots, x_s < \infty, \quad t \ge 0,$$

under initial conditions

$$u(x,0) = f(x) \tag{8.2.2}$$

and boundary conditions at $x_1 = 0, x_1 = 1$. As before, all quantities are assumed to be real, for simplicity, C^∞-smooth, and 1-periodic in $x_2, \ldots, x_s$. This assumption applies also to the boundary conditions specified below. An essential assumption is the symmetry

$$B_i(x,t) = B_i^*(x,t), \quad i = 1, \ldots, s;$$

i.e., we have a symmetric hyperbolic system. We will also discuss such systems in the half-space

$$0 \le x_1 < \infty, \quad -\infty < x_2, \ldots, x_s < \infty \quad \text{for } t \ge 0,$$

and in more general domains. Our main emphasis is to show that the problem becomes strongly well-posed under certain boundary conditions. In the next section we apply our results to the linearized inviscid compressible Navier-Stokes equations. If one of the boundaries $x_1 = 0$ or $x_1 = 1$ is a wall, then naturally the u-component of the base flow vanishes there. In this case, as we will see, the corresponding boundary is characteristic. A complete discussion of this situation would be very complicated in several space dimensions. For a characteristic boundary we will restrict ourselves to the treatment of some special cases.

8.2.1. The Basic Estimate and Bounds for Derivatives

After transforming the independent variables u, if necessary, we can assume that

$$B_1(x,t) = \Lambda(x,t)$$

is real and diagonal. First let us assume that $B_1 = \Lambda$ is nonsingular for all arguments. Then, without restriction,

$$B_1 = \Lambda = \begin{pmatrix} \Lambda_- & 0 \\ 0 & \Lambda_+ \end{pmatrix}, \quad \Lambda_- \leq -\gamma I, \quad \Lambda_+ \geq \gamma I, \quad \gamma > 0.$$

We partition the variables u correspondingly into

$$u = \begin{pmatrix} u_- \\ u_+ \end{pmatrix}.$$

As motivated by the discussion in one space dimension (see Section 7.6), we consider boundary conditions at $x_1 = 0, x_1 = 1$ of the form

$$\begin{aligned} u_-(0, y, t) &= S_0(y, t)u_+(0, y, t) + g_0(y, t), \\ u_+(1, y, t) &= S_1(y, t)u_-(1, y, t) + g_1(y, t), \quad y = (x_2, \ldots, x_s). \end{aligned} \tag{8.2.3}$$

Here S_0, S_1 are matrices of appropriate dimensions.

Difference between one and more space dimensions. There is a significant difference between the cases of one and of more space dimensions: In one dimension, we could force the matrices S_0, S_1 to become "small" by applying a suitable transformation to u, if necessary; see the proof of Lemma 7.6.2. If we were to apply such a transformation here, the matrices $B_2, \ldots, B_s$ would lose their symmetry, in general. The symmetry is essential, however, for deriving the estimates for well-posedness by the energy method. For this reason, the assumption

$$|S_0| + |S_1| \text{ "small"},$$

which we need below, constitutes a real restriction on the boundary conditions (8.2.3). The question remains whether or not it is possible to show well-posedness by other means if $|S_0| + |S_1|$ is large. We will employ the method of Laplace transform in Section 8.4 to discuss this question. It turns out that well-posedness can indeed be lost if $|S_0| + |S_1|$ is not small. In one space dimension this is not possible.

The basic estimate. We want to show first

Lemma 8.2.1. *Suppose that $u(x, t)$ is a (real) smooth solution of* (8.2.1)–(8.2.3) *which is 1-periodic in $x_2, \ldots, x_s$. If*

$$|S_0(y, t)| + |S_1(y, t)|$$

is sufficiently small for all arguments, then for any finite time interval $0 \le t \le T$ *there is a constant* K_T *with*

(8.2.4)

$$\|u(\cdot,t)\|^2 + \int_0^t \|u(\cdot,\tau)\|_\Gamma^2\, d\tau \le K_T\Big\{\|f\|^2 + \int_0^t \Big\{\|g(\cdot,\tau)\|_\Gamma^2 + \|F(\cdot,\tau)\|^2\Big\}\, d\tau\Big\}$$

for $0 \le t \le T$. *The constant* K_T *does not depend on* f, F, *or* g.

Proof. The differential equation (8.2.1) gives us that

$$\begin{aligned}\frac{d}{dt}\|u(\cdot,t)\|^2 &= 2(u,u_t)\\ &\le 2\Big(u, \sum_{i=1}^{s} B_i D_i u\Big) + c_1\big\{\|u\|^2 + \|F\|^2\big\}.\end{aligned}$$

For $i = 2, \ldots, s$,

$$\begin{aligned}(u, B_i D_i u) &= (B_i u, D_i u)\\ &= -(B_i D_i u, u) - \big((D_i B_i)u, u\big),\end{aligned}$$

and thus

$$(u, B_i D_i u) \le c_2 \|u\|^2.$$

For $i = 1$ we obtain an additional boundary term. Therefore

$$\frac{d}{dt}\|u\|^2 \le c_3\big\{\|u\|^2 + \|F\|^2\big\} + (u,\Lambda u)_\Gamma\Big|_0^1.$$

Since $|S_0| + |S_1|$ is assumed small, we can treat the boundary term in exactly the same way as in one space dimension (compare the proof of Lemma 7.6.2) and obtain that

(8.2.5)
$$(u,\Lambda u)_\Gamma\Big|_0^1 \le -\frac{\gamma}{2}\|u\|_\Gamma^2 + c_4\|g\|_\Gamma^2.$$

Integration of the resulting inequality with respect to t proves the lemma.

We will show below the *existence* of a smooth solution if the data are compatible. Then the estimate of the previous lemma states *strong well-posedness* of the hyperbolic initial-boundary value problem. In obtaining the basic estimate, we have not used that $B_1 = \Lambda$ is nonsingular. In the case that Λ has zero eigenvalues at the boundary, we can obtain a similar estimate to (8.2.4) if we replace the boundary term on the left side by

$$\int_0^t \|u_\pm(\cdot,\xi)\|_\Gamma^2 \, d\xi.$$

Zero eigenvalues of $B_1 = \Lambda$ cause difficulties, however, if we want to estimate derivatives of the solution.

Estimates of derivatives. If we apply one of the operators

$$D_t, D_2, \ldots, D_s$$

to (8.2.1) we obtain

$$\begin{aligned} D_k u_t = \sum_{i=1}^{s} B_i D_i D_k u + \sum_{i=1}^{s} (D_k B_i) D_i u \\ + C D_k u + (D_k C) u + D_k F, \end{aligned}$$

$k = t,\ k = 2, \ldots, s$. Introducing the vector

$$\underline{u} = (D_t u, D_2 u, \ldots, D_s u) \in \mathbf{R}^{ns},$$

we can write this system as

$$\begin{aligned} \underline{u}_t &= \sum_{i=1}^{s} \underline{B}_i D_i \underline{u} + \underline{C}\,\underline{u} \\ &+ \begin{pmatrix} (D_t B_1) D_1 u \\ \vdots \\ (D_s B_1) D_1 u \end{pmatrix} + \begin{pmatrix} (D_t C) u \\ \vdots \\ (D_s C) u \end{pmatrix} + \begin{pmatrix} D_t F \\ \vdots \\ D_s F \end{pmatrix}, \end{aligned}$$

where

$$\underline{B}_i = \begin{pmatrix} B_i & & 0 \\ & B_i & \\ & & \ddots \\ 0 & & B_i \end{pmatrix}.$$

Using (8.2.1) again and observing the nonsingularity of $B_1 = \Lambda$, we can express $D_1 u$ in terms of $\underline{u}$, u, and F. Boundary conditions for $\underline{u}$ are obtained if we apply $D_t, D_2, \ldots, D_s$ to (8.2.3). Since we have already estimates for $\|u\|^2$ and for $\|u\|_\Gamma^2$, we obtain estimates for $\underline{u}$ in terms of the data f, F, g and their first derivatives. Clearly, we also have an estimate for $D_1 u$ from (8.2.1). This process of estimating derivatives can be continued inductively.

8.2.2. Existence of a Solution

To show the existence of a C^∞-solution, we assume compatible data; for example, it is sufficient that the functions

$$f(x), \qquad F(x,t), \qquad g_0(y,t), \qquad g_1(y,t)$$

vanish identically in a neighborhood of all points

$$(x_1 = 0, y, t = 0) \text{ and } (x_1 = 1, y, t = 0) \text{ where } y = (x_2, \ldots, x_s).$$

We choose a gridlength $h = 1/N$, N a natural number. The space gridpoints are

$$x_\nu = (h\nu_1, \ldots, h\nu_s), \quad \nu_j \in \mathbf{Z}, \quad 0 \le \nu_1 \le N,$$

and we leave time continuous.

The differential equation (8.2.1) is replaced by a system of ordinary differential equations for the gridfunction $v(x_\nu, t) = v^h(x_\nu, t)$ of the form

$$\frac{d}{dt} v(x_\nu, t) = Q_1 v(x_\nu, t) + F(x_\nu, t), \qquad 1 \le \nu_1 \le N-1. \tag{8.2.6}$$

Here Q_1 is the difference operator*

$$Q_1 = \begin{pmatrix} \Lambda_- D_{-1} & 0 \\ 0 & \Lambda_+ D_{+1} \end{pmatrix} + \sum_{i=2}^{s} B_i D_{0i} + C. \tag{8.2.7}$$

The coefficients are to be evaluated at (x_ν, t). The boundary conditions (8.2.3) are replaced by

$$\begin{aligned} v_-(0, y, t) &= S_0(y, t) v_+(h, y, t) + g_0(y, t), \\ v_+(1, y, t) &= S_1(y, t) v_-(1 - h, y, t) + g_1(y, t), \end{aligned} \tag{8.2.8}$$

where $(0, y)$, $(1, y)$ are points of the spatial grid. There are no difficulties in discretizing the initial condition (8.2.2) by

$$v(x_\nu, 0) = f(x_\nu). \tag{8.2.9}$$

The differential equation (8.2.1) is discretized only at the interior gridpoints

$$x_\nu, \qquad 1 \le \nu_1 \le N-1,$$

of the strip. If we consider the equation (8.2.6) at a point $x_\nu = (h, y)$ close to the left boundary of the strip, say, then the difference operator

$$\Lambda_-(x_\nu, t) D_{-1}$$

*In the terminology of numerical analysis, we use "upwinding" to approximate ΛD_1 and apply a central difference formula in the periodic variables $x_2, \ldots, x_s$.

involves the value $v_{-}(0, y, t)$. We can use the discretized boundary condition (8.2.8) to eliminate this part of a v-value. Using periodicity with respect to the variables $x_2, \dots, x_s$, we have obtained a finite dimensional (linear) ordinary differential system for $v(t) = v^h(t)$ under initial conditions. Thus $v^h(t)$ is uniquely determined. As previously, by mimicking the estimates for the differential equation, one can bound all difference-differential quotients of v^h independently of h. Simple error formulae (in numerical analysis terminology, the *consistency* of the difference approximation) show that v^h converges to a smooth solution $u(x, t)$ of the initial-boundary value problem for some sequence $h \to 0$. Some of the details of the estimates of v^h are provided in the next section. We summarize the result in

Theorem 8.2.2. *Given the symmetric hyperbolic initial-boundary value problem* (8.2.1)–(8.2.3) *where* $B_1 = \Lambda$ *is nonsingular and the matrices* S_0 *and* S_1 *are sufficiently small in norm.* (*For example,* $S_0 = 0$, $S_1 = 0$.) *If the data are compatible, there is a unique smooth solution. The problem is strongly well-posed.*

The smallness-assumption for $|S_0| + |S_1|$ can be quantified as in the proof of Lemma 7.6.2. What we actually need is the estimate (8.2.5) for all functions which satisfy the boundary conditions.

8.2.3. The Finite Speed of Propagation

We shall prove that the principle of finite speed of propagation is valid for the hyperbolic initial-boundary value problem (8.2.1)–(8.2.3). As in the case of the spatially periodic Cauchy problem (see Theorem 6.2.1), we construct a difference approximation whose solutions converge to the solution $u(x, t)$ of the continuous problem. For the discrete systems the speed of propagation will be finite and bounded independently of the step-size.

The spatial meshsize $h = 1/N$ and the spatial gridpoints

$$x_\nu = (h\nu_1, \dots, h\nu_s), \qquad \nu_j \in \mathbf{Z}, \qquad 0 \le \nu_1 \le N\ ,$$

are denoted as in the last section. We choose a time step

$$k = hs_0, \qquad s_0 > 0 \ \text{ fixed},$$

where s_0 will be assumed sufficiently small, but independent of h. Discretizing in time and space, we determine a grid function

$$v(x_\nu, t), \qquad t = 0, k, 2k, \dots,$$

by

$$
\begin{aligned}
&v(x_\nu, t+k) = (I + kQ_1 + khQ_2)v(x_\nu, t) + kF(x_\nu, t), \\
&1 \le \nu_1 \le N-1, \qquad t = 0, k, 2k, \dots .
\end{aligned} \tag{8.2.10}
$$

Here Q_1 is defined in (8.2.7) and

$$Q_2 = \sum_{i=2}^{s} D_{+i} D_{-i}.$$

The gridfunction v is subject to the boundary and initial conditions (8.2.8), (8.2.9); as explained in the previous section, these conditions determine v uniquely. Clearly, the ratio

$$h/k = s_0^{-1}$$

bounds the speed of propagation in the discrete system (see Section 6.2); therefore, once we have proven convergence of $v^h(x_\nu, t)$ to $u(x,t)$ as $h \to 0$ (in the sense of (6.2.4)), the number s_0^{-1} also bounds the propagation-speed for the continuous problem.

It will be sufficient to show that we can bound v^h uniformly in terms of the data. Then convergence follows, since $v^h - u$ satisfies the above difference equations with data which tend to zero for $h \to 0$. To prove these bounds for $v = v^h$, we introduce a discrete scalar product by

$$(v, w)_h = h^s \sum_{1 \le \nu_1 \le N-1} \; \sum_{0 \le \nu_2, \dots, \nu_s < N} \langle v(x_\nu), w(x_\nu) \rangle .$$

Also, we use the notation

$$(v, w)_{\Gamma(j)} = h^{s-1} \sum_{\nu_1 = j} \langle v(x_\nu), w(x_\nu) \rangle ,$$

where the sum extends over

$$0 \le \nu_2, \dots, \nu_s < N,$$

and $\nu_1 = j$ is fixed.

In (vague) analogy to integration by parts with respect to x_1, we have the rules:

Lemma 8.2.3.

$$(v, D_{-1}w)_h = -(D_{-1}v, w)_h + h(D_{-1}v, D_{-1}w)_h + (v, w)_\Gamma \Big|_0^{N-1},$$

$$(v, D_{+1}w)_h = -(D_{+1}v, w)_h - h(D_{+1}v, D_{+1}w)_h + (v, w)_\Gamma \Big|_1^{N},$$

where

$$(v, w)_\Gamma \Big|_i^j = (v, w)_{\Gamma(j)} - (v, w)_{\Gamma(i)}.$$

Proof. Since summation over

$$0 \leq \nu_2, \ldots, \nu_s < N$$

applies to all terms, it suffices to prove the result for $s = 1$. Then we have

$$\begin{aligned}
&(v, D_- w)_h + (D_- v, w)_h - h(D_- v, D_- w)_h \\
&\quad = \sum_{\nu=1}^{N-1} \big\{ \langle v_\nu, w_\nu - w_{\nu-1} \rangle + \langle v_\nu - v_{\nu-1}, w_\nu \rangle - \langle v_\nu - v_{\nu-1}, w_\nu - w_{\nu-1} \rangle \big\} \\
&\quad = \sum_{\nu=1}^{N-1} \big\{ \langle v_\nu, w_\nu \rangle - \langle v_{\nu-1}, w_{\nu-1} \rangle \big\} \\
&\quad = \langle v_{N-1}, w_{N-1} \rangle - \langle v_0, w_0 \rangle.
\end{aligned}$$

(Here we have used the index notation v_ν for $v(x_\nu)$, etc.) This proves the first relation, and the second is shown similarly.

Recursive estimate for $v = v^h$. The difference equation (8.2.10) gives us

$$\begin{aligned}
\|v(\cdot, t+k)\|_h^2 \leq \|v\|_h^2 &+ 2k(v, Q_1 v)_h + 2kh(v, Q_2 v)_h + 2k^2 \|Q_1 v\|_h^2 \\
&+ 2k^2 h^2 \|Q_2 v\|_h^2 + c_1 \big\{ k \|v\|_h^2 + k \|F\|_h^2 \big\}.
\end{aligned}$$

The inner products and norms on the right-hand side apply to the gridfunctions $v(\,\cdot\,, t)$, etc., and we have used estimates like

$$\begin{aligned}
k^2 (Q_1 v, F)_h &\leq k^2 \|Q_1 v\|_h \|F\|_h \\
&\leq k^2 \Big\{ k \|Q_1 v\|_h^2 + \frac{1}{k} \|F\|_h^2 \Big\} \\
&\leq k c_2 \frac{k^2}{h^2} \|v\|_h^2 + k \|F\|_h^2 \\
&\leq k c_2 s_0^2 \|v\|_h^2 + k \|F\|_h^2 .
\end{aligned}$$

To bound the term

$$2k(v, Q_1 v)_h,$$

we apply Lemma 8.2.3 to treat the contribution of

$$\left(v, \begin{pmatrix} \Lambda_- D_{-1} & 0 \\ 0 & \Lambda_+ D_{+1} \end{pmatrix} v\right)_h = (v_-, \Lambda_- D_{-1} v_-)_h + (v_+, \Lambda_+ D_{+1} v_+)_h.$$

Since Λ_- and Λ_+ are bounded away from zero, we have, with some $\delta > 0$,

$$2k(v, Q_1 v) \leq -kh\delta\{\|D_{-1}v_-\|_h^2 + \|D_{+1}v_+\|_h^2\} + k(v_-, \Lambda_- v_-)_\Gamma\Big|_0^{N-1}$$
$$+ k(v_+, \Lambda_+ v_+)_\Gamma\Big|_1^N + c_3 k \|v\|_h^2.$$

Using the discrete boundary conditions (8.2.8) and the assumption that

$$|S_0| + |S_1|$$

is small, we can obtain an estimate for the boundary terms

$$(v_-, \Lambda_- v_-)_\Gamma\Big|_0^{N-1} + (v_+, \Lambda_+ v_+)_\Gamma\Big|_1^N$$

in terms of the boundary data g_0 and g_1. Therefore,

$$2k(v, Q_1 v)_h \leq -kh\delta\{\|D_{-1}v_-\|_h^2 + \|D_{+1}v_+\|_h^2\} + c_4 k\{\|v\|_h^2 + \|g\|_{h,\Gamma}^2\}.$$

The difference operator Q_2 acts only in the periodic directions $x_2, \ldots, x_s$; thus summation by parts gives us

$$2kh(v, Q_2 v)_h = -2kh \sum_{i=2}^{s} \|D_{+i}v\|_h^2.$$

Finally, it is not difficult to show that

$$2k^2 \|Q_1 v\|_h^2 \leq c_5 k^2 \Big\{\|D_{-1}v_-\|_h^2 + \|D_{+1}v_+\|_h^2 + \sum_{i=2}^{s} \|D_{+i}v\|_h^2 + \|v\|_h^2\Big\},$$

$$2k^2 h^2 \|Q_2 v\|_h^2 \leq c_5 k^2 \sum_{i=2}^{s} \|D_{+i}v\|_h^2,$$

where c_5 does not depend on s_0.

Summarizing, we have shown that

$$\|v(\cdot, t+k)\|_h^2 \leq (-kh\delta + c_5 k^2)\{\|D_{-1}v_-\|_h^2 + \|D_{+1}v_+\|_h^2\}$$
$$+ (-2kh + 2c_5 k^2) \sum_{i=2}^{s} \|D_{+i}v\|_h^2 + (1 + kc_6) \|v\|_h^2$$
$$+ kc_6\{\|g\|_{h,\Gamma}^2 + \|F\|_h^2\}.$$

Here $c_5k^2 = c_5khs_0$. If s_0 is chosen so small that

$$c_5s_0 \le \min\{\delta, 1\},$$

then we have the recursive estimate

$$\|v(\cdot, t+k)\|_h^2 \le (1+kc)\,\|v\|_h^2 + kc\{\|g\|_{h,\Gamma}^2 + \|F\|_h^2\}.$$

In any finite time interval we obtain the desired estimate of v in terms of the data. Thus we have proven

Theorem 8.2.4. *Under the assumptions of Theorem* 8.2.2 *the principle of finite speed of propagation is valid for the symmetric hyperbolic initial-boundary value problem* (8.2.1)–(8.2.3).

8.2.4. The Strip Problem vs. the Half-Space and Periodic Cauchy Problems

Thus far we have treated symmetric hyperbolic systems (8.2.1): a) under the side-condition of spatial periodicity; b) in a strip $0 \le x_1 \le 1$ with boundary conditions at $x_1 = 0$, $x_1 = 1$ and periodicity assumptions with respect to $x_2, \dots, x_s$. Only in the latter case did we need that B_1 is nonsingular.

In this section we will show how to use the strip problem to solve the hyperbolic system (8.2.1) in the half-space

$$0 \le x_1 < \infty, \qquad -\infty < x_2, \dots, x_s < \infty \;\text{ for }\; t \ge 0$$

with boundary conditions at $x_1 = 0$. Conversely, the strip problem can be reduced to left half-space, right half-space and periodic Cauchy problems. Essentially these results follow quite easily from the finite speed of propagation. The reduction of the strip problem to half-space and periodic Cauchy problems will be useful in the next section, where we remove the assumption that B_1 is nonsingular.

Solution of half-space problems. Suppose the coefficients and data in (8.2.1), (8.2.2) are defined for all $x \in \mathbf{R}^s$, are 1-periodic with respect to $x_2, \dots, x_s$, and are uniformly smooth. Furthermore, assume that the data $f(x)$ and $F(x,t)$ have compact support with respect to x_1; i.e., there is a constant M with

$$|f(x)| + |F(x,t)| = 0 \;\text{ for }\; |x_1| > M. \tag{8.2.11}$$

We always assume symmetry $B_i = B_i^*$, and also require in this section that B_1 is nonsingular. Our first result is

Theorem 8.2.5. *Under the above assumptions, consider the half-space problem*

$$u_t = \sum_{i=1}^{s} B_i(x,t)D_i u + C(x,t)u + F(x,t) =: Pu + F(x,t)$$

in $0 \le x_1 < \infty,\ -\infty < x_2, \ldots, x_s < \infty,\ t \ge 0$, *with the initial condition*

$$u(x,0) = f(x)$$

and the boundary condition at $x_1 = 0$,

$$u_-(0,y,t) = S_0(y,t)u_+(0,y,t) + g_0(y,t).$$

Assume the data are compatible and $|S_0|$ *is sufficiently small. There is a unique smooth solution which vanishes for* $0 \le t \le T$ *if the argument* $x_1 \ge c(T)$ *is sufficiently large.*

Proof. We fix a time $T > 0$ and consider the above problem in the strip

$$0 \le x_1 \le \alpha M, \qquad \alpha > 1,$$

with the additional boundary condition

$$u_+(\alpha M, y, t) = 0$$

at $x_1 = \alpha M$. According to Section 8.2.2, there is a unique smooth solution $u(x,t)$ of this strip problem.

The finite speed of propagation implies that u is identically zero in a neighborhood of the points

$$(x_1 = \alpha M, y, t), \qquad 0 \le t \le T,$$

if α is chosen sufficiently large. Setting $u = 0$ for $x_1 \ge \alpha M,\ 0 \le t \le T$, we have solved the half-space problem for $0 \le t \le T$. Uniqueness follows as previously by an energy estimate. Also, the solution constructed in $0 \le t \le T$ does not change if we choose another $T_1 > T$.

Solution of strip problems. Consider the strip problem (8.2.1)–(8.2.3) and assume, as we did previously, that the coefficients and data are defined for all x. We want to show that we can express the solution of the strip problem as

$$u = u^{(1)} + u^{(2)} + u^{(3)},$$

where $u^{(1)}$ and $u^{(2)}$ solve left and right half-space problems, respectively, whereas $u^{(3)}$ solves a periodic Cauchy problem. Such a representation is valid in a sufficiently small time interval

$$0 \le t \le \delta;$$

after time δ one can restart the process.

We choose a number* $0 < \tau \le 1/2$ and a monotone function

$$\phi_1 = \phi_1(x_1), \qquad \phi_1 \in C^\infty,$$

with

$$\phi_1(x_1) = \begin{cases} 1 & \text{for } x_1 \le \tau/4, \\ 0 & \text{for } x_1 \ge \tau/2, \end{cases}$$

and define

$$\phi_2(x_1) = \phi_1(1 - x_1), \qquad \phi_3(x_1) = 1 - \phi_1(x_1) - \phi_2(x_1).$$

(1) The right half-space problem is

$$u_t^{(1)} = Pu^{(1)} + \phi_1 F,$$
$$u^{(1)}(x, 0) = \phi_1(x_1) f(x), \quad x_1 \ge 0,$$
$$u_-^{(1)}(0, y, t) = S_0(y, t) u_+^{(1)}(0, y, t) + g_0(y, t).$$

(2) The left half-space problem is

$$u_t^{(2)} = Pu^{(2)} + \phi_2 F,$$
$$u^{(2)}(x, 0) = \phi_2(x_1) f(x), \quad x_1 \le 1\ ,$$
$$u_+^{(2)}(1, y, t) = S_1(y, t) u_-^{(2)}(1, y, t) + g_1(y, t).$$

(3) The periodic Cauchy problem is

$$u_t^{(3)} = Pu^{(3)} + \phi_3 F,$$
$$u^{(3)}(x, 0) = \phi_3(x_1) f(x).$$

To be precise, the last problem is not periodic in x_1 originally. However, using the assumption (8.2.11), we can alter the coefficients and data outside a sufficiently large interval

$$-\alpha M \le x_1 \le \alpha M$$

and make them periodic. The finite speed of propagation ensures that we do not change $u^{(3)}$ in

$$0 \le x_1 \le 1, \qquad 0 \le t \le \delta,$$

*The freedom in the choice of τ will be useful in the next section.

where $\delta > 0$ is sufficiently small. Clearly, the sum

$$u = u^{(1)} + u^{(2)} + u^{(3)}, \qquad 0 \le x_1 \le 1, \qquad 0 \le t \le \delta \,,$$

solves the given strip problem. The function u is nothing but the C^∞-solution of the strip problem which we obtained already in Section 8.2.2. We have just described another representation. Using the values $u(\,\cdot\,, \delta)$ as starting data at $t = \delta$, it is clear that there are no compatibility problems if we continue this process. This shows

Theorem 8.2.6. *The solution $u(x,t)$ of the strip problem* (8.2.1)–(8.2.3) *can be obtained by solving half-space and periodic Cauchy problems in successive time intervals*

$$t_0 \le t \le t_0 + \delta.$$

The time $\delta > 0$ depends on the finite speed of propagation.

8.2.5. Cases Where $B_1(x,t)$ Becomes Singular

For the strip problem (8.2.1)–(8.2.3) we want to remove the assumption that $B_1(x,t)$ is nonsingular for all arguments. This is particularly easy if B_1 becomes singular only in the interior $0 < x_1 < 1$. For any fixed finite time T, there is a number $0 < \tau \le 1/2$ such that

$$B_1^{-1}(x,t) \text{ exists for } 0 \le x_1 \le \tau, \quad 1 - \tau \le x_1 \le 1, \quad 0 \le t \le T.$$

Then we split the strip problem into two half-space problems and a Cauchy problem as described in the last section. For definiteness, consider the right half-space problem. We change the coefficient $B_1(x,t)$ to

$$\tilde{B}_1(x,t) = \phi_1(x_1/4)B_1(x,t) + \big(1 - \phi_1(x_1/4)\big)B_1(0,y,t),$$
$$x = (x_1, y).$$

For $0 \le x_1 \le \tau$ it holds that $\tilde{B}_1 = B_1$; furthermore, it is not difficult to check that $\tilde{B}_1$ is nonsingular throughout if $\tau > 0$ is sufficiently small. Thus the changed right half-space problem (with $\tilde{B}_1$) has a solution $\tilde{u}^{(1)}$. In some small time interval the function $\tilde{u}^{(1)}$ is zero for $x_1 > \tau$, thus $\tilde{u}^{(1)} = u^{(1)}$ solves the unchanged right half-space problem. Similar considerations apply to the left half-space problem. For the periodic Cauchy problem it was not important that the coefficient B_1 can become singular. Adding the three solutions, we obtain a smooth solution

$$u = u^{(1)} + u^{(2)} + u^{(3)}$$

of the strip problem in a small time interval $0 \le t \le \delta$. We take the constructed function $u(\,\cdot\,, \delta)$ as initial data at $t = \delta$; in this way, no compatibility problems arise when we restart the process at $t = \delta$. As the proof of Lemma 8.2.1 showed, the basic energy estimate is valid also for singular B_1; hence uniqueness of the solution is ensured. We have

Theorem 8.2.7. *The result of Theorem 8.2.2 remains valid if B_1 becomes singular in the interior $0 < x_1 < 1$ of the strip.*

Cases where B_1 is singular on the boundary. For simplicity we restrict ourselves to right half-space problems (thus $x_1 \ge 0$) and consider coefficient matrices $B_1 = B_1(x, t)$ of the following three forms.

Case 1.

$$B_1 = \begin{bmatrix} \Lambda_-(x,t) & 0 & 0 \\ 0 & x_1^p \Lambda_0(x,t) & 0 \\ 0 & 0 & \Lambda_+(x,t) \end{bmatrix}, \qquad p > 0,$$

where $\Lambda_- \le -\gamma I,\ \Lambda_+ \ge \gamma I,\ \gamma > 0$, and Λ_0 is nonsingular.

Case 2.

$$B_1 = \begin{bmatrix} \Lambda_-(x,t) & 0 & 0 \\ 0 & 0 & 0 \\ 0 & 0 & \Lambda_+(x,t) \end{bmatrix},$$

where $\Lambda_- \le -\gamma I,\ \Lambda_+ \ge \gamma I,\ \gamma > 0$.

Case 3.

$$B_1 = \begin{bmatrix} x_1^p \Lambda_0(x,t) & 0 \\ 0 & \Lambda_+(x,t) \end{bmatrix}, \qquad p > 0,$$

where $\Lambda_+ \ge \gamma I,\ \gamma > 0$.

Here Λ_-, Λ_0, Λ_+ are always real diagonal matrices. In Case 3, the matrices

$$B_1(0, y, t)$$

have no negative eigenvalues and (in contrast to Case 1) the matrix Λ_0 is allowed to become singular.

The above representations are required for $0 \le x_1 \le \tau$, i.e., in some neighborhood of the boundary $x_1 = 0$. The differential equation (8.2.1) is supplemented by an initial condition (8.2.2) and a left boundary condition.

$$u_-(0, y, t) = S(y, t)u_+(0, y, t) + g(y, t).$$

In Case 3 no boundary condition is required.

We want to show

Theorem 8.2.8. *Assume that $B_1(x, t)$ is singular on the boundary $x_1 = 0$ and has either of the three forms described above. If $|S|$ is sufficiently small, the right half-space problem has a unique smooth solution.*

Proof. There are no problems in obtaining the basic energy estimate through integration by parts; compare the proof of Lemma 8.2.1. The boundary term of the left side of (8.2.4) is here

$$\int_0^t \|u_\pm(\cdot, \xi)\|_\Gamma^2 \, d\xi.$$

Also, if we want to estimate first derivatives, we apply the operators

$$D_t, \ D_2, \ \ldots \ , \ D_s$$

to the differential equation and the boundary conditions. In the resulting system we have to express

(8.2.12) $$(D_k B_1) D_1 u, \qquad k = t, \qquad k = 2, \ldots, s \ ,$$

in terms of $D_t u, D_2 u, \ldots, D_s u$ using the differential equation (8.2.1).

In Case 1 we have, for $0 < x_1 \le \tau$,

$$D_1 u = B_1^{-1} \left\{ D_t u - \sum_{j=2}^{s} B_j D_j u - Cu - F \right\}.$$

Observing that

$$(D_k B_1) B_1^{-1}, \qquad k = t, \qquad k = 2, \ldots, s \ ,$$

remains bounded for $x_1 \to 0$, no difficulties arise. In Case 2 the situation is even easier, since the "dangerous" part of the terms (8.2.12) vanishes identically near $x_1 = 0$. In Case 3 we do not need any boundary conditions. Thus we apply all the operators

$$D_t, D_1, D_2, \ldots, D_s$$

to the differential equation (8.2.1) and obtain the a priori estimates for first derivatives. This process can be continued, and one obtains bounds for all the derivatives

$$D_t^{\nu_1} D_2^{\nu_2} \dots D_s^{\nu_s} u$$

in Case 1 and 2; in the third case *all* derivatives can be estimated.

We shall now estimate the x_1-derivatives in the first two cases. To this end, we write the differential equation in the partitioned form

$$\frac{\partial}{\partial t}\begin{pmatrix} u_- \\ u_0 \\ u_+ \end{pmatrix} = \sum_{j=1}^{s} \begin{pmatrix} B_j^{11} & B_j^{12} & B_j^{13} \\ B_j^{21} & B_j^{22} & B_j^{23} \\ B_j^{31} & B_j^{32} & B_j^{33} \end{pmatrix} D_j \begin{pmatrix} u_- \\ u_0 \\ u_+ \end{pmatrix} + Cu + F.$$

Using the differential equation, we can estimate $D_1 u_-$, $D_1 u_+$ and their y, t-derivatives. Therefore we obtain, for $v = D_1 u_0$,

$$\frac{\partial}{\partial t} v_0 = \sum_{j=1}^{s} B_j^{22} D_j v_0 + (D_1 B_1^{22}) v_0 + H.$$

For H and its y, t-derivatives we have already established a bound. The matrices B_j^{22} are symmetric and $B_1^{22}(0, y, t) \equiv 0$ by assumption. Therefore, we can estimate v_0 and all its y, t-derivatives. Then we can obtain bounds for $D_1^2 u_-$, $D_1^2 u_+$ and the y, t-derivatives of these functions. In turn, this gives us estimates for $D_1^2 u_0$. This process can be continued.

To show existence of a solution, we replace B_1 by $B_1 + \sigma I$, $\sigma > 0$. If $\sigma > 0$ is sufficiently small, there exists a solution depending on σ. The estimates are independent of σ, and the desired result follows by a limit argument.

All these considerations apply to a small neighborhood near $x_1 = 0$. With the splitting technique of Section 8.2.4 we obtain the result for the general case.

8.2.6. Results for More General Domains

For simplicity we restrict ourselves to two space dimensions and consider a bounded domain $\Omega \subset \mathbf{R}^2$ with a smooth boundary curve Γ.

Suppose we are given a symmetric hyperbolic system

$$u_t = B_1 u_x + B_2 u_y + Cu + F.$$

Along the boundary, we can write the differential equation in the form

$$u_t = B_n \frac{\partial u}{\partial n} + B_s \frac{\partial u}{\partial s} + \tilde{C} u + F,$$

where $\partial/\partial n$ and $\partial/\partial s$ denote the derivatives in the inward normal and tangential directions, respectively. Here

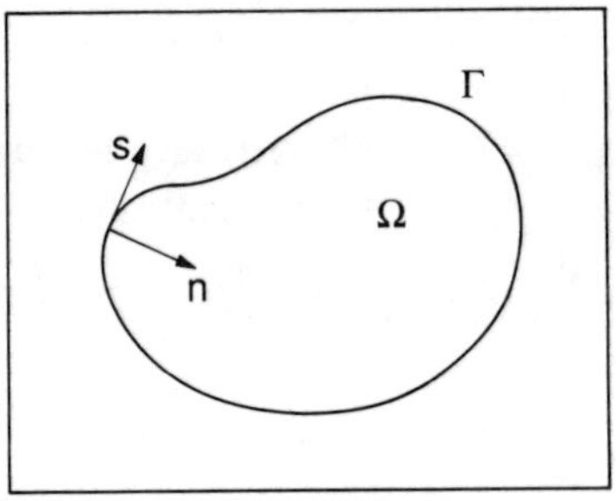

FIGURE 8.2.1. Domain Ω with boundary Γ.

$$B_n = B_1 \cos\alpha + B_2 \sin\alpha,$$

where α denotes the angle between the x-axis and the inward normal to the boundary.

We want to prove

Theorem 8.2.9. *Assume that B_n is nowhere singular on Γ and has exactly r negative eigenvalues. Suppose further that the boundary conditions consist of r linear relations*

$$Lu = g\,, \qquad \text{rank}\, L = r,$$

at each boundary point. The matrix L and the function g may vary smoothly along Γ. If at each boundary point we have an estimate

$$\langle u, B_n u\rangle \geq \delta\,|u|^2 - c\,|Lu|^2\,, \qquad u \in \mathbf{R}^n, \tag{8.2.13}$$

with $\delta > 0$, $c > 0$, then the problem is well-posed.

Proof. Integration by parts gives us the basic energy estimate

$$\begin{aligned}
\frac{d}{dt}\,\|u(\,\cdot\,,t)\|^2 &= 2(u,u_t)\\
&= 2\left\{(u, B_1 u_x) + (u, B_2 u_y) + (u, Cu) + (u, F)\right\}\\
&= -\int_\Gamma \langle u, B_n u\rangle\, ds - (u, B_{1x} u) - (u, B_{2y} u) + 2(u, Cu) + 2(u, F)\\
&\leq c_1\left\{\|u\|^2 + \|F\|^2 + \int_\Gamma |g|^2\, ds\right\} - \delta \int_\Gamma |u|^2\, ds\,.
\end{aligned}$$

Here we have used the pointwise assumption (8.2.13) to treat the boundary integral. The basic estimate follows if we integrate with respect to t.

To show the existence of a solution, we reduce the given problem to Cauchy problems and half-plane problems. To this end, let $\{\phi_j\}$ denote a (finite) partition of unity and let Ω_j denote the support of ϕ_j. We split the data functions $f = f(x, y)$, $F = F(x, y, t)$, $g = g(x, y, t)$ into

$$f_j = \phi_j f, \qquad F_j = \phi_j F, \qquad g_j = \phi_j g.$$

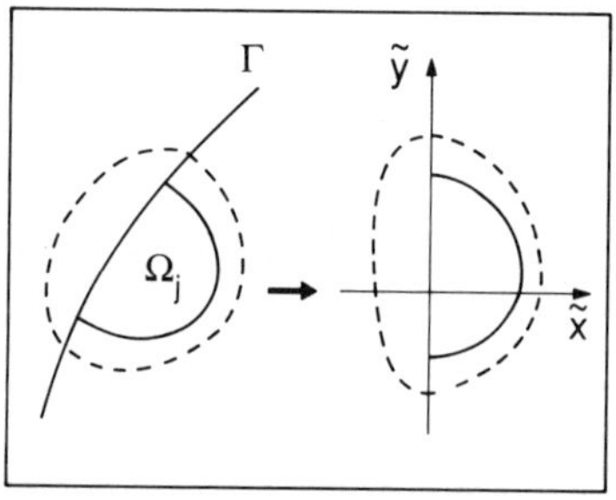

FIGURE 8.2.2. Partition of unity and transformation.

If the support Ω_j of ϕ_j lies in the interior of Ω then $g_j = 0$, and we solve the Cauchy problem with the data f_j, F_j ignoring the boundary conditions. For a sufficiently small time interval, the solution fulfills the homogeneous boundary conditions automatically. If Ω_j contains a part of the boundary Γ, then we transform a neighborhood of Ω_j in such a way that the local piece of Γ is transformed into a piece of the new $\tilde{y}$-axis and the normal directions are transformed to directions parallel to the $\tilde{x}$-axis.

The new problem can be extended to a half-plane problem of the type discussed earlier. Transforming back and adding the solutions, we have solved the problem in a small time interval. Thereafter, we can restart the process.

8.2.7. An Example in Two Dimensions

Consider the system

$$u_t = \begin{pmatrix} 1 & 0 \\ 0 & -1 \end{pmatrix} u_x + \begin{pmatrix} 0 & 1 \\ 1 & 0 \end{pmatrix} u_y + F$$

in the half-plane

$$0 \le x < \infty\,, \qquad -\infty < y < \infty$$

under our usual periodicity assumption with respect to y. The general theory suggests imposing one boundary condition at $x = 0$ which expresses u_- in terms of u_+. We consider a condition of the general form*

$$w(0,y,t) = \alpha(y)v(0,y,t) + g(y,t), \qquad u = \begin{pmatrix} v \\ w \end{pmatrix},$$

with smooth functions α and g. To obtain an energy estimate, we proceed as before:

$$\begin{aligned}\frac{d}{dt}\,\|u(\,\cdot\,,t)\|^2 &= 2(u,u_t)\\ &= 2\left\{(u, \begin{bmatrix} 1 & 0 \\ 0 & -1 \end{bmatrix} u_x) + (u, \begin{bmatrix} 0 & 1 \\ 1 & 0 \end{bmatrix} u_y) + (u,F)\right\}\\ &= \int_0^1 \left[-v^2(0,y,t) + w^2(0,y,t)\right] dy + 2(u,F).\end{aligned}$$

Now we replace w using the boundary condition. It is apparent that the desired estimate can be obtained as long as the boundary coefficient $\alpha(y)$ satisfies

$$|\alpha(y)| \le 1 - \gamma, \qquad \gamma > 0,$$

for all y. This quantifies our previous assumption of "smallness" of the coefficient matrices in the boundary condition. (Compare Lemma 8.2.1, for example.)

Now suppose that

$$|\alpha(y)| \le 1 \text{ for all } y, \qquad \alpha(y_0) = 1 \text{ for some } y_0.$$

Then our estimate breaks down since the above boundary integrand is

$$\begin{aligned}-v^2 + w^2 &= -v^2 + (\alpha v + g)^2\\ &= (-1 + \alpha^2)v^2 + 2\alpha v g + g^2\\ &= 2vg + g^2 \text{ at } y = y_0.\end{aligned}$$

*We could also treat the case $\alpha = \alpha(y,t)$.

We still can derive a basic energy estimate if we first transform to homogeneous boundary conditions: Let

$$\tilde{w}(x,y,t) = w(x,y,t) - e^{-x}g(y,t), \qquad \tilde{v} = v, \qquad \tilde{u} = \begin{pmatrix} \tilde{v} \\ \tilde{w} \end{pmatrix}.$$

The differential equation becomes

$$\tilde{u}_t = \begin{pmatrix} 1 & 0 \\ 0 & -1 \end{pmatrix} \tilde{u}_x + \begin{pmatrix} 0 & 1 \\ 1 & 0 \end{pmatrix} \tilde{u}_y + \tilde{F},$$

$$\tilde{F} = F + e^{-x} \begin{pmatrix} g_y \\ -g + g_x - g_t \end{pmatrix},$$

and the boundary condition reads

$$\tilde{w}(0,y,t) = \alpha(y)\tilde{v}(0,y,t). \tag{8.2.14}$$

Proceeding as before, the boundary term becomes

$$\int_0^1 \left[-\tilde{v}^2(0,y,t) + \tilde{w}^2(0,y,t)\right] dy \le 0,$$

and thus we can estimate $\|\tilde{u}\|$. In this way we do not obtain a bound for $\tilde{u}$ on the boundary $x = 0$, however. Difficulties arise if we want to estimate derivatives. To obtain boundary conditions for $\tilde{u}_y$, for example, we differentiate (8.2.14) with respect to y:

$$\tilde{w}_y = \alpha\tilde{v}_y + \alpha_y\tilde{v} \quad \text{at} \quad x = 0.$$

The equation has the inhomogeneous term $\alpha_y\tilde{v}$. The difficulties are avoided if we first make a unitary transformation of the dependent variable such that the boundary conditions have constant coefficients.

8.3. The Linearized Compressible Euler Equations

We linearize the inviscid compressible N-S equations at a constant state

$$U,\ V,\ W,\ R, \quad P = r(R),$$

with

$$c^2 := \frac{dr}{d\rho}(R) > 0, \quad c > 0.$$

As discussed in Section 2.4.2, we obtain the strongly hyperbolic system

$$
\frac{\partial}{\partial t}\begin{pmatrix} u \\ v \\ w \\ \rho \end{pmatrix} + \begin{pmatrix} U & 0 & 0 & c^2/R \\ 0 & U & 0 & 0 \\ 0 & 0 & U & 0 \\ R & 0 & 0 & U \end{pmatrix} \frac{\partial}{\partial x}\begin{pmatrix} u \\ v \\ w \\ \rho \end{pmatrix}
$$

$$
+ \begin{pmatrix} V & 0 & 0 & 0 \\ 0 & V & 0 & c^2/R \\ 0 & 0 & V & 0 \\ 0 & R & 0 & V \end{pmatrix} \frac{\partial}{\partial y}\begin{pmatrix} u \\ v \\ w \\ \rho \end{pmatrix}
$$

$$
+ \begin{pmatrix} W & 0 & 0 & 0 \\ 0 & W & 0 & 0 \\ 0 & 0 & W & c^2/R \\ 0 & 0 & R & W \end{pmatrix} \frac{\partial}{\partial z}\begin{pmatrix} u \\ v \\ w \\ \rho \end{pmatrix} = 0.
$$

(If the ambient flow $U = U(\mathbf{x}, t)$, etc. is not constant, the linearized equations read as above after neglection of zero-order terms and forcing functions. Compare Section 6.4.) The above system can be symmetrized by introduction of the scaled density $\tilde{\rho} = c\rho/R$; one obtains the symmetric hyperbolic system

$$
\frac{\partial}{\partial t}\begin{pmatrix} u \\ v \\ w \\ \tilde{\rho} \end{pmatrix} = \{B_1 \frac{\partial}{\partial x} + B_2 \frac{\partial}{\partial y} + B_3 \frac{\partial}{\partial z}\} \begin{pmatrix} u \\ v \\ w \\ \tilde{\rho} \end{pmatrix},
$$

$$
B_1 = -\begin{pmatrix} U & 0 & 0 & c \\ 0 & U & 0 & 0 \\ 0 & 0 & U & 0 \\ c & 0 & 0 & U \end{pmatrix}, \quad \text{etc.}
$$

We want to discuss this symmetric system in the half-space

$$
0 \le x < \infty, \quad -\infty < y, z < \infty
$$

under the usual periodicity requirement with respect to y and z, and we want to derive boundary conditions at $x = 0$ which lead to a well-posed problem. (The matrix B_1 is not diagonal; however, after a simple transformation the results of Section 8.2 apply.)

The eigenvalues of B_1 are

$$
\kappa_{1,2} = -U, \quad \kappa_{3,4} = -U \pm c.
$$

Following the general theory, we will specify as many boundary conditions at $x = 0$ as there are negative eigenvalues of B_1. Accordingly, we distinguish a number of different cases.

Case 1. **$U < 0$ *(outflow of the ambient flow)*.** In the supersonic or transonic case $|U| \geq c$, the matrix B_1 has no negative eigenvalue and no boundary condition needs to be specified. In the subsonic case $|U| < c$, there is exactly one negative eigenvalue, $\kappa_4 = -U - c$. Thus one boundary condition has to be specified. (This is necessary but not sufficient for well-posedness.) In order to obtain an energy estimate, we consider, as above, the expression

$$\frac{d}{dt} \| \begin{pmatrix} \mathbf{u} \\ \tilde{\rho} \end{pmatrix} \|^2$$

and use integration by parts with respect to x, y, and z. Assuming functions which vanish at $x = +\infty$, the critical boundary term to be estimated is

$$-\left(\begin{pmatrix} \mathbf{u} \\ \tilde{\rho} \end{pmatrix}, B_1 \begin{pmatrix} \mathbf{u} \\ \tilde{\rho} \end{pmatrix}\right)_{\Gamma_0} = \{U(u^2 + v^2 + w^2 + \tilde{\rho}^2) + 2cu\tilde{\rho}\}_{\Gamma_0}, \tag{8.3.1}$$

where the index Γ_0 indicates integration over y and z at the boundary $x = 0$. We try to obtain an estimate analogous to (8.2.5) and specify a boundary condition

$$\tilde{\rho} = \alpha u + g_0 \quad \text{at} \quad x = 0,$$

for example. The boundary term (8.3.1) becomes

$$\{U(v^2 + w^2 + \tilde{\rho}^2) + (U + 2c\alpha)u^2 + 2cug_0\}_{\Gamma_0}.$$

Since $U < 0$, it suffices to require that

$$U + 2c\alpha < 0.$$

This can be viewed as a requirement for the coefficient α in the boundary condition. With such a boundary condition the desired energy estimate is obtained, including an estimate of

$$\| \begin{pmatrix} \mathbf{u} \\ \tilde{\rho} \end{pmatrix} \|_{\Gamma_0}$$

on the boundary $x = 0$. We can also require that

$$u = 0 \quad \text{at} \quad x = 0, \tag{8.3.2}$$

or we can derive sufficient conditions for more general boundary conditions

$$\tilde{\rho} = \alpha u + \beta v + \gamma w + g_0.$$

Once the basic energy estimate can be obtained with boundary conditions of the above form, we can also bound all derivatives of the solution up to the boundary. (Note that in the homogeneous case (8.3.2), the *coefficient* in the boundary condition is constant.)

Case 2. $\boldsymbol{U > 0}$ ***(inflow of the ambient flow).*** In the supersonic case

$$U > c > 0,$$

the matrix B_1 has four negative eigenvalues, and consequently all four variables u, v, w, $\tilde{\rho}$ need to be specified. The boundary term (8.3.1) being given, there are no difficulties in obtaining the desired estimates. In the subsonic case

$$c > U > 0,$$

there are three negative eigenvalues, and one eigenvalue is positive. We specify three conditions, for example:

$$v = g_1, \quad w = g_2, \quad \tilde{\rho} = \alpha u + g_0.$$

Substituting the expression for $\tilde{\rho}$ into (8.3.1), we see that an energy estimate can be obtained if

$$U(1 + \alpha^2) + 2c\alpha < 0.$$

A possible choice is $\alpha = -1$. In the transonic case

$$c = U > 0,$$

the matrix B_1 has three negative eigenvalues, and the eigenvalue zero. As before, we need three boundary conditions to obtain an energy estimate. If the state

$$U, \ V, \ W, \ R, \quad P = r(R),$$

at which we linearize is constant, then the conditions of Theorem 8.2.8 are automatically fulfilled, and we obtain a well-posed problem. (After diagonalizing B_1, we have Case 2 of the theorem.) If the state is not constant, we need conditions as required in Case 1 or 3 of Theorem 8.2.8.

Case 3. $\boldsymbol{U = 0}$ ***(a wall).*** In this case B_1 is singular and has exactly one negative eigenvalue; hence we have to specify one boundary condition. For example, if the boundary condition takes the form

$$\tilde{\rho} = \alpha u + g_0 \quad \text{at} \quad x = 0,$$

then the integrand in (8.3.1) becomes

$$2c\alpha u^2 + 2cug_0,$$

and we obtain an energy estimate if $\alpha < 0$. Also, we can use the conditions

$$u = 0 \quad \text{at} \quad x = 0, \tag{8.3.3}$$

or

$$\tilde{\rho} = 0 \quad \text{at} \quad x = 0.$$

(Note that $\tilde{\rho}$ is a density *correction*; thus also the latter condition might be physically reasonable.) In all cases we obtain the basic energy estimate, i.e., an estimate for

$$\| \begin{pmatrix} \mathbf{u} \\ \tilde{\rho} \end{pmatrix} \|^2.$$

If the conditions of Theorem 8.2.8 are not met, there are difficulties in estimating derivatives up to the boundary.

8.4. The Laplace Transform Method for Hyperbolic Systems

Consider a strongly hyperbolic system

$$u_t = Au_x + Bu_y + F(x,y,t), \quad u = \begin{pmatrix} u_1(x,y,t) \\ \vdots \\ u_n(x,y,t) \end{pmatrix},$$

in the half-space

$$0 \le x < \infty, \quad -\infty < y < \infty, \quad t \ge 0,$$

under an initial condition at $t = 0$ and boundary conditions at $x = 0$. The energy method employed in Section 8.2 provides a tool to derive sufficient conditions for well-posedness, but if the energy method does not apply then one does not know whether or not the problem is well-posed.

The technique described here can be used – in principle – to *decide* the question of well-posedness. Employing Fourier transformation in y (or Fourier expansion in the periodic case) and Laplace transformation in t, one obtains a family of ordinary boundary value problems on the half-line $0 \le x < \infty$, which can be discussed explicitly. In the case of one space dimension, we have proceeded in a similar way in Section 7.4; however, the presence of the dual variable ω to the space variable y brings a new element into the discussion here.

There are no difficulties in generalizing from two space dimensions to three dimensions or more. A generalization to variable coefficients $A = A(x,y,t)$, etc. is more involved. Roughly speaking, it turns out that the frozen-coefficient principle is valid; i.e., if all frozen-coefficient problems are well-posed then the variable-coefficient problem is also well-posed. We will formulate a corresponding result without proof.

8.4.1. An Example

To illustrate the general theory, we will discuss a specific example, namely the symmetric hyperbolic system

$$u_t = \begin{pmatrix} -1 & 0 \\ 0 & 1 \end{pmatrix} u_x + \begin{pmatrix} 0 & 1 \\ 1 & 0 \end{pmatrix} u_y + F(x, y, t), \quad u = \begin{pmatrix} u_1(x, y, t) \\ u_2(x, y, t) \end{pmatrix}, \tag{8.4.1}$$

in the half-space

$$0 \le x < \infty, \quad -\infty < y < \infty, \quad t \ge 0,$$

with initial condition

$$u(x, y, 0) = f(x, y). \tag{8.4.2}$$

At $x = 0$ we prescribe a boundary condition

$$u_1(0, y, t) = a u_2(0, y, t) + g(y, t), \quad y \in \mathbf{R}, \quad t \ge 0, \tag{8.4.3}$$

where $a \in \mathbf{C}$ is a given constant. In contrast to the previous section, we do not request periodicity in y here. Instead, u, f, g, F, and all derivatives of these functions are assumed to lie in L_2 for each $t \ge 0$. Also, we assume the data to be compatible at $t = 0$, and require that F and g have compact support in $0 < t < \infty$.

A family of eigenvalue problems. Let $F = 0$, $g = 0$. Substitution of a function

$$u(x, y, t) = e^{\alpha(st + i\omega y)} \phi(\alpha x), \quad \phi = \begin{pmatrix} \phi_1 \\ \phi_2 \end{pmatrix}, \tag{8.4.4}$$

with parameters

$$\alpha > 0, \quad s \in \mathbf{C}, \quad \omega \in \mathbf{R}$$

into the differential equation (8.4.1) leads to

$$s\phi(x) = \begin{pmatrix} -1 & 0 \\ 0 & 1 \end{pmatrix} \phi_x(x) + i\omega \begin{pmatrix} 0 & 1 \\ 1 & 0 \end{pmatrix} \phi(x), \quad 0 \le x < \infty. \tag{8.4.5a}$$

If we add the boundary conditions

$$\phi_1(0) = a\phi_2(0), \quad \|\phi\|^2 = \int_0^\infty |\phi(x)|^2 \, dx < \infty, \tag{8.4.5b}$$

we have obtained a family of eigenvalue problems.

Definition 1. Let $\omega \in \mathbf{R}$ be fixed. A number $s \in \mathbf{C}$ is called an eigenvalue of (8.4.5) for the parameter ω if there is a nontrivial solution $\phi \in C^\infty$ of (8.4.5).

It is not difficult to show

Lemma 8.4.1. *If for some* $\omega_0 \in \mathbf{R}$ *there is an eigenvalue* $s_0 \in \mathbf{C}$ *with* $\operatorname{Re} s_0 > 0$, *then the initial-boundary value problem is ill-posed in any sense.*

Proof. The parameter $\alpha > 0$ in (8.4.4) can be chosen arbitrarily large, and consequently there is no bound on the rate of exponential growth in time. (Formally, the function u given in (8.4.4) is not a *solution* since $u(x, \cdot, t) \notin L_2$. However, we can multiply on the Fourier transform side by a function $\psi(\omega)$ having a peak at ω_0 and integrate w.r.t ω. See the proof of Theorem 2.2.2.)

Discussion of the eigenvalue condition. The previous lemma naturally leads to the following question: What is the requirement on the parameter $a \in \mathbf{C}$ appearing in the boundary condition such that there is no eigenvalue $s \in \mathbf{C}$ with $\operatorname{Re} s > 0$? This requirement on a is *necessary* for well-posedness, as follows from the previous lemma. An explicit discussion shows

Lemma 8.4.2. *The following conditions are equivalent:*

(i) *The number* $a \in \mathbf{C}$ *satisfies*

$$a \in \mathbf{R} \quad \text{or} \quad |a| \leq 1.$$

(ii) *For all* $\omega \in \mathbf{R}$, *all eigenvalues* $s \in \mathbf{C}$ *of* (8.4.5) *satisfy* $\operatorname{Re} s \leq 0$.

Proof. Equation (8.4.5a) is equivalent to

$$\phi_x(x) = \begin{pmatrix} -s & i\omega \\ -i\omega & s \end{pmatrix} \phi(x), \quad 0 \leq x < \infty.$$

If $\operatorname{Re} s > 0$ then the above matrix has the two eigenvalues

$$\kappa_1 = -\sqrt{s^2 + \omega^2}, \quad \kappa_2 = -\kappa_1$$

with

$$\operatorname{Re} \kappa_1 < 0 < \operatorname{Re} \kappa_2.$$

The eigenvector to κ_1 reads

$$\Phi_1 = \begin{pmatrix} s + \sqrt{s^2 + \omega^2} \\ i\omega \end{pmatrix},$$

and therefore the general L_2-solution of (8.4.5a) is given by

$$\phi(x) = \sigma \exp(-\sqrt{s^2 + \omega^2}\, x) \begin{pmatrix} s + \sqrt{s^2 + \omega^2} \\ i\omega \end{pmatrix}, \quad \sigma \in \mathbf{C}.$$

The boundary condition at $x = 0$ requires that

$$\sigma(s + \sqrt{s^2 + \omega^2}) = \sigma a i \omega,$$

and consequently the number $s \in \mathbf{C}$, $\operatorname{Re} s > 0$, is an eigenvalue to the parameter ω if and only if

$$s + \sqrt{s^2 + \omega^2} = a i \omega. \tag{8.4.6}$$

For $\omega = 0$, equation (8.4.6) has no solution s with $\operatorname{Re} s > 0$. Thus let $\omega \neq 0$; dividing (8.4.6) by $|\omega|$, we obtain:

$$z + \sqrt{z^2 + 1} = ai\frac{\omega}{|\omega|}, \quad z = \frac{s}{|\omega|}, \quad \operatorname{Re} z > 0.$$

The function

$$\psi(z) = z + \sqrt{z^2 + 1}, \quad \operatorname{Re} z \geq 0,$$

maps the line $\operatorname{Re} z = 0$ onto the boundary of the domain Ω in Figure 8.4.1, and $\operatorname{Re} z > 0$ is mapped onto

$$\Omega = \{\psi \in \mathbf{C} : \operatorname{Re} \psi > 0 \quad \text{and} \quad |\psi| > 1\}.$$

Therefore the equation

$$\psi(z) = ai\frac{\omega}{|\omega|}$$

has *no* solution z with $\operatorname{Re} z > 0$ if and only if

$$\pm ia \notin \Omega,$$

i.e., $a \in \mathbf{R}$ or $|a| \leq 1$.

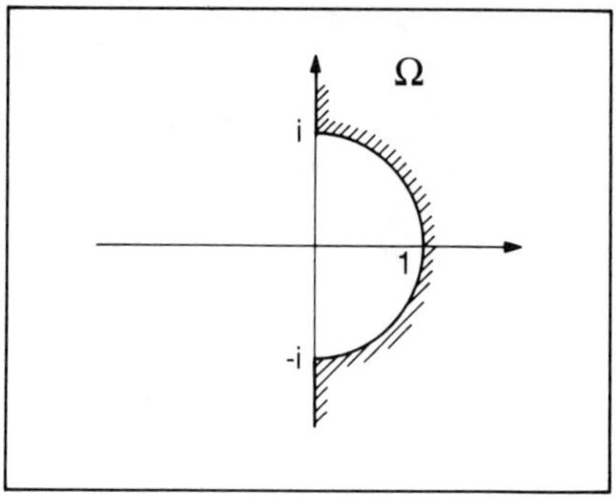

FIGURE 8.4.1. Image of $\psi(z)$, $\operatorname{Re} z \geq 0$.

8.4.2. Example Continued: Formal Solution

If there is no eigenvalue $s \in \mathbf{C}$ with $\operatorname{Re} s > 0$, then we can solve the problem *formally*. (This means, the algebraic equations and ordinary boundary value problems that appear below can be solved, but the question of convergence of integrals requires further study. As a rule, if the data are sufficiently "well-behaved", a formal solution is a genuine C^∞-solution. However, the problem is well-posed only if one can derive the proper estimates of the solution in terms of the data.) To illustrate this, we consider (8.4.1)–(8.4.3) with $f \equiv 0$, $F \equiv 0$ and first apply Fourier transformation in y. Thus we write

$$u(x,y,t) = \frac{1}{\sqrt{2\pi}} \int_{-\infty}^{\infty} e^{i\omega y}\hat{u}(x,\omega,t)\,d\omega,$$

$$g(y,t) = \frac{1}{\sqrt{2\pi}} \int_{-\infty}^{\infty} e^{i\omega y}\hat{g}(\omega,t)\,d\omega,$$

and obtain:

$$\hat{u}_t = \begin{pmatrix} -1 & 0 \\ 0 & 1 \end{pmatrix} \hat{u}_x + i\omega \begin{pmatrix} 0 & 1 \\ 1 & 0 \end{pmatrix} \hat{u} \quad \text{at } (x,\omega,t),$$

$$\hat{u}(x,\omega,0) = 0,$$

$$\hat{u}_1(0,\omega,t) = a\hat{u}_2(0,\omega,t) + \hat{g}(\omega,t).$$

Thus, for any fixed ω, we have to solve an initial-boundary value problem in one space dimension $0 \le x < \infty$. Denoting the Laplace transforms in t by

$$v(x,\omega,s) = \int_0^\infty e^{-st}\hat{u}(x,\omega,t)\,dt,$$

$$h(\omega,s) = \int_0^\infty e^{-st}\hat{g}(\omega,t)\,dt, \quad \operatorname{Re} s > 0,$$

the equations transform to

$$sv = \begin{pmatrix} -1 & 0 \\ 0 & 1 \end{pmatrix} v_x + i\omega \begin{pmatrix} 0 & 1 \\ 1 & 0 \end{pmatrix} v \quad \text{at } (x,\omega,s),$$

$$v_1(0,\omega,s) = av_2(0,\omega,s) + h(\omega,s),$$

and the requirement

$$\int_0^\infty |v(x,\omega,s)|^2\,dx < \infty$$

is naturally added as a boundary condition at $x = \infty$. The equations for $v(x,\omega,s)$ can be solved for each pair

$$(\omega, s), \quad \omega \in \mathbf{R}, \quad \operatorname{Re} s > 0,$$

separately. Using the eigenvalue- and eigenvector-results of the proof of Lemma 8.4.2, we find

$$v(x, \omega, s) = \sigma \exp(-\sqrt{s^2 + \omega^2}\, x) \begin{pmatrix} s + \sqrt{s^2 + \omega^2} \\ i\omega \end{pmatrix}.$$

Here $\sigma = \sigma(\omega, s)$ is determined by the boundary condition at $x = 0$,

$$\sigma(s + \sqrt{s^2 + \omega^2}) = \sigma a i \omega + h(\omega, s).$$

By assumption, there is no eigenvalue s with $\operatorname{Re} s > 0$, and therefore we can solve for σ,

$$\sigma = \frac{h(\omega, s)}{s + \sqrt{s^2 + \omega^2} - ai\omega}, \quad \operatorname{Re} s > 0.$$

Thus we can compute the function $v(x, \omega, s)$, and if there are no convergence problems, we can transform back and obtain the solution $u(x, y, t)$ of the initial-boundary value problem.

8.4.3. Example Continued: Estimate of the Solution on the Boundary

To investigate well-posedness, we try to estimate the solution in terms of the data. The most critical question is, as it turns out, whether one can estimate the solution on the boundary $x = 0$. To study this question, consider the example (8.4.1)–(8.4.3) with $f \equiv 0$ and assume that

(8.4.7) $$a \in \mathbf{R} \quad \text{or} \quad |a| \leq 1.$$

Thus there is no eigenvalue s with $\operatorname{Re} s > 0$, and we can solve the problem formally. Using the notations of Section 8.4.2, we have that

$$\begin{aligned} |v(0, \omega, s)|^2 &= |\sigma|^2 \{|s - \kappa_1|^2 + \omega^2\} \\ &= |h(\omega, s)|^2 \frac{|s - \kappa_1|^2 + \omega^2}{|s - \kappa_1 - ai\omega|^2}, \quad \kappa_1 = -\sqrt{s^2 + \omega^2}. \end{aligned}$$

Let us denote the factor multiplying $|h(\omega, s)|^2$ by $\rho^2(\omega, s)$; thus, for $\omega \neq 0$,

(8.4.8) $$\rho^2(\omega, s) = \frac{|z + \sqrt{z^2 + 1}|^2 + 1}{\left|z + \sqrt{z^2 + 1} - ai\omega/|\omega|\right|^2}, \quad z = \frac{s}{|\omega|}, \quad \operatorname{Re} s > 0,$$

and $\rho^2(0, s) = 1$ for $\omega = 0$, $\operatorname{Re} s > 0$. There are two fundamentally different cases:

Case 1. There is a constant c_0 with $\rho^2(\omega, s) \leq c_0$ for all $\omega \in \mathbf{R}$, $\operatorname{Re} s > 0$.

Case 2. The function $\rho^2(\omega, s)$ is not bounded in the domain $\omega \in \mathbf{R}$, $\operatorname{Re} s > 0$. For the example under consideration, it is not difficult to show that Case 1 prevails if and only if $|a| < 1$, and therefore Case 2 prevails if and only if

$$|a| = 1 \quad \text{or} \quad \big(a \in \mathbf{R} \ \text{ and } \ |a| > 1\big).$$

(Recall the assumption (8.4.7).) Concerning the question of well-posedness, the two cases of behaviour of $\rho^2(\omega, s)$ – bounded or unbounded – lead to different answers. Only in the first case can one derive estimates of the solution on the boundary which express strong well-posedness in the generalized sense. (We will prove even more below, namely strong well-posedness in the usual sense.) In Case 2 such estimates are not possible, and the problem is not strongly well-posed in the generalized sense. Nevertheless, the explicit solution formula can be used in Case 2 also, and one can derive weaker estimates. We concentrate here on Case 1 and prove

Lemma 8.4.3. *Suppose that $\rho^2(\omega, s) \leq c_0$ for all $\omega \in \mathbf{R}$ and all $s \in \mathbf{C}$ with $\operatorname{Re} s > 0$. Then, for each time T, there is a constant K_T independent of the boundary data g with*

$$\int_0^T \|u(0, \cdot, t)\|_\Gamma^2 dt \leq K_T \int_0^T \|g(\cdot, t)\|_\Gamma^2 dt.$$

The constants K_T can be chosen uniformly for $0 \leq T \leq T_0$; i.e., $K_T = K(T_0)$. In the above formula

$$\|g(\cdot, t)\|_\Gamma = \Big\{\int_{-\infty}^{\infty} |g(y, t)|^2 \, dy\Big\}^{1/2}$$

denotes the L_2-norm of the boundary data at time t, and similarly $\|u(0, \cdot, t)\|_\Gamma$ is the L_2-norm of the solution on the boundary $x = 0$ at time t.

Proof. First we recall the relation

$$v(0, \omega, s) = \int_0^\infty e^{-st} \hat{u}(0, \omega, t) \, dt, \quad \operatorname{Re} s > 0,$$

between $\hat{u}$ and its Laplace transform v. For fixed $\omega \in \mathbf{R}$ and $\eta > 0$ we apply (7.4.9) and find that

$$\int_0^T |\hat{u}(0,\omega,t)|^2\,dt \le \frac{e^{2\eta T}}{2\pi}\int_{-\infty}^{\infty} |v(0,\omega,\eta+i\xi)|^2\,d\xi$$
$$\le \frac{e^{2\eta T}}{2\pi}c_0\int_{-\infty}^{\infty} |h(\omega,\eta+i\xi)|^2\,d\xi =: I$$

The second estimate clearly follows from our assumption that $\rho^2(\omega,s) \le c_0$. Since $h(\omega,\cdot)$ is the Laplace transform of $\hat{g}(\omega,\cdot)$, equation (7.4.8) yields

$$I = e^{2\eta T}c_0\int_0^{\infty} e^{-2\eta t}|\hat{g}(\omega,t)|^2\,dt \le e^{2\eta T}c_0\int_0^{\infty} |\hat{g}(\omega,t)|^2\,dt.$$

The resulting inequality

$$\int_0^T |\hat{u}(0,\omega,t)|^2\,dt \le e^{2\eta T}c_0\int_0^{\infty} |\hat{g}(\omega,t)|^2\,dt$$

is integrated w.r.t. ω. After interchanging the order of integration, an application of Parseval's relation yields

$$\int_0^T\int_{-\infty}^{\infty} |u(0,y,t)|^2\,dy\,dt \le e^{2\eta T}c_0\int_0^{\infty}\int_{-\infty}^{\infty} |g(y,t)|^2\,dy\,dt.$$

By an argument as given in the proof of Lemma 7.4.6, we can change the data $g(y,t)$ for $t > T$ without affecting the solution $u(x,y,t)$ for $t \le T$, and the result follows.

8.4.4. Example Continued: Strong Well-Posedness in the Generalized Sense

We consider now the inhomogeneous equation

$$u_t = Au_x + Bu_y + F, \quad A = \begin{pmatrix} -1 & 0 \\ 0 & 1 \end{pmatrix}, \quad B = \begin{pmatrix} 0 & 1 \\ 1 & 0 \end{pmatrix},$$

with homogeneous initial data $u(x,y,0) = 0$ and the same boundary condition as before,

$$u_1(0,y,t) = au_2(0,y,t) + g(y,t).$$

After Fourier and Laplace transformation, we obtain:

(8.4.9)
$$sv = Av_x + i\omega Bv + H(x,\omega,s), \qquad s = i\xi + \eta, \quad \eta > 0,$$
$$v_1(0,\omega,s) = av_2(0,\omega,s) + h(\omega,s), \qquad \|v(\cdot,\omega,s)\| < \infty.$$

We assume that there is no eigenvalue s with $\operatorname{Re} s > 0$. Then, for every fixed s, ω, the ordinary boundary value problem (8.4.9) for $v(\cdot,\omega,s)$ can be solved, and we obtain a formal solution of the given initial-boundary value problem by inverting the Laplace and Fourier transform. The formal solution is a genuine solution if one can derive sufficiently strong estimates.

To derive estimates, we take the scalar product of (8.4.9) with v, integrate over $0 \le x < \infty$ and consider the real part:

$$\begin{aligned} \operatorname{Re} s(v,v) &= \operatorname{Re}(v, Av_x) + \operatorname{Re} i\omega\,(v, Bv) + \operatorname{Re}(v,H) \\ &= -\frac{1}{2}\,\langle v(0,\omega,s), Av(0,\omega,s)\rangle + \operatorname{Re}(v,H). \end{aligned} \tag{8.4.10}$$

First, assume that $|a| < 1$. Using the boundary condition for v, we obtain that

$$\begin{aligned} -\frac{1}{2}\,\langle v(0,\omega,s), Av(0,\omega,s)\rangle &= \frac{1}{2}\left(|v_1(0,\omega,s)|^2 - |v_2(0,\omega,s)|^2\right) \\ &\le -\frac{1}{4}\left(1-|a|^2\right)|v_2(0,\omega,s)|^2 + K_0\,|h(\omega,s)|^2 \\ &\le -\frac{1}{8}\left(1-|a|^2\right)|v(0,\omega,s)|^2 + K_1\,|h(\omega,s)|^2. \end{aligned}$$

Therefore, (8.4.10) yields that

$$\begin{aligned} &\frac{\eta}{2}\,\|v(\cdot,\omega,s)\|^2 + \frac{1}{8}\left(1-|a|^2\right)|v(0,\omega,s)|^2 \\ &\quad\le \frac{1}{2\eta}\,\|H(\cdot,\omega,s)\|^2 + K_1\,|h(\omega,s)|^2, \qquad \eta = \operatorname{Re} s. \end{aligned} \tag{8.4.11}$$

Here K_1 is independent of ω, s, h, and H.

By Parseval's relation, we obtain estimates of u in terms of the data F and g; see Section 7.4, in particular (7.4.11). The estimates imply that the given problem is strongly well-posed in the generalized sense. The formal solution is indeed a genuine solution.

Second, assume that $|a| = 1$ and $g \equiv 0$. In this case

$$-\frac{1}{2}\,\langle v(0,\omega,s), Av(0,\omega,s)\rangle \;\le\; 0,$$

and we obtain, instead of (8.4.11),

$$\eta\,\|v(\cdot,\omega,s)\|^2 \le \frac{1}{\eta}\,\|H(\cdot,\omega,s)\|^2.$$

Thus the problem is weakly well-posed (see Definition 4 in Section 7.3); we do not obtain an estimate of the solution u on the boundary $x = 0$.

Let us summarize the results for the example. The method of Laplace transformation shows that the condition $|a| < 1$ is necessary and sufficient for strong well-posedness in the generalized sense. In Section 8.2.7 we have treated the example already by the energy method, and could derive energy estimates under the same condition, namely $|a| < 1$. This might be misleading, because the method of Laplace transformation has generally a wider range of applicability. We have seen this already in the parabolic case. For hyperbolic systems, the energy method requires *symmetric* hyperbolicity, whereas the Laplace transform method can be applied to *strictly* hyperbolic systems, too. On the other hand, an application of the energy method is preferred, whenever possible, since intricate algebraic discussions are usually avoided.

8.4.5. Generalizations

Consider a strongly hyperbolic initial-boundary value problem

(8.4.12)

$$\begin{aligned} u_t &= Au_x + Bu_y + F(x,y,t), && 0 \le x < \infty, \quad y \in \mathbf{R}, \quad t \ge 0, \\ u(x,y,0) &= f(x,y), && 0 \le x < \infty, \quad y \in \mathbf{R}, \\ u^I(0,y,t) &= Ru^{II}(0,y,t) + g(y,t), && y \in \mathbf{R}, \quad t \ge 0. \end{aligned}$$

Here A, B, and R are constant matrices and

$$A = \begin{pmatrix} \Lambda_1 & 0 \\ O & \Lambda_2 \end{pmatrix}, \quad \Lambda_1 = \operatorname{diag}(\lambda_1, \ldots, \lambda_r), \quad \Lambda_2 = \operatorname{diag}(\lambda_{r+1}, \ldots, \lambda_n),$$

$$\lambda_1 < 0, \ldots, \lambda_r < 0, \quad \lambda_{r+1} > 0, \ldots, \lambda_n > 0.$$

The unknown vector-function u is partitioned accordingly,

$$u = \begin{pmatrix} u^I \\ u^{II} \end{pmatrix}, \quad u^I \text{ of length } r.$$

If $f \equiv 0$, then we apply Fourier transformation in y, Laplace transformation in t, and obtain, with the same notations as before,

(8.4.13)

$$\begin{aligned} sv &= Av_x + i\omega Bv + H(x,\omega,s), && 0 \le x < \infty, \quad \omega \in \mathbf{R}, \quad \operatorname{Re} s > 0, \\ v^I(0,\omega,s) &= Rv^{II}(0,\omega,s) + h(\omega,s), && \|v(\cdot,\omega,s)\| < \infty. \end{aligned}$$

The eigenvalue condition. Suppose that for some $\omega \in \mathbf{R}$ and some $s \in \mathbf{C}$ with $\operatorname{Re} s > 0$ there is a nontrivial solution $\phi(x)$ of

$$\begin{aligned} s\phi(x) &= A\phi_x(x) + i\omega B\phi(x), \\ \phi^I(0) &= R\phi^{II}(0), \quad \int_0^\infty |\phi(x)|^2 dx = \|\phi\|^2 < \infty. \end{aligned}$$

Then the problem (8.4.12) is ill-posed, because the functions

$$u(x, y, t) = e^{\alpha(st+i\omega y)}\phi(\alpha x), \quad \alpha > 0,$$

are solutions for $F \equiv 0$, $g \equiv 0$, and hence there is no bound on the exponential growth rate in time. To discuss the above eigenvalue condition, we note that the general solution of

$$\phi_x = A^{-1}(sI - i\omega B)\phi, \quad \|\phi\| < \infty,$$

can be written in the form

$$\phi(x) = \sum_{\text{Re}\,\kappa_j<0} \sigma_j \Phi_j(x) \exp(\kappa_j x), \quad \sigma_j \in \mathbf{C}. \tag{8.4.14}$$

Here $\kappa_j = \kappa_j(\omega, s)$ are the roots of

$$\det\Big(A^{-1}(sI - i\omega B) - \kappa I\Big) = 0 \tag{8.4.15}$$

and $\Phi_j(x) = \Phi_j(x, \omega, s)$ are the corresponding vector-functions; these are polynomials in x of degree $\le m_j - 1$ where m_j is the algebraic multiplicity of κ_j. For a further discussion, the result of the next lemma is important. Recall first the assumption of hyperbolicity, which implies that all matrices

$$i\omega_1 A + i\omega_2 B, \quad \omega_1, \omega_2 \in \mathbf{R},$$

have only purely imaginary eigenvalues.

Lemma 8.4.4. *Let $\omega \in \mathbf{R}$ and $s \in \mathbf{C}$ with $\text{Re}\, s > 0$ be given. The characteristic equation* (8.4.15) *has exactly r roots κ_j with negative real parts and exactly $n - r$ roots with positive real parts. (The roots are counted according to their algebraic multiplicity.)*

Proof. There is no purely imaginary root $\kappa_1 = i\omega_1$ since otherwise

$$0 = \det(sI - i\omega B - i\omega_1 A), \quad \text{Re}\, s > 0,$$

in contradiction to the hyperbolicity assumption. Now fix ω and let

$$s \in \mathbf{R}, \quad s \gg 1.$$

The roots μ_j of

$$\det\Big(A^{-1}(I - \frac{i\omega}{s}B) - \mu I\Big) = 0, \quad \mu = \frac{\kappa}{s},$$

are

$$\mu_j = \frac{1}{\lambda_j} + O(\frac{1}{s^{1/n}}), \quad j = 1, \ldots, n.$$

Hence the assertion follows, since the κ_j depend continuously on s, and no κ_j can cross the imaginary axis for $\text{Re}\, s > 0$.

According to the lemma, the sum in (8.4.14) consists of r terms. In particular, we obtain at $x = 0$ a representation of ϕ,

$$\phi(0, \omega, s) = \Phi(\omega, s)\sigma, \quad \sigma \in \mathbf{C}^r,$$

with a matrix $\Phi(\omega, s)$ of size $n \times r$. Introducing this representation into the boundary condition, we obtain

Lemma 8.4.5. *A number $s \in \mathbf{C}$ with* $\text{Re}\, s > 0$ *is an eigenvalue for a given ω if and only if the $r \times r$ matrix*

$$\Phi^I(\omega, s) - R\Phi^{II}(\omega, s)$$

is singular. Here Φ^I consists of the first r and Φ^{II} of the last $n - r$ rows of Φ.

Formal solution and estimates. Suppose there are no eigenvalues s with positive real parts; i.e., the matrices $\Phi^I - R\Phi^{II}$ are all nonsingular. As in the example, we can solve (8.4.13) and obtain a *formal* solution of (8.4.12) (with $f \equiv 0$) by inverting the Fourier-Laplace transform. To derive estimates, consider first the case $F \equiv 0$, $f \equiv 0$. We have $H \equiv 0$ in (8.4.13), and therefore

$$v(0, \omega, s) = \Phi(\omega, s)\sigma \quad \text{where} \quad \{\Phi^I(\omega, s) - R\Phi^{II}(\omega, s)\}\sigma = h(\omega, s).$$

One can derive an estimate

$$\int_0^t \|u(0, \cdot, \tau)\|_\Gamma^2 \, d\tau \le K_T \int_0^t \|g(\cdot, \tau)\|_\Gamma^2 \, d\tau$$

for $0 \le t \le T$, with K_T independent of g, if and only if the function

$$\left|\Phi(\omega, s)\{\Phi^I(\omega, s) - R\,\Phi^{II}(\omega, s)\}^{-1}\right|, \quad \omega \in \mathbf{R}, \quad \text{Re}\, s > 0, \tag{8.4.16}$$

is bounded.

Inhomogeneous differential equations and the symmetrizer. Now consider the case with an inhomogeneous term F in the differential equation (8.4.12). As before, we assume that the function (8.4.16) is defined and bounded. The simple process of Section 8.4.4 for deriving estimates of u in terms of g and F

does not work, in general. However, if the system (8.4.12) is *strictly* hyperbolic, one can construct a symmetrizer, which allows us to derive the estimates.

Let us define

$$s' = \frac{s}{\sqrt{|s|^2 + \omega^2}}, \qquad \omega' = \frac{\omega}{\sqrt{|s|^2 + \omega^2}}, \qquad s = i\xi + \eta, \quad \eta > 0.$$

Then (8.4.13) can be written as follows:

$$s'v = \frac{1}{\sqrt{|s|^2 + \omega^2}} Av_x \ + \ i\omega' Bv \ + \ \frac{1}{\sqrt{|s|^2 + \omega^2}} H. \tag{8.4.17}$$

One can show

Lemma 8.4.6. *Suppose that* (8.4.12) *is strictly hyperbolic and the function* (8.4.16) *is defined and bounded. For every fixed* $\eta > 0$ *there is a matrix function* $\hat{S} = \hat{S}(\omega', \xi')$ *which is defined for the arguments* $-1 \le \omega', \ \xi' \le 1$ *and which has the following properties:*

1) *The function* $\hat{S}(\omega', \xi')$ *is* C^∞*-smooth.*
2) *For all arguments* ω', ξ' *the matrix* $\hat{S}A$ *is Hermitian.*
3) *If* $y \in \mathbf{C}^n$, $h \in \mathbf{C}^r$ *are vectors with* $y^I = Ry^{II} + h$, *then*

$$\langle y, \hat{S}Ay \rangle \ge \delta_1 |y|^2 - C|h|^2.$$

Here $\delta_1 > 0$ *and* C *are constants independent of* y, h, ω', $\xi', \eta > 0$.

4) *With a constant* δ_2 *independent of* ω', ξ', $\eta > 0$, *it holds that*

$$\hat{S}(s'I - i\omega' B) + (s'I - i\omega' B)^* \hat{S}^* \ge \delta_2 \eta' I, \quad \eta' = \operatorname{Re} s'.$$

For a proof we refer the reader to Kreiss (1970). Let us assume here that the symmetrizer is constructed. Then one can derive estimates of the solution $v = v(x, \omega, s)$ of (8.4.13) as follows: We multiply (8.4.17) with $\hat{S}$, take the inner product with v, and integrate over $0 \le x < \infty$. As a result we obtain

(8.4.18)
$$-\frac{1}{\sqrt{|s|^2 + \omega^2}} \Big(v, \hat{S}Av_x\Big) + \Big(v, \hat{S}(s'I - i\omega' B)v\Big) = \frac{1}{\sqrt{|s|^2 + \omega^2}} \Big(v, \hat{S}H\Big).$$

Also, using integration by parts and the properties 2. and 3. of $\hat{S}$, we find that

$$2(v, \hat{S}Av_x) = \langle v, \hat{S}Av \rangle \Big|_{x=0} \ge \delta_1 |v(0, \omega, s)|^2 - C|h(\omega, s)|^2.$$

Now we take the real part in (8.4.18), use property 4. of $\hat{S}$, and multiply with $2(|s|^2 + \omega^2)^{1/2}$. Then it follows that

$$\delta_1 |v(0, \omega, s)|^2 + \delta_2 \eta \|v(\cdot, \omega, s)\|^2 \le \text{const} \, \|v\| \, \|H\| + C|h(\omega, s)|^2.$$

This gives us the desired estimate for the transformed function v. By Parseval's relation we obtain

Theorem 8.4.7. *Suppose that the system* $u_t = Au_x + Bu_y$ *is strictly hyperbolic. Then the initial-boundary value problem* (8.4.12) (*with* $f \equiv 0$) *is strongly well-posed in the generalized sense if and only if the function* (8.4.16) *is defined and bounded.*

Remark. The functions $\Phi_j(x) = \Phi_j(x, \omega, s)$ in (8.4.14), and consequently the matrix $\Phi(\omega, s)$, are not unique. However, if the Φ_j are suitably normalized, then – for each fixed ω – the matrix function $\Phi(\omega, s)$ can be extended continuously to $\mathrm{Re}\, s = 0$ in such a way that rank $\Phi(\omega, s) = r$ for all $\omega \in \mathbf{R}$ and all $s \in \mathbf{C}$ with $\mathrm{Re}\, s \geq 0$. With such an extension, the expression (8.4.16) is defined and bounded if and only if the matrix

$$\Phi^I(\omega, s) - R\Phi^{II}(\omega, s) \tag{8.4.19}$$

is nonsingular for all ω and all $\mathrm{Re}\, s \geq 0$. This observation is helpful for a discussion of the boundedness of (8.4.16). If a matrix (8.4.19) is singular at ω_0, $s_0 = i\xi_0$, then s_0 is called a generalized eigenvalue for ω_0. The corresponding roots

$$\kappa_j(\omega_0, i\xi_0) = \lim_{\eta \to 0+} \kappa_j(\omega_0, i\xi_0 + \eta)$$

of (8.4.15) might be purely imaginary, and the corresponding nontrivial solution $\phi(x)$ is generally not in L_2.

The example (8.4.1), which we discussed in the previous sections of this chapter, is even strongly well-posed in the sense of Definition 1, Section 7.3, provided that $|a| < 1$. Thus, under the assumption that $|a| < 1$, we can also estimate the solution for inhomogeneous initial data. This has been shown already in Section 8.2.7 by a direct energy estimate. For the general case discussed here, one cannot show an energy estimate directly. Nevertheless, one can prove

Theorem 8.4.8. *Suppose that the system* $u_t = Au_x + Bu_y$ *is strictly hyperbolic. If the initial-boundary value problem* (8.4.12) *is strongly well-posed in the generalized sense, then it is also strongly well-posed in the sense of Definition* 1, *Section* 7.3.

A proof of this result is contained in Rauch (1972 a,b, 1973). The arguments are rather involved.

All results generalize to half-space problems in more than two space dimensions. We Fourier transform in all space variables tangential to the boundary $x = 0$, and obtain a dual variable ω which is a vector instead of a scalar. The other considerations remain the same.

Variable coefficients. Consider a strictly hyperbolic problem (8.4.12) with smoothly varying coefficients

$$A = A(x, y, t), \quad B = B(x, y, t), \quad R = R(y, t).$$

We assume A to be diagonal* for all arguments, and nonsingular at $x = 0$. The (constant) block-structure of $A(0, y, t)$ with $\Lambda^I < 0$, $\Lambda^{II} > 0$, and the corresponding partitioning of u into u^I, u^{II} are the same as in the case of constant coefficients. In particular, $R = R(y, t)$ has constant size $r \times (n - r)$. For each boundary point $(0, y_0, t_0)$ we obtain a problem with frozen coefficients,

$$\begin{aligned} u_t &= A(0, y_0, t_0)u_x + B(0, y_0, t_0)u_y + F(x, y, t), \\ u(x, y, 0) &= 0, \\ u^I(0, y, t) &= R(y_0, t_0)u^{II}(0, y, t) + g(y, t). \end{aligned} \tag{8.4.20}$$

For these problems we can decide the question of strong well-posedness by Theorems 8.4.7 and 8.4.8. For the variable-coefficient problem we state

Theorem 8.4.9. *The variable-coefficient problem is strongly well-posed if the differential equation is strictly hyperbolic and all frozen-coefficient problems* (8.4.20) *are strongly well-posed.*

Using the theory of pseudodifferential operators (see, for example, Eskin (1973) or Nierenberg (1970)) and the symmetrizer $\hat{S}$, we want to outline the proof of strong well-posedness in the generalized sense. As before, the main problem is the estimate. Let $\eta > 0$ be fixed, and transform the differential equation to

$$\tilde{u}_t = (P - \eta I)\tilde{u} + \tilde{F}, \quad P = A\frac{\partial}{\partial x} + B\frac{\partial}{\partial y}, \quad \tilde{u} = e^{-\eta t}u.$$

We construct the symmetrizer $\hat{S} = \hat{S}(x, y, t, \omega', s')$ *pointwise*: at each point (x, y, t) the properties 1., 2., and 4. of Lemma 8.4.6 hold, and at each boundary point $(x = 0, y, t)$ property 3. also holds. (One can construct $\hat{S}$ as a smooth function of all arguments.) The symmetrizer $\hat{S}$ is used as the symbol of a

*If A is not diagonal, we apply a transformation $\tilde{u} = S^{-1}u$, $S = S(x, y, t)$, such that $S^{-1}AS$ is diagonal. This introduces lower-order terms, but these do not influence well-posedness.

pseudodifferential operator S. For sufficiently large η we obtain the desired estimate from

$$\begin{aligned}\delta_1 \int_{-\infty}^{\infty} \|\tilde{u}(\cdot, y, t)\|_\Gamma^2 \, dt + \delta_2 \eta \int_{-\infty}^{\infty} \|\tilde{u}(\cdot, \cdot, t)\|^2 \, dt - C \int_{-\infty}^{\infty} \|\tilde{g}(\cdot, y, t)\|_\Gamma^2 \, dt \\ \leq 2 \operatorname{Re} \int_{-\infty}^{\infty} \Big(\tilde{u}, S\big(\tilde{u}_t - (P - \eta I)\tilde{u}\big)\Big) \, dt \\ = 2 \operatorname{Re} \int_{-\infty}^{\infty} \big(\tilde{u}, S\tilde{F}\big) \, dt.\end{aligned}$$

Here the data functions are set to zero for negative t. The above estimate yields strong well-posedness in the generalized sense. Again, the results Rauch (1972 a,b, 1973) show that the problem is even strongly well-posed in the sense of Definition 1, Section 7.3.

Strictly hyperbolic systems are not common in applications. For example, the linearized compressible Euler equations in 3D are not strictly hyperbolic. In Agranovich (1972) it is shown that the previous theorem remains valid for strongly hyperbolic systems if there is a transformation $\hat{S}(x, y, t, \omega_1, \omega_2)$ which is analytic in ω_1, ω_2 and smooth in x, y, t, and which diagonalizes the symbol

$$\hat{P} = \hat{P}(x, y, t, i\omega_1, i\omega_2) = i\Big(A(x, y, t)\omega_1 + B(x, y, t)\omega_2\Big).$$

In other words, for all arguments we have

$$\hat{S}^{-1}\hat{P}\hat{S} = i \operatorname{diag}(\lambda_1, \ldots, \lambda_n), \quad \lambda_j \text{ real.}$$

A corresponding result holds for strongly hyperbolic half-space problems in any number of space dimensions. One can show that a transformation $\hat{S}$ with the above properties exists for the linearized compressible Euler equations; therefore, the theory outlined above does apply. In Oliger and Sundström (1978) a rather complete discussion of boundary conditions for the Euler equations is given. The existence of the transformation $\hat{S}$ seems to be typical for hyperbolic systems which appear in applications.

Let us note further that problems in more general domains (with smooth boundaries) can be reduced to half-space problems and pure Cauchy problems by partition of unity arguments and local transformations. For this reason the study of half-space problems is most important. The main restriction we made throughout Section 8.4 is the assumption of a *noncharacteristic* boundary. General results for a characteristic boundary are not known. However, Case 2 of Theorem 8.2.7 has been treated in Majda and Osher (1975) in the Fourier-Laplace transform framework.

8.5. Remarks on Mixed Systems and Nonlinear Problems

Mixed systems. As in the one-dimensional case, we can extend our results to mixed hyperbolic-parabolic systems. Initial-boundary value problems for parabolic systems were discussed in Section 8.1, for hyperbolic systems in Section 8.2. Now couple two such systems – in the same way as in the Cauchy problem – by adding lower-order terms. (See Section 6.3.1 or Theorem 2.5.1.) The initial-boundary value problem for the coupled system is well-posed with the same boundary conditions provided that the boundary is *noncharacteristic* for the hyperbolic part. If the boundary is characteristic and the conditions of Section 8.2.5 are met, then the problem is still well-posed if we use a Dirichlet condition for the parabolic part.

Much more general boundary conditions can be discussed in the framework of Laplace-Fourier transforms. Here we refer to Strikwerda (1977).

For the treatment of the N-S equations, one is also interested in the behavior of solutions near the boundary when the viscosity converges to zero. We refer to Gustafsson and Sundström (1978) and to Michelson (1988).

Nonlinear problems. No new difficulties arise for short-time existence of solutions of quasilinear problems. Consider, for example, a half-plane problem for a quasilinear symmetric hyperbolic system of the form

$$\begin{aligned} u_t = &\Big\{A(x,y,t) + \epsilon A_1(x,y,t,u)\Big\}u_x + \Big\{B(x,y,t) + \epsilon B_1(x,y,t,u)\Big\}u_y \\ &+ C(x,y,t)u + F(x,y,t) + \epsilon F_1(x,y,t,u), \\ &\qquad 0 \le x < \infty, \quad -\infty < y < \infty, \quad t \ge 0, \end{aligned}$$

with initial and boundary conditions

$$\begin{aligned} &u(x,y,0) = f(x,y), \\ &u^I(0,y,t) = \Big\{S_0(y,t) + \epsilon S_1(y,t,u)\Big\}u^{II}(0,y,t) + g(y,t) + \epsilon g_1(y,t,u). \end{aligned}$$

We assume that all coefficients and data are C^∞-smooth and that A, A_1, B, B_1 are Hermitian. The matrix $A(0,y,t)$ is assumed to be nonsingular and diagonal. The variables $u = (u^I, u^{II})$ are partitioned according to the signs of the diagonal entries of this matrix; see Section 8.2.1. One can show a result of the following type:

Theorem. *Assume that the linear problem which is obtained for $\epsilon = 0$ is strongly well-posed. Then – for sufficiently small ϵ – the nonlinear problem has* *a unique smooth solution in a time interval* $0 \leq t \leq T_\epsilon$. *Here* $T_\epsilon \to \infty$ *for* $\epsilon \to 0$. *As in Section* 5.3, *one can also derive asymptotic expansions of the solution.*

Clearly, nonlinear systems usually do not occur directly in the form described above. Nevertheless, local existence results can frequently be derived by a theorem of the above type. To illustrate the principle, we consider a quasilinear system

$$\begin{aligned} u_t &= P(x,t,u,\partial/\partial x)u + F(x,t,u), \\ u(x,0) &= f(x), \end{aligned}$$

together with quasilinear boundary conditions. Now x can be a vector of any dimension. Then we solve the linear problem

$$\begin{aligned} w_t &= P(x,t,f,\partial/\partial x)w + F(x,t,f), \\ w(x,0) &= f(x); \end{aligned}$$

i.e., we evaluate the nonlinear coefficients at the initial data. The same process applies to the boundary conditions for w. We substitute

$$u = \epsilon\tilde{u} + w$$

into the original equations. For the new variables $\tilde{u}$ we obtain a system which has the above structure in a time interval of length $0 \leq t \leq K\epsilon$; i.e., the nonlinear terms are multiplied by ϵ. Results of the type sketched above are valid for parabolic, for hyperbolic, and for mixed hyperbolic-parabolic problems. The systems need not be symmetric. Estimates can be derived which yield strong well-posedness in the generalized sense. We refer to Strikwerda (1977) and to Michelson (1988).

Notes on Chapter 8

Energy estimates for initial-boundary value problems have a long history. Some references are Friedrichs (1958), Ladyzhenskaya (1984), and Lions (1961). For systems $\partial u/\partial t = P(\partial/\partial x)u$ with constant coefficients, Hersch (1963) derived necessary and sufficient conditions for weak well-posedness of half-space problems. As in Section 8.5, after Fourier and Laplace transform, he obtained an eigenvalue problem. The half-space problem is weakly well-posed if and only if the eigenvalues s satisfy Re $s \leq$ const. For a homogenous operator $P(\partial/\partial x)$ this condition becomes: There are no eigenvalues s with Re $s > 0$.

For parabolic systems the corresponding condition leads to weakly well-posed problems in the sense of Chapter 2, even for problems with variable coefficients.

This was proved by Agranovich and Vishik (1964), Eidelman (1964), and Solonnikov (1965) using generalizations of Levy's parametrix.

For hyperbolic first-order systems with variable coefficients we required the stronger condition: there are no eigenvalues or generalized eigenvalues s with Re $s \geq 0$. A first result in this direction is due to Agmon (1962). The general theory for first-order systems was developed by Kreiss (1970) with extensions by Ralston (1971), Rauch (1972a,b, 1973) and Agranovich (1971, 1972). For single higher-order equations the theory is due to Sakamoto (1970, 1982). There is no general theory for equations with variable coefficients which satisfy the weaker eigenvalue condition (Lopatinskii condition) pointwise. For more details we refer to Eskin (1983) and Volevich and Gindinkin (1980).

For difference approximations a similar theory has been developed. See Kreiss (1968), Osher (1972), and Gustafsson, Kreiss, and Sundström (1972) for results in one dimension, and Michelson (1983, 1987) for results in more dimensions.

9 The Incompressible Navier-Stokes Equations: The Spatially Periodic Case

In the next two chapters we apply the general theory to the Navier-Stokes equations. For references and deeper insight we refer the reader to von Wahl (1985), Temam (1984, 1983), and Ladyzhenskaya (1969).

9.1. The Spatially Periodic Case in Two Dimensions

Without further mentioning, all functions in this section are assumed to be real, C^∞-smooth and 2π-periodic in x and y. We consider the viscous incompressible N-S equations

(9.1.1) $$u_t + uu_x + vu_y + p_x = \nu\Delta u + F(x, y, t),$$

(9.1.2) $$v_t + uv_x + vv_y + p_y = \nu\Delta v + G(x, y, t), \quad \nu > 0,$$

$$u_x + v_y = 0,$$

and give initial conditions

(9.1.3) $$u(x, y, 0) = f(x, y), \quad v(x, y, 0) = g(x, y).$$

Clearly, consistency of (9.1.2) and (9.1.3) requires us to assume that

(9.1.4) $$f_x + g_y = 0.$$

Notations. We define an inner product and a norm for scalar functions by

$$(u,v) = \int_0^{2\pi}\int_0^{2\pi} u(x,y)v(x,y)\,dx\,dy, \quad \|u\|^2 = (u,u).$$

With 1 we denote the function which is identically one.

Without changing the terms p_x, p_y in (9.1.1), an arbitrary function $p_0(t)$ could be added to the pressure. Thus p is not uniquely determined by the incompressible equations. We use the normalization*

(9.1.5) $$(1, p(\cdot,t)) = 0 \quad \text{for all} \quad t.$$

Our aim is to prove

Theorem 9.1.1. *Assume the compatibility condition* (9.1.4) *to be satisfied. The spatially periodic Cauchy problem* (9.1.1)–(9.1.3) *with side-condition* (9.1.5) *has a unique solution* (u,v,p). *The solution is* C^∞*-smooth and exists for all time* $t \geq 0$.

9.1.1. Outline of the Proof; Vorticity Equation

A simplifying assumption. Suppose that (u,v,p) is a solution. The first momentum equation (9.1.1) gives us, for the spatial average of u,

$$\begin{aligned}\frac{d}{dt}(1,u) = (1,u_t) &= -(1,uu_x) - (1,vu_y) + (1,F)\\ &= (u_x+v_y,u) + (1,F) = (1,F).\end{aligned}$$

Therefore

$$(1,u(\cdot,t)) = (1,f) + \int_0^t (1,F(\cdot,\tau))\,d\tau$$

is known in terms of the data. If we introduce

$$\tilde{u} = u - \frac{1}{4\pi^2}(1,u), \quad \tilde{v} = v - \frac{1}{4\pi^2}(1,v),$$

then $(1,\tilde{u}) = (1,\tilde{v}) = 0$ and

$$\tilde{u}_t + \tilde{u}\tilde{u}_x + \tilde{v}\tilde{u}_y + p_x = \nu\Delta\tilde{u} + F - \frac{1}{4\pi^2}(1,u)\tilde{u}_x - \frac{1}{4\pi^2}(1,v)\tilde{u}_y,$$

*The ambiguity disappears for the compressible equations and the normalization (9.1.5) can be justified by a limit process.

and similarly for $\tilde{v}$. Since the coefficients $(1,u)$, $(1,v)$ can be considered as known, the study of the system for $(\tilde{u},\tilde{v})$ is very similar to that of the system (9.1.1)–(9.1.3) with

$$F = G = 0, \quad (1,f) = (1,g) = 0. \tag{9.1.6}$$

Henceforth we assume (9.1.6), for simplicity, and leave the extensions to the reader. The average flow quantities vanish:

$$(1,u) = (1,v) = 0 \quad \text{for all} \quad t.$$

Vorticity. The expression

$$\xi = v_x - u_y$$

is called the vorticity. Differentiation of the first momentum equation (9.1.1) with respect to y, of the second with respect to x, and subtraction gives us the *vorticity equation*

$$\xi_t + u\xi_x + v\xi_y = \nu\Delta\xi. \tag{9.1.7}$$

Clearly, ξ satisfies the initial condition

$$\xi(\cdot,0) = g_x - f_y. \tag{9.1.8}$$

If u, v were known, then the vorticity could be determined by the linear parabolic problem (9.1.7), (9.1.8).

Conversely, as we will show, if ξ is known, then at each time t the velocities are uniquely determined by the inhomogeneous Cauchy-Riemann system

$$u_x + v_y = 0, \quad v_x - u_y = \xi,$$

with side-condition

$$(1,u) = (1,v) = 0.$$

This suggests the iteration

$$\begin{aligned}
&\xi_t^{n+1} + u^n\xi_x^{n+1} + v^n\xi_y^{n+1} = \nu\Delta\xi^{n+1}, && u^0 \equiv v^0 \equiv 0,\\
&\xi^{n+1}(\cdot,0) = g_x - f_y, &&\\
&u_x^{n+1} + v_y^{n+1} = 0, && v_x^{n+1} - u_y^{n+1} = \xi^{n+1},\\
&(1,u^{n+1}) = (1,v^{n+1}) = 0, && n = 0,1\ldots.
\end{aligned} \tag{9.1.9}$$

We will use this iteration to prove *global* existence. Once u, v are determined, the pressure p can be obtained from

$$\Delta p + 2(v_x u_y - u_x v_y) = 0, \quad (1, p) = 0.$$

9.1.2. Some Results for Elliptic Equations

We need some elementary results about elliptic equations, which we prove by Fourier expansions. Recall that all functions are assumed to be C^∞-smooth and 2π-periodic in x and y.

Lemma 9.1.2. *If $F = F(x, y)$ is a function with $(1, F) = 0$, then*

$$\Delta w = F, \quad (1, w) = 0$$

has a unique solution w. There is a constant K independent of F with

$$\|w\|_{H^2}^2 \leq K\|F\|^2.$$

Thus we "gain" two derivatives.

Proof. We can write

$$F(x, y) = \sum_{\mathbf{k}\neq 0} \hat{F}(\mathbf{k})e^{i\langle\mathbf{k},\mathbf{x}\rangle}, \quad w(x, y) = \sum_{\mathbf{k}\neq 0} \hat{w}(\mathbf{k})e^{i\langle\mathbf{k},\mathbf{x}\rangle},$$

$$\mathbf{k} = (k_1, k_2), \quad \mathbf{x} = (x, y), \quad \langle\mathbf{k}, \mathbf{x}\rangle = k_1 x + k_2 y.$$

The sums extend over all wave-vectors $\mathbf{k} \neq 0$ with integer components k_1, k_2. In terms of Fourier coefficients, the equation $\Delta w = F$ is equivalent to

$$-(k_1^2 + k_2^2)\hat{w}(\mathbf{k}) = \hat{F}(\mathbf{k}).$$

This shows existence, uniqueness, and the estimate.

Lemma 9.1.3. *Let F, G denote functions with*

$$G_x - F_y = 0, \quad (1, F) = (1, G) = 0.$$

There is a unique solution φ of

$$\varphi_x = F, \quad \varphi_y = G, \quad (1, \varphi) = 0.$$

This function φ is also the unique solution of

$$\Delta\varphi = F_x + G_y, \quad (1, \varphi) = 0.$$

Proof. As in the proof of the previous lemma, we write F, G, and φ as Fourier series. By assumption,

$$k_1\hat{G}(\mathbf{k}) - k_2\hat{F}(\mathbf{k}) = 0 \quad \text{for all} \quad \mathbf{k}. \tag{9.1.10}$$

The equations $\varphi_x = F$, $\varphi_y = G$ are equivalent to

$$ik_1\hat{\varphi}(\mathbf{k}) = \hat{F}(\mathbf{k}), \quad ik_2\hat{\varphi}(\mathbf{k}) = \hat{G}(\mathbf{k}). \tag{9.1.11}$$

If $\hat{\varphi}(\mathbf{k})$ satisfies these equations then

$$\hat{\varphi}(\mathbf{k}) = -i\frac{k_1\hat{F}(\mathbf{k}) + k_2\hat{G}(\mathbf{k})}{k_1^2 + k_2^2}. \tag{9.1.12}$$

Conversely, if $\hat{\varphi}(\mathbf{k})$ is defined by (9.1.12), then (9.1.11) follows from (9.1.10). The remaining statements are shown in the same way.

Lemma 9.1.4. *Suppose that* $(1, F) = (1, G) = 0$. *Then the system*

$$u_x + v_y = F, \quad v_x - u_y = G, \quad (1, u) = (1, v) = 0$$

has a unique solution (u, v). *For each* $j = 1, 2, \ldots,$ *there is a constant* K_j *independent of* F, G *with*

$$\|u\|_{H^j}^2 + \|v\|_{H^j}^2 \;\le\; K_j(|F|_{H^{j-1}}^2 + |G|_{H^{j-1}}^2).$$

If F *and* G *dependent smoothly on* t, *then so do* u *and* v.

Proof. If the system has a solution then

$$\Delta u = F_x - G_y, \quad \Delta v = F_y + G_x,$$

and therefore

$$\hat{u}(\mathbf{k}) = -i\frac{k_1\hat{F}(\mathbf{k}) - k_2\hat{G}(\mathbf{k})}{k_1^2 + k_2^2},$$

$$\hat{v}(\mathbf{k}) = -i\frac{k_2\hat{F}(\mathbf{k}) + k_1\hat{G}(\mathbf{k})}{k_1^2 + k_2^2}.$$

Conversely, if we define u and v by the above formulae, the system is solved and the estimates follow from

$$|\hat{u}(\mathbf{k})|^2 + |\hat{v}(\mathbf{k})|^2 \le \frac{4}{|\mathbf{k}|^2}\{|\hat{F}(\mathbf{k})|^2 + |\hat{G}(\mathbf{k})|^2\}.$$

Smooth dependence on t holds, because we can differentiate the Fourier series termwise.

9.1.3. Vorticity Formulation of the Navier-Stokes Equations

The Navier-Stokes equations for the primitive variables u, v, p are equivalent to the vorticity equation together with the inhomogeneous Cauchy-Riemann system for u, v and the elliptic pressure equation

$$\Delta p + J = 0, \quad J = 2(v_x u_y - u_x v_y), \quad (1, p) = 0. \tag{9.1.13}$$

More precisely,

Theorem 9.1.5. *Suppose that $f_x + g_y = 0$, $(1, f) = (1, g) = 0$. If u, v, p solve*

$$\begin{aligned} u_t + uu_x + vu_y + p_x &= \nu \Delta u, \\ v_t + uv_x + vv_y + p_y &= \nu \Delta v, \qquad (1, p) = 0 \\ u_x + v_y &= 0, \\ u(x, y, 0) = f(x, y), &\quad v(x, y, 0) = g(x, y), \end{aligned} \tag{9.1.14}$$

and ξ is defined as $\xi = v_x - u_y$, then

$$\begin{aligned} \xi_t + u\xi_x + v\xi_y &= \nu \Delta \xi, \\ \xi(x, y, 0) &= g_x(x, y) - f_y(x, y), \\ u_x + v_y = 0, \quad v_x - u_y &= \xi, \quad (1, u) = (1, v) = 0, \end{aligned} \tag{9.1.15}$$

and p satisfies (9.1.13). *Conversely, if u, v, ξ solve* (9.1.15), *and p is defined as the solution of* (9.1.13), *then u, v, p solve the first set of equations.*

Proof. First suppose that u, v, p solve the equations (9.1.14). The system (9.1.15) has been derived in Section 9.1.1. Also, differentiation of the first momentum equation with respect to x and of the second with respect to y yields, for the pressure,

$$\Delta p + J = 0, \quad J = u_x u_x + 2v_x u_y + v_y v_y = 2(v_x u_y - u_x v_y).$$

This proves the first part of the equivalence.

Conversely, assume that u, v, ξ solve the equations (9.1.15). We define F and G by

$$\begin{aligned} u_t + uu_x + vu_y - \nu \Delta u &= F, \\ v_t + uv_x + vv_y - \nu \Delta v &= G, \end{aligned}$$

and obtain

$$G_x - F_y = \xi_t + u\xi_x + v\xi_y + (u_x + v_y)\xi - \nu \Delta \xi = 0, \quad (1, F) = (1, G) = 0.$$

By Lemma 9.1.3 there is a unique solution φ of

$$\varphi_x = F, \quad \varphi_y = G, \quad (1, \varphi) = 0.$$

Clearly, if we set $p = -\varphi$, the momentum equations are satisfied by u, v, p. Also,

$$\Delta\varphi = F_x + G_y = -2(v_x u_y - u_x v_y),$$

and therefore $-\varphi = p$ is the solution of (9.1.13). The initial conditions $u = f$, $v = g$ at $t = 0$ follow from the *uniqueness* of the solution u, v of

$$u_x + v_y = 0, \quad v_x - u_y = \xi, \quad (1, u) = (1, v) = 0,$$

and the initial condition for ξ, $\xi = g_x - f_y$ at $t = 0$.

9.1.4. Uniqueness and Existence

To begin with, we show *uniqueness*; we assume that u_j, v_j, p_j, $j = 1, 2$, solve (9.1.14). Subtraction yields for the differences $\bar{u} = u_1 - u_2$, etc.

$$\begin{aligned} \bar{u}_t + u_1\bar{u}_x + v_1\bar{u}_y + \bar{u}u_{2x} + \bar{v}u_{2y} + \bar{p}_x &= \nu\Delta\bar{u}, \\ \bar{v}_t + u_1\bar{v}_x + v_1\bar{v}_y + \bar{u}v_{2x} + \bar{v}v_{2y} + \bar{p}_y &= \nu\Delta\bar{v}, \\ \bar{u}_x + \bar{v}_y &= 0. \end{aligned}$$

Thus, using integration by parts,

$$\begin{aligned} \frac{1}{2}\frac{d}{dt}\{\|\bar{u}\|^2 + \|\bar{v}\|^2\} &= (\bar{u}, \bar{u}_t) + (\bar{v}, \bar{v}_t) \\ &\le K_T\{\|\bar{u}\|^2 + \|\bar{v}\|^2\}, \quad 0 \le t \le T, \end{aligned}$$

and therefore the initial conditions imply that

$$\bar{u} = \bar{v} = 0.$$

Furthermore, we obtain from (9.1.13) that $\Delta\bar{p} = 0$, $(1, \bar{p}) = 0$, and thus $\bar{p} = 0$. This shows the *uniqueness* of the solution of (9.1.14).

To prove *existence* we use the iteration (9.1.9) starting with $u^0 \equiv v^0 \equiv 0$. For every fixed n we have linear equations with C^∞-coefficients; thus u^n, v^n, ξ^n exist for all time. (See Lemma 9.1.4 for the determination of u^{n+1}, v^{n+1} in terms of ξ^{n+1}.) We first show the following estimate for the sequence ξ^n.

Lemma 9.1.6. *Let $h = g_x - f_y$ denote the initial data of ξ^n. There is a constant K depending on ν and $\|h\|_{H^1}$, but independent of n and t with*

$$\|\xi^n(\cdot, t)\|_{H^1}^2 \le K, \quad n = 1, 2, \ldots, \quad t \ge 0. \qquad (9.1.16)$$

Proof. a) We write $\xi = \xi^{n+1}, u = u^n, v = v^n$; thus

$$\xi_t = -u\xi_x - v\xi_y + \nu\Delta\xi, \quad \xi(\cdot, 0) = h. \tag{9.1.17}$$

Therefore,

$$\begin{aligned}\frac{1}{2}\frac{d}{dt}\|\xi(\cdot,t)\|^2 &= (\xi, \xi_t) \\ &= -(\xi, u\xi_x) - (\xi, v\xi_y) - \nu\{\|\xi_x\|^2 + \|\xi_y\|^2\}.\end{aligned}$$

Here

$$-(\xi, u\xi_x) = (\xi_x u, \xi) + (\xi u_x, \xi),$$

and thus

$$-(\xi, u\xi_x) = \frac{1}{2}(\xi u_x, \xi).$$

A similar computation for $(\xi, v\xi_y)$ and the relation $u_x + v_y = 0$ yield

$$\frac{d}{dt}\|\xi\|^2 = -2\nu\{\|\xi_x\|^2 + \|\xi_y\|^2\} \leq 0.$$

Hence we have, for all n,

$$\|\xi^n(\cdot,t)\|^2 \leq \|h\|^2, \quad 2\nu\int_0^t |\xi|_{H^1}^2\, d\tau \leq \|h\|^2. \tag{9.1.18}$$

b) Integration by parts and (9.1.17) imply that

$$\begin{aligned}\frac{1}{2}\frac{d}{dt}\|\xi_x\|^2 &= (\xi_x, \xi_{xt}) = -(\xi_{xx}, \xi_t) \\ &\leq (\xi_{xx}, u\xi_x) + (\xi_{xx}, v\xi_y) - \nu\|\xi_{xx}\|^2.\end{aligned}$$

Here

$$\begin{aligned}(\xi_{xx}, u\xi_x) &\leq |u|_\infty \|\xi_{xx}\|\,\|\xi_x\|, \\ (\xi_{xx}, v\xi_y) &\leq |v|_\infty \|\xi_{xx}\|\,\|\xi_y\|.\end{aligned}$$

Adding a similar estimate for $(d/dt)\|\xi_y\|^2$, one finds that

$$\begin{aligned}\frac{d}{dt}|\xi(\cdot,t)|_{H^1}^2 &= \frac{d}{dt}\{\|\xi_x\|^2 + \|\xi_y\|^2\} \\ &\leq 2|\mathbf{u}|_\infty\{\|\xi_{xx}\| + \|\xi_{yy}\|\}\{\|\xi_x\| + \|\xi_y\|\} \\ &\quad - 2\nu\{\|\xi_{xx}\|^2 + \|\xi_{yy}\|^2\}, \quad |\mathbf{u}(\cdot,t)|_\infty = \max_{x,y}|\mathbf{u}(x,y,t)|,\end{aligned}$$

and therefore

$$\frac{d}{dt}|\xi|_{H^1}^2 \le c_1|\mathbf{u}|_\infty^2|\xi|_{H^1}^2, \quad c_1 = c_1(\nu). \tag{9.1.19}$$

c) A Sobolev inequality (Theorem A.3.6) and Lemma 9.1.4 give us that

$$|\mathbf{u}^n|_\infty^2 \le c_2\|\mathbf{u}^n\|_{H^2}^2 \le c_3|\xi^n|_{H^1}^2.$$

d) To summarize, if we add

$$\frac{d}{dt}\|\xi^{n+1}\|^2 \le 0$$

and (9.1.19), we obtain that

$$\frac{d}{dt}\|\xi^{n+1}\|_{H^1}^2 \le c|\xi^n|_{H^1}^2\|\xi^{n+1}\|_{H^1}^2.$$

Here c is a constant depending only on ν. By (9.1.18),

$$\alpha := c\int_0^t |\xi^n|_{H^1}^2\, d\tau \le \text{const}\,\|h\|^2.$$

Therefore,

$$\|\xi^{n+1}(\cdot,t)\|_{H^1}^2 \le e^\alpha\,\|h\|_{H^1}^2$$

is bounded independently of n and t.

Next we show that *all* space derivatives of ξ^n can be estimated independently of n in any given time interval. It is convenient to note that *mixed* space derivatives can always be estimated by *pure* space derivatives. More precisely, for all functions w,

$$|w|_{H^j}^2 \le \gamma_j\{\|D_1^j w\|^2 + \|D_2^j w\|^2\} \tag{9.1.20}$$

where γ_j is a numerical constant independent of w. This can be shown by Fourier expansion. (See formula (A.3.2) in Appendix 3.)

Lemma 9.1.7. *Suppose that the sequence ξ^n is defined by the iteration* (9.1.9), *and $T > 0$ is arbitrary but fixed. For each $j = 1, 2, \ldots$, there exists K_j with*

$$\|\xi^n(\cdot,t)\|_{H^j} \le K_j, \quad 0 \le t \le T.$$

The constant K_j depends on ν, T, and $\|h\|_{H^j}$, but is independent of n.

Proof. We abbreviate, as before, $\xi = \xi^{n+1}$, $u = u^n$, $v = v^n$, and let $j \ge 2$. Using induction on j, we can assume the above estimate to hold up to $j - 1$; thus

$$\|\xi^n\|_{H^{j-1}} \le K_{j-1}, \quad \|\mathbf{u}\|_{H^j} \le c_1 K_{j-1}.$$

Applying the operator D_1^j to (9.1.17), we obtain that

$$D_1^j \xi_t = -D_1^j(u\xi_x) - D_1^j(v\xi_y) + \nu \Delta D_1^j \xi.$$

Therefore,

$$\begin{aligned}\frac{1}{2}\frac{d}{dt}\|D_1^j \xi\|^2 &= (D_1^j \xi, D_1^j \xi_t)\\ &\le (D_1^{j+1}\xi, D_1^{j-1}(u\xi_x)) + (D_1^{j+1}\xi, D_1^{j-1}(v\xi_y)) - \nu\|D_1^{j+1}\xi\|^2\\ &\le c_2\{\|D_1^{j-1}(u\xi_x)\|^2 + \|D_1^{j-1}(v\xi_y)\|^2\}, \quad c_2 = c_2(\nu, j).\end{aligned}$$

The differentiated products in the last formula are evaluated by Leibniz' rule. One obtains an estimate in terms of

(9.1.21) $\quad \|(D_1^k u)(D_1^{j-k}\xi)\|^2, \quad \|(D_1^k v)(D_1^{j-1-k} D_2 \xi)\|^2, \quad 0 \le k \le j-1.$

Consider the u-terms for definiteness,

$$\|(D_1^k u)(D_1^{j-k}\xi)\|^2 \le |D_1^k u|_\infty^2 \|D_1^{j-k}\xi\|^2.$$

For $0 \le k \le j-2$:

$$|D_1^k u|_\infty^2 \le c_3 \|u\|_{H^j}^2 \le c_4 K_{j-1}^2,$$

and

$$\|D_1^{j-k}\xi\|^2 \le \|\xi\|_{H^j}^2.$$

For $k = j-1$:

$$|D_1^k u|_\infty^2 = |D_1^{j-1} u|_\infty^2 \le c_5 \|\xi^n\|_{H^j}^2,$$

and

$$\|D_1^{j-k}\xi\|^2 = \|D_1 \xi\|^2 \le K_1^2.$$

A similar estimate applies to the v-terms in (9.1.21). Thus we have shown that

$$\frac{d}{dt}\|D_1^j \xi^{n+1}\|^2 \le c_6 \Big\{1 + |\xi^n|_{H^j}^2 + |\xi^{n+1}|_{H^j}^2\Big\}.$$

Clearly, the term

$$\frac{d}{dt}\|D_2^j \xi^{n+1}\|^2$$

can be bounded in the same way. Introducing the functions

$$G_j^n(t) = \|D_1^j \xi^n(\cdot, t)\|^2 + \|D_2^j \xi^n(\cdot, t)\|^2$$

and observing (9.1.20), we have that

$$\frac{d}{dt}G_j^{n+1}(t) \le c\{1 + G_j^n(t) + G_j^{n+1}(t)\}, \quad 0 \le t \le T,$$
$$G_j^n(0) \le |h|_{H^j}^2.$$

Here c does not depend on n. Therefore, by Gronwall's Lemma 3.1.1,

$$G_j^{n+1}(t) \le K\Big\{1 + \int_0^t G_j^n(\tau)\,d\tau\Big\}, \quad K = K(T),$$

in $0 \le t \le T$. By Picard's Lemma 3.3.4 we obtain the desired bound independent of n. This completes the proof of the lemma.

The above considerations (and Sobolev inequalities) show that we have bounds independent of n for *all* derivatives of the functions

$$u^n, \ v^n, \ \xi^n$$

in $0 \le t \le T$. We can now argue in exactly the same way as in the proof of local existence for Burgers' equation; see Theorem 4.1.6. Thus – in any time interval $0 \le t \le T$ – the sequence $u^n, \ v^n, \ \xi^n$ converges along with all its derivatives to the solution $u, \ v, \ \xi$ of (9.1.15). The Equivalence Theorem 9.1.5 shows existence of a solution $u, \ v, \ p$ of the N-S equations.

Theorem 9.1.8. *Consider the system of equations* (9.1.15) *with initial data* $h = g_x - f_y$ *where*

$$f_x + g_y = 0, \quad (1, f) = (1, g) = 0.$$

There is a unique C^∞-solution. The solutions exists for all time.

9.1.5. Maximum Principle and Smallest Scale

It is an important property of the vorticity (in 2D) that it obeys a maximum principle. We formulate this next.

Lemma 9.1.9. *For all $t \ge 0$ the vorticity ξ satisfies*

$$|\xi(\cdot, t)|_\infty \le |\xi(\cdot, 0)|_\infty.$$

Proof. Let $\alpha > 0$ be a constant and set

$$\tilde{\xi}(x, y, t) = e^{-\alpha t}\xi(x, y, t).$$

The vorticity equation yields

$$\tilde{\xi}_t + u\tilde{\xi}_x + v\tilde{\xi}_y = \nu\Delta\tilde{\xi} - \alpha\tilde{\xi}.$$

Suppose $\tilde{\xi}$ attains its maximum over

$$x, y \in \mathbf{R}, \quad 0 \le t \le T$$

in (x_0, y_0, t_0) with $\tilde{\xi}(x_0, y_0, t_0) > 0$. If $0 < t_0 \le T$ then

$$\tilde{\xi}_x = \tilde{\xi}_y = 0, \quad \tilde{\xi}_t \ge 0, \quad \Delta\tilde{\xi} \le 0$$

at (x_0, y_0, t_0); this contradicts the differential equation for $\tilde{\xi}$. Thus we have $t_0 = 0$; i.e., a positive maximum can only be attained at $t = 0$. Similarly, a negative minimum can only occur at $t = 0$. If we send $\alpha \to 0$, the estimate for ξ follows.

The maximum principle allows one to improve the estimates of the previous section considerably. In Henshaw, Kreiss, and Reyna (1988) it is proved that

$$J_j^2(t) = \|D_1^j \xi(\cdot, t)\|^2 + \|D_2^j \xi(\cdot, t)\|^2$$

can essentially be bounded in terms of

$$\left(\frac{1}{\nu} |\xi(\cdot, 0)|_\infty\right)^j.$$

If

$$\xi(\mathbf{x}, t) = \sum_{\mathbf{k}} \hat{\xi}(\mathbf{k}, t) e^{i\langle \mathbf{k}, \mathbf{x}\rangle}$$

denotes the Fourier expansion of the vorticity, these results imply that

$$|\hat{\xi}(\mathbf{k}, t)| \sim e^{-\nu^{1/2}|\mathbf{k}|}$$

provided that $|\xi(\cdot, 0)|_\infty = O(1)$. Thus, there is very little energy in the high wave-numbers $|\mathbf{k}| \gg \nu^{-1/2}$, and therefore the smallest spatial length scale is $\sim \nu^{1/2}$.

We have restricted ourselves to smooth initial data. However, in the same way as for Burgers' equation, we can show a smoothing principle for the vorticity: We can estimate

$$t^j \|\xi(\cdot, t)\|_{H^j}^2$$

in terms of $\|\xi(\cdot, 0)\|^2$. Therefore we can define generalized solutions for nonsmooth initial data.

9.2. The Spatially Periodic Case in Three Dimensions

As in the previous section, we restrict ourselves for simplicity to the case with zero forcing. The equations have the form

$$\mathbf{u}_t + u\mathbf{u}_x + v\mathbf{u}_y + w\mathbf{u}_z + \operatorname{grad} p = \nu\Delta\mathbf{u}, \quad \nu > 0, \tag{9.2.1a}$$

$$\operatorname{div}\mathbf{u} = 0, \tag{9.2.1b}$$

$$\mathbf{u}(\mathbf{x}, 0) = \mathbf{f}(\mathbf{x}), \quad \operatorname{div}\mathbf{f} = 0. \tag{9.2.1c}$$

Without further mentioning, all functions in this section are assumed to be real, C^∞-smooth and 2π-periodic in x, y, z . We require that

$$(1, f_i) = 0, \quad i = 1, 2, 3,$$

and obtain, as in the 2D case,

$$(1, u(\cdot, t)) = (1, v(\cdot, t)) = (1, w(\cdot, t)) = 0, \quad t \geq 0,$$

as long as the solution exists.

9.2.1. Vorticity Formulation; Local Existence

In three space dimensions the vorticity is defined as the vector

$$\xi = \operatorname{curl}\mathbf{u} = \begin{pmatrix} w_y - v_z \\ u_z - w_x \\ v_x - u_y \end{pmatrix}. \tag{9.2.2}$$

If one applies the curl-operator to the momentum equations and observes that $\operatorname{div}\mathbf{u} = 0$, then one finds after some computation the vorticity equation

$$\xi_t + u\xi_x + v\xi_y + w\xi_z - B(\mathbf{u})\xi = \nu\Delta\xi. \tag{9.2.3}$$

Here $B(\mathbf{u})$ is the Jacobian matrix of the velocity field,

$$B(\mathbf{u}) = \begin{pmatrix} u_x & u_y & u_z \\ v_x & v_y & v_z \\ w_x & w_y & w_z \end{pmatrix}.$$

To show a local existence result, we want to use an iteration like (9.1.9) in the 2D case. To this end, we first note that

$$\operatorname{div}\xi = 0,$$

as follows directly from the definition (9.2.2). Also, integration by parts implies the vanishing of the spatial averages

$$(1, \xi_i(\cdot, t)) = 0, \quad i = 1, 2, 3.$$

This allows us to determine **u** in terms of ξ by a generalized Cauchy-Riemann system. We show by Fourier expansion:

Lemma 9.2.1. *Suppose that $\xi = \xi(\mathbf{x})$ satisfies*

$$\operatorname{div} \xi = 0, \quad (1, \xi_i) = 0, \quad i = 1, 2, 3.$$

Then the system

$$L\mathbf{u} := \begin{pmatrix} w_y - v_z \\ u_z - w_x \\ v_x - u_y \\ u_x + v_y + w_z \end{pmatrix} = \begin{pmatrix} \xi_1 \\ \xi_2 \\ \xi_3 \\ 0 \end{pmatrix} \tag{9.2.4}$$

with side-condition

$$(1, u) = (1, v) = (1, w) = 0$$

has a unique solution **u**. *For each $j = 1, 2, \ldots$, there is a constant K_j independent of ξ with*

$$\|\mathbf{u}\|^2_{H^j} \le K_j |\xi|^2_{H^{j-1}}.$$

Proof. Suppose that the system has a solution. We can combine the equations to obtain

$$\Delta u = \xi_{2z} - \xi_{3y}, \quad \Delta v = \xi_{3x} - \xi_{1z}, \quad \Delta w = \xi_{1y} - \xi_{2x}.$$

These equations allow us to express the Fourier coefficients of u, v, w in terms of the Fourier coefficients of ξ_1, ξ_2, ξ_3. For example,

$$\hat{u}(\mathbf{k}) = -i\, \frac{k_3 \hat{\xi}_2(\mathbf{k}) - k_2 \hat{\xi}_3(\mathbf{k})}{k_1^2 + k_2^2 + k_3^2}.$$

This shows uniqueness and the estimates. Conversely, defining u, v, w in terms of its Fourier coefficients by formulae as above, we obtain existence.

To prove local existence for the N-S equations, we consider the sequence

$$\mathbf{u}^n, \xi^n, \quad n = 1, 2, \ldots,$$

defined by

$$\xi_t^{n+1} + u^n \xi_x^{n+1} + v^n \xi_y^{n+1} + w^n \xi_z^{n+1} - B(\mathbf{u}^n)\xi^{n+1} = \nu \Delta \xi^{n+1},$$

$$\xi^{n+1}(\cdot, 0) = \operatorname{curl} \mathbf{f},$$

$$L\mathbf{u}^{n+1} = \begin{pmatrix} \xi^{n+1} \\ 0 \end{pmatrix},$$

$$(1, u^{n+1}) = (1, v^{n+1}) = (1, w^{n+1}) = 0,$$

starting with $\mathbf{u}^0 \equiv 0$. In contrast to the 2D case, one obtains here an estimate of $\|\xi\|_{H^1}$ only in a finite (small) time interval $0 \le t \le T$, where T depends on the H_1-norm of the initial vorticity, i.e., on $\|\text{curl}\,\mathbf{f}\|_{H^1}$. Then one can show – as in the 2D case – that the functions $\mathbf{u}^n$, ξ^n are uniformly smooth in this interval $0 \le t \le T$. As before, it follows that the functions $\mathbf{u}^n$, ξ^n converge along with all their derivatives to C^∞-functions $\mathbf{u}$, ξ which solve the vorticity equation (9.2.3) and the generalized Cauchy-Riemann system (9.2.4) at each time t , $0 \le t \le T$.

To obtain a solution $\mathbf{u}$, p of the N-S equations, it remains to determine the pressure. We take the divergence of the momentum equations to find

$$\Delta p + J = 0,$$

$$\begin{aligned} J &= u_x u_x + v_y v_y + w_z w_z + 2\{v_x u_y + w_x u_z + w_y v_z\} \\ &= 2\{v_x u_y - u_x v_y + w_x u_z - u_x w_z + w_y v_z - v_y w_z\}. \end{aligned}$$

If we impose the side-condition $(1, p) = 0$, the pressure is uniquely determined. (Note that $(1, J) = 0$.) Except for the analog of Lemma 9.1.6 (which is replaced by a local result), these details are more or less the same as in the 2D case. Also, the uniquenes of a solution follows as in Section 9.1.4. We summarize the result in

Theorem 9.2.2. *The spatially periodic Cauchy problem for the 3D viscous incompressible Navier-Stokes equations with C^∞-data has a unique C^∞-solution existing in some time interval $0 \le t \le T$. The time $T > 0$ depends only on the viscosity ν and on $\|\mathbf{h}\|_{H^1}$. Here $\mathbf{h} = \text{curl}\,\mathbf{f}$ is the vorticity at time $t = 0$.*

9.2.2. Extension of the Solution Interval for Bounded Velocities

It is not known whether the solutions $\mathbf{u}$, p of the viscous incompressible N-S equations (9.2.1) exist for all time $0 \le t < \infty$ or whether it is possible (for some initial data) that the solution ceases to exist due to the development of a singularity in finite time. If one assumes a solution in a time interval

$$0 \le t < T,$$

one can ask for conditions which allow the extension of the solution beyond T. For example, the Local Theorem 9.2.2 implies: If

$$\sup_{0 \le t < T} \|\xi(\cdot, t)\|_{H^1}, \quad \xi = \operatorname{curl} \mathbf{u},$$

is finite, then the solution can be extended beyond T. In other words, if the solution ceases to exist at $t = T$, then the function

$$\|\xi(\cdot, t)\|_{H^1}, \quad 0 \le t < T,$$

must have a singularity at $t = T$. The boundedness of $\|\xi\|_{H^1}$ implies boundedness of the velocity:

$$|\mathbf{u}|_\infty \le c_1 \|\mathbf{u}\|_{H^2} \le c_2 \|\xi\|_{H^1}.$$

In this section we want to show that boundedness of the velocity is also sufficient to extend the solution beyond T. One can prove even more, namely estimates of all derivatives of the solution in terms of the initial derivatives and the maximal velocity.

Notations. For any $j = 0, 1, \ldots,$ let

(9.2.5) $$J_j^2(t) = \|D_1^j \mathbf{u}(\cdot, t)\|^2 + \|D_2^j \mathbf{u}(\cdot, t)\|^2 + \|D_3^j \mathbf{u}(\cdot, t)\|^2,$$

$$D_1 = \partial/\partial x, \quad D_2 = \partial/\partial y, \quad D_3 = \partial/\partial z, \quad |\mathbf{u}(\cdot, t)|_\infty = \max_{\mathbf{x}} |\mathbf{u}(\mathbf{x}, t)|.$$

We assume throughout that $\mathbf{u}$, p is a smooth solution of (9.2.1) defined in $0 \le t < T$; we start with the basic energy estimate. This estimate is valid without any assumption on $|\mathbf{u}|_\infty$.

Lemma 9.2.3. *The energy-change satisfies*

$$\frac{d}{dt} \|\mathbf{u}(\cdot, t)\|^2 = -2\nu J_1^2(t), \quad 0 \le t < T,$$

and therefore

$$\|\mathbf{u}(\cdot, t)\|^2 \le \|\mathbf{f}\|^2, \quad 2\nu \int_0^T J_1^2 \, dt \le \|\mathbf{f}\|^2.$$

Proof. From (9.2.1a),

$$\frac{1}{2} \frac{d}{dt} \|\mathbf{u}\|^2 = (\mathbf{u}, \mathbf{u}_t) = -S - (\mathbf{u}, \operatorname{grad} p) + \nu(\mathbf{u}, \Delta \mathbf{u}).$$

Here S consists of nine terms,

$$\begin{aligned} S = (u, uu_x) + (u, vu_y) + (u, wu_z) \\ + (v, uv_x) + (v, vv_y) + (v, wv_z) \\ + (w, uw_x) + (w, vw_y) + (w, ww_z). \end{aligned}$$

Integration by parts yields that

$$(u, uu_x) = (u^2, u_x) = -2(uu_x, u) = 0,$$

and similarly

$$(u, vu_y) = -\frac{1}{2}(u^2, v_y).$$

Therefore,

$$\begin{aligned} S &= -\frac{1}{2}\Big\{(u^2, v_y) + (u^2, w_z) + (v^2, u_x) + (v^2, w_z) + (w^2, u_x) + (w^2, v_y)\Big\} \\ &= -\frac{1}{2}\Big\{(u^2, -u_x) + (v^2, -v_y) + (w^2, -w_z)\Big\} = 0. \end{aligned}$$

Furthermore,

$$(\mathbf{u}, \Delta\mathbf{u}) = (\mathbf{u}, \mathbf{u}_{xx}) + (\mathbf{u}, \mathbf{u}_{yy}) + (\mathbf{u}, \mathbf{u}_{zz}) = -J_1^2,$$

and the lemma is proved.

We now proceed to estimate the first space derivaties of $\mathbf{u}$ and assume the velocity to be bounded,

$$|\mathbf{u}|_{\infty,T} := \sup_{0\le t<T} |\mathbf{u}(\cdot, t)|_\infty < \infty.$$

Since the dependence of the estimates on the viscosity ν is of some interest, we will make it explicit below. The constants c, c_1, c_2, etc. are numbers independent of ν, T and the solution $\mathbf{u}$ under consideration.

Lemma 9.2.4. *For the function $J_1(t)$ measuring the first space derivatives of $\mathbf{u}$ we have that*

$$J_1^2(t) \le J_1^2(0) + c\frac{|\mathbf{u}|_{\infty,T}^2}{\nu^2}\|\mathbf{f}\|^2.$$

The right-hand side also is a bound for

$$\nu\int_0^T J_2^2\,dt.$$

Proof. Application of any of the operators

$$D = D_1,\ D_2,\ D_3$$

to the momentum equations yields

$$D\mathbf{u}_t + D\big\{u\mathbf{u}_x + v\mathbf{u}_y + w\mathbf{u}_z\big\} + \text{grad}\, Dp = \nu\Delta D\mathbf{u}.$$

Therefore,

$$\begin{aligned}\frac{1}{2}\frac{d}{dt}\|D\mathbf{u}\|^2 &= (D\mathbf{u}, D\mathbf{u}_t)\\ &= \left(D^2\mathbf{u}, u\mathbf{u}_x + v\mathbf{u}_y + w\mathbf{u}_z\right) - \nu\{\|D\mathbf{u}_x\|^2 + \|D\mathbf{u}_y\|^2 + \|D\mathbf{u}_z\|^2\}\\ &\le c_1|\mathbf{u}|_\infty J_1 J_2 - \nu\|D^2\mathbf{u}\|^2.\end{aligned}$$

Summing these three estimates for $D = D_1,\ D_2,\ D_3$, we find that

$$\begin{aligned}\frac{d}{dt}J_1^2 &\le c_2\,|\mathbf{u}|_\infty J_1 J_2 - 2\nu J_2^2\\ &\le \frac{c_2^2}{\nu}\,|\mathbf{u}|_\infty^2 J_1^2 - \nu J_2^2.\end{aligned}$$

Integration of the inequality – together with the estimate of $\int J_1^2\,dt$ by Lemma 9.2.3 – proves the result.

We want to show now a similar estimate for $J_2(t)$; more precisely,

$$\frac{d}{dt}J_2^2 \le c|\mathbf{u}|_\infty J_2 J_3 - 2\nu J_3^2. \tag{9.2.6}$$

Then we can proceed exactly as in the proof of the previous lemma and obtain an estimate of $J_2(t)$ in terms of

$$J_2(0),\ J_1(0),\ \|\mathbf{f}\|,\ \nu,\ |\mathbf{u}|_{\infty,T}\,.$$

To show (9.2.6), we apply any of the operators

$$D^2 = D_1^2,\ D_2^2,\ D_3^2$$

to the momentum equations:

$$D^2\mathbf{u}_t + D^2\{u\mathbf{u}_x + v\mathbf{u}_y + w\mathbf{u}_z\} + \operatorname{grad} D^2 p = \nu\Delta D^2\mathbf{u}.$$

Therefore,

$$\begin{aligned}\frac{1}{2}\frac{d}{dt}\|D^2\mathbf{u}\|^2 &= (D^2\mathbf{u}, D^2\mathbf{u}_t)\\ &\le -\left(D^2\mathbf{u}, D^2\{u\mathbf{u}_x + v\mathbf{u}_y + w\mathbf{u}_z\}\right) - \nu\|D^3\mathbf{u}\|^2.\end{aligned}$$

Now Leibniz' rule is applied to the term

$$D^2\{u\mathbf{u}_x + v\mathbf{u}_y + w\mathbf{u}_z\};$$

for example,

$$D^2(v\mathbf{u}_y) = (D^2v)\mathbf{u}_y + 2(Dv)(D\mathbf{u}_y) + vD^2\mathbf{u}_y.$$

Typical terms which need to be estimated are

$$\big(D^2\mathbf{u}, (D^2v)\mathbf{u}_y\big), \quad \big(D^2\mathbf{u}, (Dv)(D\mathbf{u}_y)\big), \quad \big(D^2\mathbf{u}, vD^2\mathbf{u}_y\big).$$

There is no difficulty in bounding the third term by

$$|\mathbf{u}|_\infty J_2 J_3.$$

(Actually, the three terms of this form add up to zero since $\operatorname{div}\mathbf{u} = 0$). Also, we can use integration by parts to remove the first derivative from $\mathbf{u}_y$ and from Dv in the first and in the second term, respectively. This shows that (9.2.6) holds.

To summarize, if the velocity is bounded in $0 \le t < T$, then $J_1(t)$, $J_2(t)$ are also bounded. Therefore,

$$\|\xi(\cdot, t)\|_{H^1}, \quad 0 \le t < T,$$

is bounded, and our previous considerations show that the existence interval can be extended beyond T.

Theorem 9.2.5. *If the solution* **u** *to* (9.2.1) *satisfies a bound*

$$\sup\{|\mathbf{u}(\cdot, t)|_\infty : 0 \le t < T\} < \infty, \tag{9.2.7}$$

then **u** *can be extended to a solution in an interval* $0 \le t < T + \Delta T, \ \Delta T > 0$.

Remark. Instead of (9.2.7) one only needs to assume

$$\sup\big\{\big(\mathbf{u}(\cdot, t), |\mathbf{u}(\cdot, t)|\mathbf{u}(\cdot, t)\big) : 0 \le t < T\big\} < \infty.$$

Then the solution can be extended beyond T; see von Wahl (1985).

10 The Incompressible Navier-Stokes Equations under Initial and Boundary Conditions

In two space dimensions the nonlinear equations have the form

$$\mathbf{u}_t + u\mathbf{u}_x + v\mathbf{u}_y + \operatorname{grad} p = \nu\Delta\mathbf{u} + \mathbf{F}, \quad u_x + v_y = 0, \tag{10.1.1}$$

$$\mathbf{u} = (u, v), \quad \mathbf{F} = (F, G), \quad \mathbf{x} = (x, y).$$

We consider the equations in the region

$$0 \le x \le 1, \quad -\infty < y < \infty, \quad t \ge 0,$$

and prescribe initial conditions

$$u(x, y, 0) = u_0(x, y), \; v(x, y, 0) = v_0(x, y), \quad u_{0x} + v_{0y} = 0,$$

and boundary conditions at $x = 0$, $x = 1$. All data are assumed to be C^∞-smooth and 2π-periodic in y; we seek a C^∞-solution which is 2π-periodic in y.

10.1. The Linearized Equations in 2D

In this section we start our discussion of the linearized equations under boundary conditions. To this end, let

$$U,\ V,\ P,\quad U_x + V_y = 0,$$

denote smooth functions of $x,\ y,\ t$ which are 2π-periodic in y, and substitute

$$u = U + u',\ v = V + v',\ p = P + p'$$

into the N-S equations. Neglecting terms quadratic in the corrections u', etc. and dropping the prime $'$ in our notation, we find the linear equations

$$\text{(10.1.2)}\qquad \mathbf{u}_t + U\mathbf{u}_x + V\mathbf{u}_y + A\mathbf{u} + \operatorname{grad} p = \nu\Delta\mathbf{u} + \mathbf{F},\quad u_x + v_y = 0,$$

where

$$A = \begin{pmatrix} U_x & U_y \\ V_x & V_y \end{pmatrix}.$$

The forcing function $\mathbf{F}$ in (10.1.2) is the defect of (U, V, P) in (10.1.1). It is convenient to study (10.1.2) for a general smooth matrix function $A(x, y, t)$ which is 2π-periodic in y.

Our plan is to approximate (10.1.2) by the evolutionary system

$$\text{(10.1.3}a\text{)}\qquad \mathbf{u}_t + U\mathbf{u}_x + V\mathbf{u}_y + A\mathbf{u} + \operatorname{grad} p = \nu\Delta\mathbf{u} + \mathbf{F},$$

$$\text{(10.1.3}b\text{)}\qquad \epsilon p_t + u_x + v_y = 0,\quad \epsilon > 0.$$

We shall choose boundary conditions for (u, v, p) in such a way that our theory for mixed parabolic-hyperbolic equations applies for every fixed $\epsilon > 0$. Then, if we can estimate the solutions of (10.1.3) and their derivatives independently of ϵ, we obtain a smooth solution of (10.1.2) in the limit $\epsilon \to 0$. We start with the basic estimate.

Lemma 10.1.1. *Suppose that* $\mathbf{u}, p$ *solve* (10.1.3). *There is a constant* $c > 0$, *depending on the coefficients* $U,\ V,$ *in the differential equation* (10.1.3) *but independent of* ϵ *and* $\mathbf{F}$, *with*

(10.1.4)

$$\frac{1}{2}\frac{d}{dt}\{\|\mathbf{u}\|^2 + \epsilon\|p\|^2\} \le c\|\mathbf{u}\|^2 + (\mathbf{u}, \mathbf{F}) - \nu\left\{\|\mathbf{u}_x\|^2 + \|\mathbf{u}_y\|^2\right\} + \left\{-\frac{1}{2}U(u^2 + v^2) - up + \nu(uu_x + vv_x)\right\}\Bigg|_{x=0}^{x=1}.$$

(Here we use the notation

$$\{\, b \,\}\Big|_{x=0}^{x=1} = \int_0^{2\pi} b(1, y, t)\, dy - \int_0^{2\pi} b(0, y, t)\, dy$$

for boundary terms.)

Proof. The differential equations (10.1.3) yield

$$\begin{aligned}\frac{1}{2}\frac{d}{dt}\{\|\mathbf{u}\|^2+\epsilon\|p\|^2\} &= (\mathbf{u},\mathbf{u}_t)+(p,\epsilon p_t)\\ &= -(\mathbf{u},U\mathbf{u}_x)-(\mathbf{u},V\mathbf{u}_y)-(\mathbf{u},A\mathbf{u})-(\mathbf{u},\operatorname{grad}p)\\ &\quad+\nu(\mathbf{u},\Delta\mathbf{u})+(\mathbf{u},\mathbf{F})-(p,u_x+v_y).\end{aligned}$$

The result follows from integration by parts and estimates of zero-order terms. The boundary contributions appear from integration by parts in the x-variable.

Boundary conditions providing the basic estimate. The boundary conditions – together with the term

$$-\nu\{\|\mathbf{u}_x\|^2+\|\mathbf{u}_y\|^2\} =: -\nu J_1^2(\mathbf{u})$$

on the right-hand side of (10.1.4) – have to provide bounds for the boundary term

$$B.T. = \left\{-\frac{1}{2}U(u^2+v^2)-up+\nu(uu_x+vv_x)\right\}\Big|_{x=0}^{x=1}. \tag{10.1.5}$$

We consider different possibilities.

Case 1. The lines $x=0$, $x=1$ represent walls. In this case it is reasonable to prescribe the boundary velocities

$$u=v=0 \quad \text{at} \quad x=0,\ x=1$$

for the nonlinear system (10.1.1). (This is obvious for u and confirmed by observation for v.) Thus one chooses a base flow with

$$U=V=0 \quad \text{at} \quad x=0,\ x=1$$

and obtains

$$u=v=0 \quad \text{at} \quad x=0,\ x=1 \tag{10.1.6}$$

for the linearized system. Clearly, the boundary term (10.1.5) vanishes, and the basic estimate follows.

Case 2. Inflow at $x=0$, outflow at $x=1$. Now one chooses

$$U(0,y,t)>0, \quad U(1,y,t)>0.$$

Again, we can use the Dirichlet conditions (10.1.6). Also, the following conditions of Neumann type may be used:

$$
\begin{aligned}
-\nu u_x(j,y,t) + p(j,y,t) &= \alpha_j(y,t)u(j,y,t) + g_j(y,t), \\
-\nu v_x(j,y,t) &= \beta_j(y,t)v(j,y,t) + h_j(y,t), \quad j = 0,1.
\end{aligned}
\tag{10.1.7}
$$

Here α_j, β_j, g_j, h_j are given smooth functions which are 2π-periodic in y. For the boundary term (10.1.5) we obtain

$$
\begin{aligned}
B.T. &= \left\{ -\frac{1}{2}U(u^2+v^2) - up + u(p - \alpha u - g) - v(\beta v + h) \right\}\Big|_{x=0}^{x=1} \\
&\le c\left\{ \|\mathbf{u}\|_\Gamma^2 + \|g\|_\Gamma^2 + \|h\|_\Gamma^2 \right\}.
\end{aligned}
$$

Here

$$
\|\mathbf{u}\|_\Gamma^2 = \int_0^{2\pi} \{|\mathbf{u}(0,y,t)|^2 + |\mathbf{u}(1,y,t)|^2\}\,dy,
$$

etc. Using a (one-dimensional) Sobolev inequality for each fixed y and integrating over y, we have

$$
c\|\mathbf{u}\|_\Gamma^2 \le \frac{\nu}{2}\,J_1^2(\mathbf{u}) + \frac{c_1}{\nu}\|\mathbf{u}\|^2. \tag{10.1.8}
$$

Thus the basic estimate follows from Lemma 10.1.1.

Remarks.

1. The above boundary conditions are of the general form discussed for mixed parabolic-hyperbolic systems in Section 8.5. Therefore, if smooth initial data are given for u, v, p and provided that suitable compatibility conditions are fulfilled at $t = 0$, the equations (10.1.3) determine a unique smooth solution satisfying these boundary conditions.
2. In many applications the viscosity constant $\nu > 0$ is very small. Then, if estimate (10.1.8) is employed, rapid exponential growth of the energy is allowed. However, if the coefficients α_j, β_j in the boundary conditions (10.1.7) are suitably chosen, one obtains an estimate of the boundary term (10.1.5) by $c\{\|g\|_\Gamma^2 + \|h\|_\Gamma^2\}$. In this case, inequality (10.1.8) need not be employed, and the rate of exponential growth becomes independent of ν. The same is true, of course, for Dirichlet boundary conditions.
3. It is no restriction to assume the Dirichlet conditions (10.1.6) to be homogeneous. If inhomogeneous data are given, we can introduce new variables $\tilde{\mathbf{u}} = \mathbf{u} - \phi$ which satisfy the homogeneous condition, and if we differentiate with respect to y and t, the condition remains homogeneous. In contrast to this, consider (10.1.7). Of course we could also transform to the homogeneous case, but if we want to estimate y- and t-derivatives, new inhomogeneous terms are introduced, and nothing is gained.

4. We have assumed conditions of the same type at $x = 0$ and $x = 1$, either of Dirichlet or of Neumann type. Clearly, this assumption is not necessary. In applications one often specifies $u = v = 0$ at inflow and (10.1.7) at outflow. If $\nu > 0$ is small and α, β are suitably chosen (see Remark 2), such a Neumann condition at outflow leads to a "smoother" solution than the simple requirement $u = v = 0$ at outflow.

10.2. Auxiliary Results for Poisson's Equation

In this section we prove some elementary estimates for solutions of Laplace's equation and Poisson's equation in the strip

$$0 \le x \le 1, \quad -\infty < y < \infty.$$

As in the last section, all functions are assumed to be C^∞-smooth and 2π-periodic in y. We apply Fourier expansion in y-direction and use the notation

$$u(x, y) = \sum_{k=-\infty}^{\infty} \hat{u}(x, k)e^{iky}, \quad \hat{u}(x, k) = \frac{1}{2\pi} \int_0^{2\pi} u(x, y)e^{-iky}\, dy.$$

Then Parseval's relation reads

$$\int_0^{2\pi} |u(x, y)|^2\, dy = 2\pi \sum_{k=-\infty}^{\infty} |\hat{u}(x, k)|^2, \quad 0 \le x \le 1,$$

and integration over x gives us

$$\|u\|^2 = 2\pi \sum_{k=-\infty}^{\infty} \int_0^1 |\hat{u}(x, k)|^2\, dx. \tag{10.2.1}$$

Lemma 10.2.1. *Suppose that $u \in C^\infty$ solves*

$$\Delta u = 0 \quad \text{in} \quad 0 \le x \le 1, \quad -\infty < y < \infty,$$

and u is 2π-periodic in y. There is a constant K_1 independent of u with

$$\|u_y\|^2 \le K_1 \|u_x\|^2.$$

If the spatial average of u vanishes, i.e.,

$$0 = \int_0^1 \int_0^{2\pi} u(x, y)\, dy\, dx = 2\pi \int_0^1 \hat{u}(x, 0)\, dx,$$

then

$$\|u\|^2 \le K_2\|u_x\|^2$$

with K_2 independent of u.

Proof. Since $\Delta u = 0$ it follows that

$$\hat{u}_{xx}(x,k) - k^2\hat{u}(x,k) = 0, \quad k = 0, \pm 1, \ldots.$$

Thus we have a representation

$$\text{(10.2.2)} \qquad \hat{u}(x,k) = \begin{cases} a_k e^{-|k|x} + b_k e^{|k|(x-1)} & k \ne 0, \\ a_0 + b_0 x & k = 0. \end{cases}$$

Using (10.2.1) for u_y, we find that

$$\begin{aligned} \|u_y\|^2 &= 2\pi \sum_{k=-\infty}^{\infty} k^2 \int_0^1 |\hat{u}(x,k)|^2\, dx \\ &= 2\pi \sum_{k=-\infty}^{\infty} k^2 \int_0^1 \left| a_k e^{-|k|x} + b_k e^{|k|(x-1)} \right|^2 dx. \end{aligned}$$

Similarly,

$$\begin{aligned} \|u_x\|^2 &= 2\pi \sum_{k=-\infty}^{\infty} \int_0^1 |\hat{u}_x(x,k)|^2\, dx \\ &= 2\pi \sum_{k=-\infty}^{\infty} k^2 \int_0^1 \left| -a_k e^{-|k|x} + b_k e^{|k|(x-1)} \right|^2 dx \; + \; 2\pi |b_0|^2. \end{aligned}$$

Elementary computation shows that (for $k \ne 0$)

$$\int_0^1 \left| \pm a_k e^{-|k|x} + b_k e^{|k|(x-1)} \right|^2 dx = \frac{|a_k|^2 + |b_k|^2}{2|k|}(1 - e^{-2|k|}) \pm 2\,\mathrm{Re}\,(a_k b_k)\, e^{-|k|}.$$

Using

$$|2\,\mathrm{Re}\,(a_k b_k)| \le 2|a_k b_k| \le |a_k|^2 + |b_k|^2,$$

we find that the estimate $\|u_y\|^2 \le K_1 \|u_x\|^2$ is valid if K_1 satisfies

$$\frac{1}{2|k|}\left(1 - e^{-2|k|}\right) + e^{-|k|} \le K_1\left(\frac{1}{2|k|}(1 - e^{-2|k|}) - e^{-|k|}\right)$$

for all integers $k \ne 0$.

To show the estimate for $\|u\|^2$, we first observe that

$$0 = \int_0^1 \hat{u}(x,0)\,dx = a_0 + \frac{1}{2} b_0,$$

and therefore

$$\hat{u}(x,0) = b_0(x - \frac{1}{2}).$$

Thus (10.2.1) yields

$$\begin{aligned} \|u\|^2 &\leq \|u_y\|^2 + 2\pi \int_0^1 |\hat{u}(x,0)|^2\,dx \\ &\leq \|u_y\|^2 + 2\pi |b_0|^2 \leq \|u_y\|^2 + \|u_x\|^2. \end{aligned}$$

This proves the lemma.

An easy generalization of the last lemma is

Lemma 10.2.2. *Suppose that $u \in C^\infty$ solves*

$$\Delta u = 0 \quad \text{in} \quad 0 \leq x \leq 1, \quad -\infty < y < \infty,$$

and u is 2π-periodic in y. For every $j \geq 1$ there is a constant K_j independent of u such that

$$|u|^2_{H^j} \leq K_j \| \frac{\partial^j u}{\partial x\, \partial y^{j-1}} \|^2.$$

Proof. The equation $\Delta u = 0$ allows us to express even order x-derivatives in terms of y-derivatives:

$$\frac{\partial^{2l} u}{\partial x^{2l}} = (-1)^l \frac{\partial^{2l} u}{\partial y^{2l}}.$$

Thus, any derivative term appearing in $|u|_{H^j}$ reduces to

$$\pm \frac{\partial^j u}{\partial x\, \partial y^{j-1}} \quad \text{or} \quad \pm \frac{\partial^j u}{\partial y^j}.$$

The estimate of the last lemma finishes the proof.

Next we want to show estimates of u in terms of boundary data on the lines $x = 0,\ x = 1$.

Lemma 10.2.3. *Let 2π-periodic C^∞-functions $g_0(y)$, $g_1(y)$ be given. The boundary value problem*

$$\Delta u = 0, \quad u(0, y) = g_0(y), \quad u(1, y) = g_1(y)$$

has a unique C^∞-solution $u(x, y)$ which is 2π-periodic in y. For any $j = 0, 1, \ldots$ there is K_j independent of g with

$$\|u\|_{H^j}^2 \leq K_j \|g\|_{H^j(\Gamma)}^2.$$

Here

$$\|g\|_{H^j(\Gamma)}^2 = \|g_0\|_{H^j}^2 + \|g_1\|_{H^j}^2$$

denotes the H^j-norm of the boundary data.

Proof. First assume that there is a smooth solution u. We use (10.2.1) and the representation (10.2.2) to obtain

$$\|u\|^2 = 2\pi \sum_{k \neq 0} \int_0^1 \left|a_k e^{-|k|x} + b_k e^{|k|(x-1)}\right|^2 dx \; + \; 2\pi \int_0^1 |a_0 + b_0 x|^2 \, dx$$

$$\leq c_1 \sum_{k=-\infty}^{\infty} \left(|a_k|^2 + |b_k|^2\right).$$

On the other hand,

$$\|g_0\|^2 + \|g_1\|^2 = 2\pi \sum_{k=-\infty}^{\infty} \left(|\hat{g}_0(k)|^2 + |\hat{g}_1(k)|^2\right).$$

The formula (10.2.2) gives us the relations

$$\begin{pmatrix} 1 & e^{-|k|} \\ e^{-|k|} & 1 \end{pmatrix} \begin{pmatrix} a_k \\ b_k \end{pmatrix} = \begin{pmatrix} \hat{g}_0(k) \\ \hat{g}_1(k) \end{pmatrix}, \quad k \neq 0,$$

$$\begin{pmatrix} 1 & 0 \\ 1 & 1 \end{pmatrix} \begin{pmatrix} a_0 \\ b_0 \end{pmatrix} = \begin{pmatrix} \hat{g}_0(k) \\ \hat{g}_1(k) \end{pmatrix}, \quad k = 0.$$

Thus we find that

$$|a_k|^2 + |b_k|^2 \leq c_2 \left(|\hat{g}_0(k)|^2 + |\hat{g}_1(k)|^2\right),$$

and the bound $\|u\|^2 \leq K_0 \|g\|_\Gamma^2$ follows. Similarly,

$$\|u_x\|^2 \leq c_1 \sum_{k \neq 0} k^2 \left(|a_k|^2 + |b_k|^2\right) \; + \; 2\pi |b_0|^2$$

$$\leq c\left\{\|dg/dy\|^2 + \|g\|^2\right\}.$$

If we apply this estimate to

$$\frac{\partial^{j-1} u}{\partial y^{j-1}}$$

instead of u, and use Lemma 10.2.2, then we obtain the desired bound of $\|u\|^2_{H^j}$. The bound clearly implies uniqueness. *Existence* of a solution follows , because we can define u in terms of its Fourier expansion.

Now consider the inhomogeneous differential equation

$$\Delta u = F(x, y) \tag{10.2.3}$$

where $F \in C^\infty$ is 2π-periodic in y. We ask for estimates of a C^∞-solution with homogeneous boundary data

$$u(0, y) = u(1, y) = 0, \quad -\infty < y < \infty, \tag{10.2.4}$$

and require, as before, that u is 2π-periodic in y. We show

Lemma 10.2.4. *The above strip problem has a unique solution u. For any $j = 0, 1, \ldots$, there is a constant K_j independent of F with*

$$\|u\|^2_{H^{j+2}} \le K_j \|F\|^2_{H^j}.$$

In this sense, we "gain" two derivatives.

Proof. First assume that the problem has a solution u. Fourier expansion of (10.2.3), (10.2.4) yields

$$\hat{u}_{xx}(x, k) - k^2 \hat{u}(x, k) = \hat{F}(x, k), \tag{10.2.5a}$$

$$\hat{u}(0, k) = \hat{u}(1, k) = 0. \tag{10.2.5b}$$

Thus each function

$$\hat{u}(\cdot, k), \quad k = 0, \pm 1, \ldots,$$

is determined by an ordinary boundary value problem. By

$$\|\hat{u}(\cdot, k)\| = \left\{ \int_0^1 |\hat{u}(x, k)|^2 \, dx \right\}^{1/2}$$

we denote the L_2-norm of such a function of x, and obtain through integration by parts (for $k \neq 0$)

$$\begin{aligned} \|\hat{u}_x(\cdot,k)\|^2 + k^2\|\hat{u}(\cdot,k)\|^2 &= -\Big(\hat{u}(\cdot,k), \hat{F}(\cdot,k)\Big) \\ &\leq \frac{1}{2k^2}\|\hat{F}(\cdot,k)\|^2 + \frac{k^2}{2}\|\hat{u}(\cdot,k)\|^2. \end{aligned}$$

Thus

$$k^4\|\hat{u}(\cdot,k)\|^2 \leq \|\hat{F}(\cdot,k)\|^2,$$

and with (10.2.5a)

$$\begin{aligned} \|\hat{u}_{xx}(\cdot,k)\| &\leq k^2\|\hat{u}(\cdot,k)\| + \|\hat{F}(\cdot,k)\| \\ &\leq 2\|\hat{F}(\cdot,k)\|. \end{aligned}$$

For $k = 0$ an explicit integration of the boundary value problem (10.2.5) shows that

$$\|\hat{u}(\cdot,0)\| \leq \|\hat{F}(\cdot,0)\|.$$

Thus we have obtained that

$$(1+k^4)\|\hat{u}(\cdot,k)\|^2 + (1+k^2)\|\hat{u}_x(\cdot,k)\|^2 + \|\hat{u}_{xx}(\cdot,k)\|^2 \leq c\|\hat{F}(\cdot,k)\|^2,$$

and Parseval's relation (10.2.1) yields

$$\|u\|^2_{H^2} \leq K_0\|F\|^2.$$

This shows the desired bound for $j = 0$. There are no difficulties in estimating higher y-derivatives of

$$u, \quad u_x, \quad u_{xx}$$

because (10.2.3), (10.2.4) can be differentiated with respect to y. If we want to estimate higher x-derivatives, we use (10.2.3) and replace them by y-derivatives, e.g.,

$$u_{xxx} = F_x - u_{xyy}.$$

Thus we obtain the estimate of the lemma. Existence of a C^∞-solution follows as before since we can define u by its Fourier expansion.

If both – the boundary data and the differential equation – are inhomogeneous, i.e.,

$$\Delta u = F(x,y), \quad u(0,y) = g_0(y), \quad u(1,y) = g_1(y),$$

then we decompose the problem in an obvious way, and obtain the solution as a sum,

$$u = u^{(1)} + u^{(2)}.$$

Here Lemma 10.2.3 applies to $u^{(1)}$ and Lemma 10.2.4 applies to $u^{(2)}$. One obtains the estimate

$$\|u\|^2_{H^{j+2}} \leq K_j \Big\{ \|F\|^2_{H^j} + \|g\|^2_{H^{j+2}(\Gamma)} \Big\}.$$

10.3. The Linearized Navier-Stokes Equations under Boundary Conditions

The linearized system (10.1.2) is not of the standard form of an evolution equation since we do not have an equation for p_t. To show the existence of a solution of (10.1.2) under initial and boundary conditions, we go over to (10.1.3) for $\epsilon > 0$ and send $\epsilon \to 0$.

10.3.1. Convergence Theorem

We consider the system (10.1.3) and restrict ourselves, for simplicity, to the no-slip condition

(10.3.1) $$u = v = 0 \quad \text{at} \quad x = 0,\ x = 1$$

and the initial condition

(10.3.2) $$u(x, y, 0) = v(x, y, 0) = p(x, y, 0) = 0.$$

As previously, the forcing function **F** and the coefficients U, V, A in (10.1.3) are assumed to be C^∞-smooth and 2π-periodic in y. Also, we require that

(10.3.3) $$\partial^j F(x, y, 0)/\partial t^j = \partial^j G(x, y, 0)/\partial t^j = 0, \quad j = 0, 1, 2, \ldots.$$

This ensures compatibility of the data at $t = 0$. For each $\epsilon > 0$ the initial-boundary value problem (10.1.3), (10.3.1), (10.3.2) has a unique C^∞-solution

$$\mathbf{u} = \mathbf{u}^\epsilon(x, y, t), \quad p = p^\epsilon(x, y, t)$$

which is 2π-periodic in y. To begin with, note that at $t = $ *all* derivatives of **u**, p are $\equiv 0$, as follows from (10.3.3). Our aim is to show that all derivatives of **u**, p can be bounded independently of ϵ in any interval $0 \leq t \leq T$. Then, as $\epsilon \to 0$, we obtain a smooth solution of the linearized Navier-Stokes equations. As remarked earlier, the solution is only unique if one fixes a (time-dependent) constant to determine the pressure, e.g.,

(10.3.4) $$(1, p(\cdot, t)) = 0.$$

We will show

Theorem 10.3.1. *Consider the linearized viscous incompressible equations* (10.1.2) *in*

$$0 \le x \le 1, \quad -\infty < y < \infty, \quad t \ge 0$$

with boundary conditions $u = v = 0$ *at* $x = 0,\ x = 1$, *initial condition* $u = v = 0$ *at* $t = 0$, *and side-condition* (10.3.4). *Assume that the data and coefficients are* C^∞*-smooth,* 2π*-periodic in* y, *and fulfill* (10.3.3). *Then there is a unique* C^∞*-solution which is* 2π*-periodic in* y. *In any interval* $0 \le t \le T$, *the solution is the limit of the corresponding solutions of* (10.1.3) *as* $\epsilon \to 0$.

The assumptions of the theorem can be relaxed: One can allow, for example, certain inhomogeneous initial conditions and different boundary conditions; see Section 10.3.2.

The proof of the estimates of all derivatives of $\mathbf{u} = \mathbf{u}^\epsilon$, $p = p^\epsilon$ proceeds in several steps. We fix a time interval $0 \le t \le T$. For brevity, we will call a function $\omega = \omega^\epsilon(x, y, t)$ *estimated*, if we have established a bound for

$$\max\{\|\omega^\epsilon(\cdot, t)\| \, : \, 0 \le t \le T\},$$

which is uniform in $0 < \epsilon \le 1$. As previously, the main difficulty is estimating x-derivatives, because one cannot differentiate the boundary conditions in x-direction.

Step 1. The basic estimate (see Lemma 10.1.1) gives us bounds for $\mathbf{u}$, ϵp. We can differentiate the differential equation and the boundary conditions arbitrarily often w.r.t. y and t and obtain systems of the same form where the lower-order terms are already estimated. Thus we get bounds for all y, t-derivatives of $\mathbf{u}$, ϵp. In particular, we have bounded

$$\frac{d}{dt}\{\|\mathbf{u}\|^2 + \epsilon\|p\|^2\},$$

and the basic estimate (10.1.4) gives us a bound for $\mathbf{u}_x$. From the differentiated systems we obtain, with the same argument, bounds for all y, t-derivatives of $\mathbf{u}_x$. To summarize, we have estimates for

$$u, \ v, \ u_x, \ v_x, \ \epsilon p$$

and all y, t-derivatives of these functions.

Step 2. We differentiate the equation $\epsilon p_t + u_x + v_y = 0$ w.r.t. x and replace u_{xx} using the first equation (10.1.3a). Then we obtain

$$\epsilon\nu p_{tx} + p_x = H_1,$$

$$H_1 = F - u_t - Uu_x - Vu_y - \nu v_{xy} - a_{11}u - a_{12}v + \nu u_{yy},$$

where the function H_1 is already estimated. Integrating the above ordinary differential equation for $p_x(x, y, t)$ in time and observing that

$$\frac{1}{\epsilon\nu}\int_0^t e^{-(t-s)/(\epsilon\nu)}\,ds \leq 1,$$

we have bounded p_x. Also, since all y, t-derivatives of H_1 are estimated, we have bounds for all y, t-derivatives of p_x.

Step 3. Differentiating the u-equation (10.1.3a) w.r.t. x, the v-equation w.r.t. y, and summing the two results, we obtain that

$$(u_x + v_y)_t + U(u_x + v_y)_x + V(u_x + v_y)_y + \Delta p + L(\mathbf{u}, \mathbf{u}_x, \mathbf{u}_y)$$
$$= \nu\Delta(u_x + v_y) + F_x + G_y,$$

where $L(\mathbf{u}, \mathbf{u}_x, \mathbf{u}_y)$ is a linear expression in $\mathbf{u}, \mathbf{u}_x, \mathbf{u}_y$. We replace $u_x + v_y$ by $-\epsilon p_t$ and find that

$$\epsilon\nu\Delta p_t + \Delta p = H_2,$$

$$H_2 = F_x + G_y + \epsilon\Big(p_{tt} + Up_{xt} + Vp_{yt}\Big) - L(\mathbf{u}, \mathbf{u}_x, \mathbf{u}_y).$$

The function H_2 is already estimated, and using the same argument as in Step 2, we have established a bound for the function

$$\Delta p = H$$

and all its y, t-derivatives.

Step 4. At each time $t \geq 0$ the pressure p can be written as a sum

$$p = p_1 + p_2$$

where

$$\Delta p_1 = 0, \quad p_1(0, y, t) = p(0, y, t), \quad p_1(1, y, t) = p(1, y, t),$$

and

$$\Delta p_2 = H, \quad p_2(0, y, t) = p_2(1, y, t) = 0. \tag{10.3.5}$$

From Lemma 10.2.4 we obtain bounds for p_2, p_{2x}, p_{2xx} and all y, t-derivatives of these functions. In Step 2 we have bounded the function p_x and all its y, t-derivatives; thus we can estimate the difference p_{1x} and all its y, t-derivatives.

Step 5. We want to derive a bound for p_1 itself. To this end note that

$$(1, p(\cdot, t)) = 0,$$

as follows from $\epsilon p_t + u_x + v_y = 0$ and the boundary conditions. Therefore, the function

$$\tilde{p}_1(x, y, t) = p_1(x, y, t) + \frac{1}{2\pi}(1, p_2)$$

has spatial average zero and, using Lemma 10.2.1, we can bound $\tilde{p}_1$. (Note that $\tilde{p}_{1x} = p_{1x}$.) Since p_2 is already bounded, we obtain an estimate for p_1 itself. In the same way we obtain also a bound for each time derivative p_{1t}, p_{1tt}, etc., and Lemma 10.2.2 yields bounds for all other derivatives of p_1. Thus we have established estimates for *all* derivatives of p_1.

Step 6. We have shown estimates for p, p_x, p_{xx}, $\mathbf{u}$, $\mathbf{u}_x$ and all y, t-derivatives of these functions. It remains to bound higher x-derivatives (and their y, t-derivatives); this can now be done recursively as follows: The equations (10.1.3a) can be solved for u_{xx}, v_{xx}, and therefore these functions are also bounded. Thus all first derivatives of H_2 are estimated, and (10.3.5) gives us bounds for all third derivatives of p. Then (10.1.3a) yields bounds for $\mathbf{u}_{xxx}$, $\mathbf{u}_{xxxx}$. A simple induction argument finishes the proof.

10.3.2. Remark on Initialization

In the last section we have assumed that the initial conditions are *homogeneous*. Let us explain the reason for this. Suppose that we would give general nonhomogeneous initial data

$$u(x, y, 0) = f(x, y), \quad v(x, y, 0) = g(x, y), \quad p(x, y, 0) = h(x, y). \tag{10.3.6}$$

For every fixed $\epsilon > 0$ the corresponding initial-boundary value problem (10.1.3) has a solution. However, we need bounds of the derivatives which are *independent* of ϵ. Our estimates of the previous section show these bounds if the y, t-derivatives are bounded independently of ϵ at $t = 0$. For the y-derivatives we have no difficulties, provided the boundary conditions are compatible with the initial conditions. Therefore, the crucial question is *if one can bound the time derivatives independently of* ϵ *at* $t = 0$.

By (10.1.3b) the first time derivatives are bounded independently of ϵ at $t = 0$ if and only if

$$u_x(\cdot, 0) + v_y(\cdot, 0) = \epsilon d_1, \quad d_1 = O(1). \tag{10.3.7}$$

Thus, to first approximation, the initial data have to satisfy the divergence relation. Differentiation of (10.1.3*b*) with respect to t gives us

$$\epsilon p_{tt} = -(u_{xt} + v_{yt}) = \Delta p + H(u, v) - F_x - G_y.$$

Here $H(u, v)$ is an expression in u, v and its space derivatives. Thus, the second time derivatives are bounded independently of ϵ if we have, at $t = 0$,

$$\Delta p + H(u, v) - F_x - G_y = \epsilon p_1, \quad p_1 = O(1). \tag{10.3.8}$$

We can consider (10.3.8) as an equation for the pressure. For the third time derivative of p we obtain

$$\begin{aligned} \epsilon^2 p_{ttt} &= \epsilon \Delta p_t + \epsilon \{H_t(u, v) - F_{xt} - G_{yt}\} \\ &= -\Delta(u_x + v_y) + \epsilon\{H_t(u, v) - F_{xt} - G_{yt}\} \\ &= \epsilon\{-\Delta d_1 + H_t(u, v) - F_{xt} - G_{yt}\}. \end{aligned} \tag{10.3.9}$$

Thus we have to choose d_1 so that

$$-\Delta d_1 + H_t(u, v) - F_{xt} - G_{yt} = \epsilon d_2, \quad d_2 = O(1).$$

Note that we can use (10.1.3*a*) to express $H_t(u, v)$ in terms of u, v and their space derivatives. Further differentiation of (10.3.9) with respect to t gives us an equation for p_1.

This process can be continued. We can ensure that more and more time derivatives are bounded independently of ϵ by choosing the initial data so that the above relations are satisfied. The easiest way to comply with all requirements is to give *homogeneous* initial data and to request (10.3.3). This case has been treated above. However, more general data are allowed as long as they satisfy the above requirements and lead to bounded time derivatives. To choose initial data in this way is called *initialization by the bounded derivative principle*.

In applications, one often solves compressible equations which are almost incompressible. Here a similar problem occurs: To ensure that the solution is close to a solution of the incompressible equation, one has to "prepare" the initial data. The principle is the same, one chooses the initial data so that a number of time derivatives at $t = 0$ remain bounded independent of the compressibility. For details, see Kreiss (1985).

10.4. Remarks on the Passage from the Compressible to the Incompressible Equations

Consider the nonlinear system (10.1.1) and choose functions (an approximate solution)

$$U,\ V,\ P,\quad U_x + V_y = 0.$$

For example, the functions may satisfy inhomogeneous boundary and initial conditions. We substitute

$$u = U + u',\quad v = V + v',\quad p = P + p'$$

into (10.1.1) and obtain equivalent (nonlinear) equations for the corrections u', v', p'. If we drop the prime $'$ in our notations, the equations read

$$\text{(10.4.1)}\qquad \mathbf{u}_t + (U+u)\mathbf{u}_x + (V+v)\mathbf{u}_y + \operatorname{grad} p = \nu\,\Delta\mathbf{u} + \mathbf{F},$$
$$u_x + v_y = 0.$$

Here $\mathbf{F}$ is a new forcing which is determined by U, V, P. This system can be approximated by (10.4.1) together with

$$\text{(10.4.2)}\qquad \epsilon p_t + u_x + v_y = 0,\quad \epsilon > 0.$$

If we have boundary conditions (10.3.1), (10.3.2) and assume (10.3.3), then we can proceed in the same way as above for the linearized equation. As $\epsilon \to 0$, we obtain convergence to a solution of the nonlinear incompressible equations.

The process which has been chosen here to obtain a solution of the incompressible equations does *not* describe the passage of the compressible to the incompressible N-S equations. If one wants to study this limit, one should replace (10.4.2) by

$$\text{(10.4.3)}\qquad \epsilon\Big\{p_t + (U+u)p_x + (V+v)p_y\Big\} + u_x + v_y = 0.$$

(We recall the continuity equation $\rho_t + \operatorname{div}(\rho\mathbf{u}) = 0$ and formally substitute $\rho = 1 + \epsilon p$.) In both cases, (10.4.1) with (10.4.2) or (10.4.1) with (10.4.3), we have a coupled hyperbolic-parabolic system. To illustrate the differences between (10.4.2) and (10.4.3), we assume inflow ($U > 0$) at $x = 0$ and require the boundary condition $u = v = 0$ at $x = 0$. Whereas the boundary $x = 0$ is *characteristic* for the equation (10.4.2), the variable p is an *ingoing characteristic variable* for (10.4.3). Therefore, no boundary condition for p is needed with (10.4.2), but (10.4.3) requires a boundary condition. For example, one can prescribe

$$p(0, y, t) = 0.$$

For the limit-equations ($\epsilon = 0$), no boundary condition for p is allowed. In Kreiss, Lorenz, and Naughton (1988) it is shown that convergence for $\epsilon \to 0$ is also obtained with (10.4.3), at least away from a boundary layer at $x = 0$.

APPENDIX

1 Notations and Results from Linear Algebra

Vectors and matrices. Let $\mathbf{C}$ denote the field of complex numbers and let $\mathbf{C}^n$ be the vector space of column vectors

$$u = \begin{pmatrix} u_1 \\ \vdots \\ u_n \end{pmatrix}, \quad u_j \in \mathbf{C}, \quad j = 1, \ldots, n.$$

We define an inner product and a norm by

$$\langle u, v \rangle = \sum_{j=1}^{n} \overline{u}_j v_j, \quad |u| = \langle u, u \rangle^{1/2}, \quad u, v \in \mathbf{C}^n.$$

Here $\overline{u}_j$ is the complex conjugate of u_j. If $A = (a_{jk}) \in \mathbf{C}^{n,n}$ is a complex n by n matrix, then

$$A^* = \overline{A}^T = (\overline{a}_{kj})$$

denotes the complex conjugate transpose of A.

It holds that

$$\langle u, Av \rangle = \langle A^* u, v \rangle \quad \text{for all} \quad u, v \in \mathbf{C}^n.$$

The spectral-norm of A is

$$|A| = \max\{|Au| : u \in \mathbf{C}^n, \quad |u| = 1\}.$$

One can show that

$$|A|^2 = \rho(A^* A)$$

where $\rho\ (B)$ denotes the spectral radius of a matrix B, i.e., the largest absolute value of all eigenvalues of B. A matrix $U \in \mathbf{C}^{n,n}$ is called unitary if

$$U^* U = I, \quad \text{i.e.} \quad U^{-1} = U^*.$$

If U is unitary then, for any $u \in \mathbf{C}^n$,

$$|Uu|^2 = \langle Uu, Uu \rangle = \langle U^* Uu, u \rangle = |u|^2,$$

and therefore

$$|U| = |U^{-1}| = |U^*| = 1.$$

A matrix U with columns $u^1, \ldots, u^n \in \mathbf{C}^n$ is unitary if and only if

$$\langle u^j, u^k \rangle = \delta_{jk} = \begin{cases} 1 & \text{for } j = k, \\ 0 & \text{for } j \neq k; \end{cases}$$

i.e., the columns form an orthonormal system.

An important result from linear algebra is

Schur's Theorem. *Let $A \in \mathbf{C}^{n,n}$ denote a matrix with eigenvalues $\lambda_1, \ldots, \lambda_n$ in any prescribed order. There is a unitary matrix U such that*

$$U^* AU = R = (r_{jk})$$

is upper triangular with diagonal entries $r_{jj} = \lambda_j,\ \ j = 1, \ldots, n$.

Proof. We use induction on n; the case $n = 1$ is trivial. Let $Au^1 = \lambda_1 u^1, |u^1| = 1$. One can choose an orthonormal basis

$$u^1,\ u^2,\ \ldots, u^n$$

of $\mathbf{C}^n$ starting with u^1. The matrix U_1 with columns $u^1, \ldots, u^n$ is unitary, and

$$AU_1 = U_1 \begin{pmatrix} \lambda_1 & * \\ 0 & A_2 \end{pmatrix}.$$

The matrix A_2 has the eigenvalues $\lambda_2, \ldots, \lambda_n$. There is a unitary matrix U_2 such that

$$A_2 U_2 = U_2 R_2,$$

where R_2 is upper triangular and $\lambda_2, \ldots, \lambda_n$ appear on the diagonal. If we set

$$U = U_1 \begin{pmatrix} 1 & 0 \\ 0 & U_2 \end{pmatrix},$$

then

$$AU = U \begin{pmatrix} \lambda_1 & * \\ 0 & R_2 \end{pmatrix},$$

and the induction is completed.

APPENDIX

2 Interpolation

Fourier interpolation in 1D. Let $m \in \{1, 2, \ldots\}$, $h = (2m+1)^{-1}$, $x_\nu = \nu h$, $\nu = 0, \pm 1, \pm 2, \ldots$. Suppose that $v(x)$ is a 1-periodic function, $v(x) \equiv v(x+1)$, defined at all gridpoints $x = x_\nu$. We want to interpolate $v(x)$ by a Fourier polynomial

$$w(x) = \sum_{k=-m}^{m} a_k e^{2\pi i k x} \tag{A.2.1}$$

at all grid points:

$$w(x_\nu) = v(x_\nu), \quad \nu = 0, \pm 1, \pm 2, \ldots. \tag{A.2.2}$$

Theorem A.2.1. *The interpolation problem* (A.2.2) *has a unique solution* (A.2.1).

Proof. If we introduce the inner product

$$\big(u(x), v(x)\big)_h = h \sum_{\nu=0}^{2m} \overline{u(x_\nu)} v(x_\nu),$$

then the gridfunctions

$$e^{2\pi i k x}, \quad k = -m, -m+1, \ldots, m,$$

form an orthonormal system, i.e.,

$$\left(e^{2\pi ilx}, e^{2\pi ikx}\right)_h = \begin{cases} 0 & \text{if } 0 < |k-l| \le 2m\ ; \\ 1 & \text{if } k = l\ . \end{cases}$$

The above result is obvious for $k = l$ and follows from the geometric-sum formula otherwise. Consequently, if $w(x)$ is an interpolant of the form (A.2.1), then

$$\left(e^{2\pi ilx}, v(x)\right)_h = \left(e^{2\pi ilx}, w(x)\right)_h = \sum_{k=-m}^{m} a_k \left(e^{2\pi ilx}, e^{2\pi ikx}\right)_h = a_l.$$

Thus the coefficients a_l are uniquely determined by the data function $v(x)$. Also, we can consider the interpolation-condition

$$\sum_{k=-m}^{m} a_k e^{2\pi ik\nu h} = v(x_\nu), \quad \nu = 0, \ldots, 2m,$$

as a linear system of $2m + 1$ equations for the $2m + 1$ unknown a_k. Since the solution vector is unique (if it exists), the system-matrix is nonsingular, and existence follows.

Fourier interpolation is very useful since the interpolant $w(x)$ inherits "smoothness" from the data $v(x)$, with estimates independent of h. More precisely, if we introduce norms by

$$\|v\|_h^2 = (v, v)_h^2, \quad \|w\|^2 = \int_0^1 |w(x)|^2\, dx,$$

then the following result holds:

Theorem A.2.2. *If $w(x)$ denotes the Fourier interpolant of the data $v(x)$ then*

$$\|w\|^2 = \|v\|_h^2 = \sum_{k=-m}^{m} |a_k|^2$$

and

$$\|D^p w\|^2 \le \left(\frac{\pi}{2}\right)^{2p} \|D_+^p v\|_h^2, \quad p = 1, 2, \ldots, \quad D = \frac{d}{dx}\ .$$

It is important that the factor $(\pi/2)^{2p}$ does not depend on h.

Proof. The first equation is nothing but Parseval's relation; it applies since the functions

$$e^{2\pi ikx}, \quad |k| \le m,$$

are orthonormal w.r.t. the continuous and discrete inner products. Therefore, differentiation of $w(x)$ gives us

$$\|D^p w\|^2 = \sum_{k=-m}^{m} |a_k|^2 (2\pi k)^{2p}.$$

Since

$$(D_+ v)(x) = \sum_{k=-m}^{m} a_k \frac{e^{2\pi ikh} - 1}{h} e^{2\pi ikx},$$

we find that

$$\|D_+^p v\|_h^2 = \sum_{k=-m}^{m} |a_k|^2 \left|\frac{e^{2\pi ikh} - 1}{h}\right|^{2p}.$$

Here

$$\left|\frac{e^{2\pi ikh} - 1}{h}\right|^2 = \left|\frac{2 \sin \pi kh}{h}\right|^2 = \left|\frac{\sin \pi kh}{\pi kh}\right|^2 (2\pi k)^2 \geq \left(\frac{2}{\pi}\right)^2 (2\pi k)^2,$$

and the result follows.

Fourier interpolation in more space dimensions. The generalization to s space dimensions is straightforward. Suppose $v = v(x)$, $x = (x_1, \ldots, x_s)$, is a function of s variables which is 1-periodic in each variable x_l and which is defined at all gridpoints $(\nu_1 h, \nu_2 h, \ldots, \nu_s h)$, $\nu_l = 0, \pm 1, \pm 2, \ldots$. Then there is a unique Fourier polynomial

$$w(x) = \sum_{l=1}^{s} \sum_{k_l=-m}^{m} a_k e^{2\pi i\langle k, x\rangle}, \quad k = (k_1, \ldots, k_s),$$

which interpolates v at all gridpoints,

$$w(\nu_1 h, \ldots, \nu_s h) = v(\nu_1 h, \ldots, \nu_s h), \quad \nu_l \in \mathbf{Z}.$$

As in Theorem A.2.2, derivatives of w can be estimated in terms of corresponding divided differences of v; the constants in the estimates are independent of h.

Interpolation of nonperiodic functions in 1D. s Consider a discrete function f defined on a grid $x_\nu = \nu h$, $h = 1/N$, $\nu = 0, 1, 2, \ldots, N$. We want to interpolate f so that the derivatives of the interpolant can be estimated in terms of the divided differences of f, with constants independent of h. There are many ways to do this. Here we first extend f to a *periodic* gridfunction so that bounds of divided differences are preserved. Then we use Fourier interpolation to obtain the desired interpolant. For transparency, we describe the corresponding extension of smooth functions first.

Consider a function $f \in C^p[0,1]$, and let $0 < \delta < \frac{1}{2}$ be given. We will extend the domain of definition of f to $-\infty < x < \infty$. The extended function, again denoted by f, will have its support in $-\delta \leq x \leq 1+\delta$. Furthermore,

$$\sum_{j=0}^{p} \|D^j f\|_E^2 \leq C \sum_{j=0}^{p} \|D^j f\|^2, \quad D = \frac{d}{dx}. \tag{A.2.3}$$

Here C depends on δ and p but not on f, and $\|\cdot\|_E$, $\|\cdot\|$ denote the L_2-norms over the whole line and the original interval, respectively.

To begin with, let us preliminarily define $f(x)$ for $1 \leq x \leq 3/2$ as the solution of the differential equation

$$D^p u = 0, \quad 1 \leq x \leq 3/2,$$

with initial conditions

$$D^j u(1) = D^j f(1), \quad j = 0, 1, 2, \ldots, p-1.$$

The solution is

$$f(x) = \sum_{j=0}^{p-1} \frac{1}{j!} D^j f(1)\,(x-1)^j.$$

(Note that the p-th derivative of the extended function jumps at $x = 1$, in general. This is not essential, however.) For $x \leq 0$ we use the corresponding construction. Now we modify the function so that its support becomes compact. To this end, let $\varphi \in C^\infty$ denote a cut-off function with

$$\varphi(x) = \begin{cases} 1 & \text{if } -\frac{\delta}{3} \leq x \leq 1 + \frac{\delta}{3}, \\ 0 & \text{if } x \leq -\frac{2\delta}{3} \text{ or } x \geq 1 + \frac{2\delta}{3}. \end{cases}$$

Then the desired extension is given by φf. By the Sobolev inequality stated in Lemma A.3.10, we can estimate $|D^j f(1)|$, $|D^j f(0)|$, $j = 0, 1, 2, \ldots, p-1$, in terms of $\|D^p f\|$ and $\|f\|$, and therefore we have proved

Lemma A.2.3. *Let $f \in C^p$ be defined for $0 \leq x \leq 1$, and let $0 < \delta < \frac{1}{2}$. One can extend the definition of f to $-\infty < x < \infty$ so that the extended function f has its support in $-\delta \leq x \leq 1+\delta$ and fulfills* (A.2.3). *The constant in* (A.2.3) *is independent of f.*

We consider now a discrete function. Let $x_\nu = \nu h,\ \nu = 0, \pm 1, \pm 2, \ldots,\ h = 1/N$, denote a grid and let $f_\nu = f(x_\nu),\ \nu = 0, 1, 2, \ldots, N$, be defined at the gridpoints $0 \le x_\nu \le 1$. We prove

Lemma A.2.4. *For any p we can extend the definition of f_ν to the whole discrete line so that the support of f_ν is contained in the interval $-\frac{2}{3} \le x \le 1 + \frac{2}{3}$ and*

$$\sum_{j=0}^{p} \|D_+^j f\|^2_{-\infty,\infty} \le \text{const} \sum_{j=0}^{p} \|D_+^j f\|^2_{0,N-j}\,. \tag{A.2.4}$$

The extension and the constant depend on p, but the constant does not depend on f or h.

Proof. The proof proceeds similarly to the continous case. For $1 \le x_\nu \le \frac{3}{2}$ we define $f_\nu = f(x_\nu)$ preliminarily as the solution of the difference equation

$$D_-^p u_{N+\mu} = 0, \quad \mu = 1, 2, \ldots\,. \tag{A.2.5}$$

As in the continous case, the solution can be written down. It has the form

$$u_{N+\mu} = P_{p-1}(x_\mu),$$

where $P_{p-1}(x)$ is a polynomial of degree $\le p-1$, whose coefficients are divided differences $D_-^j f_\nu,\ N - p \le \nu \le N,\ j = 0, 1, 2, \ldots, p-1$. For $x < 0$ we make the corresponding construction. Then φf is the desired extension. This extension is again denoted by f. By the discrete Sobolev inequality stated in Lemma A.3.18, we can estimate $D_-^j f_\nu$ in terms of $\|D_+^p f\|^2_{0,N-p}$ and $\|f\|^2_{0,N}$, and therefore we obtain the estimate (A.2.4).

We can consider the extended gridfunction f as a periodic function with period 2. If w denotes its Fourier interpolant, then

$$\sum_{j=0}^{p} \|D^j w\|^2_{[0,\,1]} \le \sum_{j=0}^{p} \|D^j w\|^2_{[-1/2,\,3/2]} \le \text{const} \sum_{j=0}^{p} \|D_+^j f\|^2_{0,N-j}\,.$$

The latter estimate follows from Theorem A.2.2 and formula (A.2.4).

We have proved

Theorem A.2.5. *Let $f_\nu = f(x_\nu)$ denote a discrete function, where $x_\nu = \nu h,\ h = 1/N,\ \nu = 0, 1, 2, \ldots, N$. Let p denote a positive integer. We can interpolate f by a function $w \in C^\infty[0, 1]$ so that*

$$\sum_{j=0}^{p} \|D^j w\|^2 \leq \text{ const } \sum_{j=0}^{p} \|D_+^j f\|_{0,N-j}^2 \, .$$

The interpolant and the constant depend on p; the constant is independent of f and h, however.

Extensions of functions in a strip. We shall now generalize the results to a strip

$$I: \; 0 \leq x \leq 1, \; -\infty < y < \infty,$$

and start again with the continous case. Let $f = f(x,y)$, $f \in C^{2p}(I)$, and assume that f is 1-periodic in y. For every fixed y we use the construction given above. For $x > 1$ we obtain

$$f(x,y) = \varphi(x) \sum_{j=0}^{p-1} \frac{1}{j!} \frac{\partial^j f}{\partial x^j}(1,y)\,(x-1)^j.$$

Therefore,

$$\frac{\partial^p f}{dy^p}(x,y) = \varphi(x) \sum_{j=0}^{p-1} \frac{1}{j!} \frac{\partial^{j+p} f}{\partial x^j \, \partial y^p}(1,y)\,(x-1)^j.$$

Thus, we need the L_2-norms of the derivatives of order $2p$ to estimate $\|\partial^p f/\partial y^p\|_E$.

Lemma A.2.6. *We can extend the definition of f so that*

$$\sum_{j=0}^{p} \Big(\| \frac{\partial^j f}{\partial x^j} \|_E^2 + \| \frac{\partial^j f}{\partial y^j} \|_E^2 \Big) \leq \text{ const } \sum_{j=0}^{2p} \Big(\| \frac{\partial^j f}{\partial x^j} \|^2 + \| \frac{\partial^j f}{\partial y^j} \|^2 \Big).$$

The corresponding interpolation result is also clear: For every p we can find an interpolant whose derivatives up to order p can be estimated in terms of divided differences of order $2p$. All these results can be extended in a straightforward way to functions in a strip of s dimensions.

APPENDIX

3 Sobolev Inequalities

We show here some frequently used inequalities which express bounds of one norm of a function (or of a derivative of a function) in terms of other norms of the same function and its derivatives. Typical examples are so-called Sobolev inequalities which give bounds of the *maximum norm* of a function in terms of L_2-norms of derivatives. It is important that the constants entering these estimates do not depend on any specific function under consideration, but are uniform for a certain class of functions.

To derive such estimates, we may always assume the functions to be C^∞-smooth, for convenience. (The functions take values in $\mathbf{C}^n$.) Then the estimate extends to all (less smooth) functions which can be approximated by C^∞-functions w.r.t. the norms entering the estimate.

L_2-estimates of periodic functions. Let $u \in C^\infty(\mathbf{R})$, $u(x) \equiv u(x+L)$; i.e., u is L-periodic. As usual,

$$\|u\|^2 = \int_0^L |u(x)|^2 \, dx.$$

We first show

Lemma A.3.1. *For all $\epsilon > 0$ and all integers $j, k \geq 1$ it holds that*

$$\|D^j u\|^2 \leq \epsilon \|D^{j+k} u\|^2 + \epsilon^{-j/k} \|u\|^2, \quad D = \frac{d}{dx}.$$

Proof. By Fourier expansion,
(A.3.1)

$$u(x) = \frac{1}{\sqrt{L}} \sum_{\omega=-\infty}^{\infty} \hat{u}(\omega) e^{2\pi i \omega x/L}, \quad \hat{u}(\omega) = \frac{1}{\sqrt{L}} \int_0^L u(x) e^{-2\pi i \omega x/L}\, dx$$

$$D^j u(x) = \frac{1}{\sqrt{L}} \sum_{\omega=-\infty}^{\infty} \left(\frac{2\pi i \omega}{L}\right)^j \hat{u}(\omega) e^{2\pi i \omega x/L},$$

and Parseval's relation yields that

$$\|D^j u\|^2 = \sum_{\omega=-\infty}^{\infty} \left(\frac{2\pi\omega}{L}\right)^{2j} |\hat{u}(\omega)|^2.$$

Therefore, the lemma is proved if we can show that

$$\left(\frac{2\pi\omega}{L}\right)^{2j} \leq \epsilon \left(\frac{2\pi\omega}{L}\right)^{2(j+k)} + \epsilon^{-j/k}, \quad \omega \in \mathbf{Z}.$$

The function

$$y(s) = \epsilon s^{2(j+k)} + \epsilon^{-j/k} - s^{2j}, \quad s \in \mathbf{R},$$

has its minimum at

$$s_0 = \left(\frac{j}{j+k}\right)^{1/(2k)} \epsilon^{-1/(2k)},$$

and

$$y(s_0) = \epsilon^{-j/k} \left\{1 - \frac{k}{j+k}\left(\frac{j}{j+k}\right)^{j/k}\right\} > 0.$$

Hence the assertion follows.

A generalization to any number of space dimensions is straightforward. Let $u \in C^\infty(\mathbf{R}^s)$ be L-periodic in each variable x_l, and denote

$$\|u\|^2 = \int_0^L \dots \int_0^L |u(x)|^2\, dx_1 \dots dx_s.$$

We use the notations

$$J_p^2(u) = \sum_{l=1}^{s} \|D_l^p u\|^2, \qquad D_l = \frac{\partial}{\partial x_l},$$

$$|u|_{H^p}^2 = \sum_{|\nu|=p} \|D^\nu u\|^2, \qquad D^\nu = D_1^{\nu_1} \dots D_s^{\nu_s};$$

i.e., J_p^2 measures the "clean" p-th derivatives whereas $|\cdot|_{H^p}$ measures all p-th derivatives. For any multi-index ν,

$$\|D^\nu u\|^2 = \sum_\omega \left(\frac{2\pi\omega_1}{L}\right)^{2\nu_1} \cdots \left(\frac{2\pi\omega_s}{L}\right)^{2\nu_s} |\hat{u}(\omega)|^2$$

where the sum extends over all vectors $\omega = (\omega_1, \ldots, \omega_s)$ with integer components. If $|\nu| = p$ then

$$\omega_1^{2\nu_1} \ldots \omega_s^{2\nu_s} \leq \omega_1^{2p} + \ldots + \omega_s^{2p},$$

and therefore

$$\|D^\nu u\|^2 \leq J_p^2(u), \quad p = |\nu|.$$

Thus one obtains the inclusion

$$J_p^2(u) \leq |u|_{H^p}^2 \leq K J_p^2(u), \tag{A.3.2}$$

where $K = K(p, s)$ is the number of multi-indices ν of order p.

A direct generalization of the previous lemma to s space dimensions is

Lemma A.3.2. *For all $\epsilon > 0$ and all integers j, $k \geq 1$ it holds that*

$$J_j^2(u) \leq \epsilon J_{j+k}^2(u) + s\epsilon^{-j/k}\|u\|^2. \tag{A.3.3}$$

The proof follows by Parseval's relation and the argument given in the proof of Lemma A.3.1. Because of the inclusion (A.3.2), we also have

Lemma A.3.3. *For all $\epsilon > 0$ and all integers j, $k \geq 1$ it holds that*

$$|u|_{H^j}^2 \leq \epsilon |u|_{H^{j+k}}^2 + C\epsilon^{-j/k}\|u\|^2, \tag{A.3.4}$$

where C is independent of ϵ, u, and L.

Functions with compact support. Let $u \in C_0^\infty(\mathbf{R}^s)$ and denote

$$\|u\|^2 = \int_{\mathbf{R}^s} |u(x)|^2 \, dx_1 \ldots dx_s.$$

If $L > 0$ is so large that the support of u is contained in

$$[-L/2\,,\, L/2]^s,$$

then we can identify u with an L-periodic function and apply the previous results. Since the estimates are independent of L, we obtain the inequalities (A.3.3), (A.3.4) for all $u \in C_0^\infty$.

Maximum norm estimates of periodic functions. We start again with one space dimension and consider an L-periodic function

$$u \in C^{\infty}(\mathbf{R}), \quad u(x) \equiv u(x+L).$$

For convenience we assume throughout that $L \geq 1$; we first show

Lemma A.3.4. *For each integer $k \geq 1$ there is a constant C_k independent of u, $0 < \epsilon \leq 1$, and $L \geq 1$ with*

$$|u|_{\infty}^2 \leq \epsilon \|D^k u\|^2 + C_k \epsilon^{-1/(2k-1)} \|u\|^2.$$

Proof. From the representation of $u(x)$ as a Fourier sum and the Cauchy-Schwarz inequality we obtain

$$\begin{aligned} L|u(x)|^2 &\leq \Big\{ \sum_{\omega=-\infty}^{\infty} |\hat{u}(\omega)| \Big\}^2 \\ &= \Big\{ \sum_{\omega} |\hat{u}(\omega)| (\epsilon |2\pi\omega/L|^k + 1)(\epsilon |2\pi\omega/L|^k + 1)^{-1} \Big\}^2 \\ &\leq \Big\{ \sum_{\omega} |\hat{u}(\omega)|^2 (\epsilon |2\pi\omega/L|^k + 1)^2 \Big\} \Big\{ \sum_{\omega} (\epsilon |2\pi\omega/L|^k + 1)^{-2} \Big\} \\ &=: P_1 P_2. \end{aligned}$$

To estimate P_1, we observe that

$$(\epsilon |2\pi\omega/L|^k + 1)^2 \leq 2\epsilon^2 |2\pi\omega/L|^{2k} + 2$$

and find, by Parseval's relation,

$$P_1 \leq 2\epsilon^2 \|D^k u\|^2 + 2\|u\|^2.$$

To estimate P_2, we write

$$\begin{aligned} P_2 &= 1 + 2\sum_{\omega=1}^{\infty} \big\{ \epsilon (2\pi\omega/L)^k + 1 \big\}^{-2} \\ &\leq 1 + 2\int_0^{\infty} \big\{ \epsilon (2\pi\omega/L)^k + 1 \big\}^{-2} d\omega \\ &= 1 + 2\epsilon^{-1/k} \frac{L}{2\pi} \int_0^{\infty} \big\{ y^k + 1 \big\}^{-2} dy \\ &\leq c_k (1 + \epsilon^{-1/k} L) \leq 2 c_k \epsilon^{-1/k} L. \end{aligned}$$

For the last estimate we used the restriction

$$0 < \epsilon \leq 1 \leq L.$$

Hence it follows that

$$|u|_\infty^2 \leq L^{-1} P_1 P_2$$
$$\leq 4c_k \epsilon^{2-1/k} \|D^k u\|^2 + 4c_k \epsilon^{-1/k} \|u\|^2,$$

and if we call*

$$4c_k \epsilon^{2-1/k} = \tilde{\epsilon},$$

then the assertion follows.

A key result in obtaining Sobolev inequalities in s space dimensions is formulated next; a proof can be given by transformation to polar coordinates. This result is responsible for the requirement $k > s/2$ in Theorem A.3.6 and it shows why maximum norm estimates depend on the number s of space dimensions.

Lemma A.3.5. *The integral*

$$\int_{\mathbf{R}^s} \left\{ |y|^{2k} + 1 \right\}^{-1} dy_1 \ldots dy_s$$

is finite if and only if $k > s/2$.

Now let $u \in C^\infty(\mathbf{R}^s)$ be L-periodic in each variable x_l. The following result is (a version of) Sobolev's inequality.

Theorem A.3.6. *For each integer* $k \geq [s/2] + 1$ *there is a constant* $C_{s,k}$ *independent of* u, $0 < \epsilon \leq 1$, *and* $L \geq 1$ *with*

$$|u|_\infty^2 \leq \epsilon J_k^2(u) + C_{s,k} \epsilon^{-s/(2k-s)} \|u\|^2.$$

Proof. From the representation of u as a Fourier sum* we have that

*We can assume $0 < \tilde{\epsilon} \leq 1$ to be given and can solve for $\epsilon > 0$. Then $0 < \epsilon \leq 1$, if $4c_k \geq 1$.
*The sums extend over all vectors $\omega = (\omega_1, \ldots, \omega_s)$ with integer components ω_l.

$$
\begin{aligned}
L^s|u(x)|^2 &\le \{\sum_\omega |\hat{u}(\omega)|\}^2 \\
&\le \Big\{\sum_\omega |\hat{u}(\omega)|^2\{\epsilon|2\pi\omega_1/L|^k + \ldots + \epsilon|2\pi\omega_s/L|^k + 1\}^2\Big\} \\
&\quad \cdot \Big\{\sum_\omega \{\epsilon|2\pi\omega_1/L|^k + \ldots + \epsilon|2\pi\omega_s/L|^k + 1\}^{-2}\Big\} \\
&=: P_1 P_2.
\end{aligned}
$$

Here

$$
\begin{aligned}
P_1 &\le \sum_\omega |\hat{u}(\omega)|^2(s+1)\{\epsilon^2(2\pi\omega_1/L)^{2k} + \ldots + \epsilon^2(2\pi\omega_s/L)^{2k} + 1\} \\
&= (s+1)\{\epsilon^2 J_k^2(u) + \|u\|^2\}.
\end{aligned}
$$

To estimate P_2, we first sum only over those ω for which all components $\omega_l \ne 0$. Then the sum can be bounded by an integral:

$$
\begin{aligned}
\tilde{P}_2 &:= \sum_{\text{all } \omega_l \ne 0} \{\epsilon|2\pi\omega_1/L|^k + \ldots + \epsilon|2\pi\omega_s/L|^k + 1\}^{-2} \\
&\le \int_{\mathbf{R}^s} \{\epsilon|2\pi\omega_1/L|^k + \ldots + \epsilon|2\pi\omega_s/L|^k + 1\}^{-2}\, d\omega \qquad \text{(A.3.5)} \\
&= (\epsilon^{-1/k}\frac{L}{2\pi})^s \int_{\mathbf{R}^s} \{|y_1|^k + \ldots + |y_s|^k + 1\}^{-2}\, dy.
\end{aligned}
$$

It is not difficult to show that

$$
\{|y_1|^k + \ldots + |y_s|^k + 1\}^2 \ge s^{-k}|y|^{2k} + 1,
$$

and using the previous lemma, we obtain that the above integral is finite. Therefore,

$$
\tilde{P}_2 \le c_{s,k}(\epsilon^{-1/k}L)^s.
$$

If we sum the expressions in (A.3.5) over all ω with $\omega_s = 0$, $\omega_l \ne 0$ for $l = 1, \ldots, s-1$, say, then we obtain in the same way a bound of the corresponding sum by

$$
c_{s-1,k}(\epsilon^{-1/k}L)^{s-1}.
$$

Therefore,

$$
\begin{aligned}
P_2 &\le c'_{s,k}\big(1 + (\epsilon^{-1/k}L)^s\big) \\
&\le 2c'_{s,k}(\epsilon^{-1/k}L)^s.
\end{aligned}
$$

Only in the latter estimate have we used that $0 < \epsilon \leq 1 \leq L$. To summarize,

$$\begin{aligned} |u|_\infty^2 &\leq L^{-s} P_1 P_2 \\ &\leq c''_{s,k}\{\epsilon^{2-s/k} J_k^2(u) + \epsilon^{-s/k}\|u\|^2\}, \end{aligned}$$

and if we call

$$c''_{s,k}\epsilon^{2-s/k} = \tilde{\epsilon},$$

then the assertion follows.

One can also estimate *derivatives* of a function in maximum norm. The following result is a simple implication of the previous theorem and Lemma A.3.3. Again, we assume $u \in C^\infty(\mathbf{R}^s)$ to be L-periodic in each variable.

Lemma A.3.7. *Let* $|\nu| = j \geq 1$, $k \geq [s/2] + 1$. *There is a constant* C *independent of* u, $0 < \epsilon \leq 1$, *and* $L \geq 1$ *with*

$$|D^\nu u|_\infty^2 \leq \epsilon J_{j+k}^2(u) + C\epsilon^{-(2j+s)/(2k-s)}\|u\|^2, \quad |\nu| = j.$$

Proof. We apply the previous theorem to $D^\nu u$ instead of u and obtain

$$\begin{aligned} |D^\nu u|_\infty^2 &\leq \epsilon J_k^2(D^\nu u) + c_1\epsilon^{-s/(2k-s)}\|D^\nu u\|^2 \\ &\leq \epsilon|u|_{H^{j+k}}^2 + c_1\epsilon^{-s/(2k-s)}\{\epsilon_1|u|_{H^{j+k}}^2 + c_2\epsilon_1^{-j/k}\|u\|^2\} \\ &\leq \{\epsilon + c_1\epsilon_1\epsilon^{-s/(2k-s)}\}|u|_{H^{j+k}}^2 + c_1c_2\epsilon^{-s/(2k-s)}\epsilon_1^{-j/k}\|u\|^2. \end{aligned}$$

If we choose

$$\epsilon_1 \sim \epsilon\epsilon^{s/(2k-s)},$$

then the result follows.

Estimates by a product. It is sometimes convenient to have estimates by a product. For example, if we use Theorem A.3.6 for $k = s$, we obtain

$$|u|_\infty^2 \leq \epsilon J_s^2(u) + C\epsilon^{-1}\|u\|^2. \tag{A.3.6}$$

If $\epsilon = \|u\|/J_s(u) \leq 1$, then

$$|u|_\infty^2 \leq (1 + C)\|u\| J_s(u).$$

In the other case, $\|u\| > J_s(u)$, we apply (A.3.6) with $\epsilon = 1$:

$$|u|_\infty^2 \leq (1 + C)\|u\|^2.$$

Thus we have proved

$$|u|_\infty^2 \le (1+C)\|u\|\Big(J_s(u) + \|u\|\Big). \tag{A.3.7}$$

We apply this result to $D^\nu u$ instead of u and find

Lemma A.3.8. *There is a constant $C = C(j, s)$ independent of u and $L \ge 1$ with*

$$|D^\nu u|_\infty^2 \le C J_j(u)\Big(J_{s+j}(u) + J_j(u)\Big), \quad j = |\nu|.$$

Nonperiodic functions on an interval. Let $u = u(x)$, $0 \le x \le L$, denote a C^∞-function with values in $\mathbf{C}^n$. There are points x_0, x_1 with

$$|u(x_0)| = \min_{0\le x\le L} |u(x)|, \quad |u(x_1)| = \max_{0\le x\le L} |u(x)| = |u|_\infty.$$

From

$$|u(x_1)|^2 - |u(x_0)|^2 = \int_{x_0}^{x_1} \frac{d}{dx}\langle u(x), u(x)\rangle\, dx \le 2\,\|u\|\,\|u_x\|$$

we find that

$$\begin{aligned} |u|_\infty^2 &\le \frac{1}{L}\|u\|^2 + 2\,\|u\|\,\|u_x\| \\ &\le \epsilon\,\|u_x\|^2 + (\frac{1}{\epsilon} + \frac{1}{L})\,\|u\|^2 \end{aligned} \tag{A.3.8}$$

for any $\epsilon > 0$. Similarly,

$$\begin{aligned} \langle u, u_x\rangle\Big|_0^L &= \int_0^L \frac{d}{dx}\langle u, u_x\rangle\, dx \\ &= (u, u_{xx}) + \|u_x\|^2, \end{aligned}$$

and therefore

$$\begin{aligned} \|u_x\|^2 &\le \|u\|\,\|u_{xx}\| + 2\,|u|_\infty\,|u_x|_\infty \\ &\le \frac{\epsilon}{2}\|u_{xx}\|^2 + \frac{1}{2\epsilon}\|u\|^2 + \eta|u|_\infty^2 + \frac{1}{\eta}|u_x|_\infty^2 \end{aligned} \tag{A.3.9}$$

for any $\epsilon > 0, \eta > 0$. We can use these formulae to show

Lemma A.3.9. *There is $C > 0$ independent of $0 < \epsilon \le 1$, $L \ge 1$, and $u \in C^\infty[0, L]$ with*

$$\|u_x\|^2 \le \epsilon\|u_{xx}\|^2 + \frac{C}{\epsilon}\|u\|^2. \tag{A.3.10}$$

Proof. From (A.3.8) follows

$$|u|_\infty^2 \le \alpha\|u_x\|^2 + (\frac{1}{\alpha} + \frac{1}{L})\|u\|^2,$$

$$|u_x|_\infty^2 \le \beta\|u_{xx}\|^2 + (\frac{1}{\beta} + \frac{1}{L})\|u_x\|^2.$$

In (A.3.9) we choose $\eta = c/\sqrt{\epsilon}$, and estimate $|u|_\infty^2$ and $|u_x|_\infty^2$ using the above formulae with

$$\alpha = \frac{\sqrt{\epsilon}}{2c}, \quad \beta = \sqrt{\epsilon}.$$

Then the choice

$$c = 4(1 + \frac{\sqrt{\epsilon}}{L})$$

proves the result.

In the following lemma we generalize the formulae (A.3.8), (A.3.10). The dependence of the constants C_i on ϵ will be discussed below.

Lemma A.3.10. *Let $j \ge 0$, $k \ge 1$ denote integers, and let $0 < \epsilon \le 1$. There are constants $C_i = C_i(j,k,\epsilon)$, $i = 1, 2$, independent of $L \ge 1$ and $u \in C^\infty[0, L]$ with*

$$|D^j u|_\infty^2 \le \epsilon\|D^{j+k}u\|^2 + C_1\|u\|^2, \quad D = \frac{d}{dx},$$

$$\|D^j u\|^2 \le \epsilon\|D^{j+k}u\|^2 + C_2\|u\|^2.$$

Proof. First let $k = 1$. The estimates are shown for $j = 0$, and we assume that they are valid for some j. Then (by (A.3.10))

$$\begin{aligned}\|D^{j+1}u\|^2 &\le \epsilon\|D^{j+2}u\|^2 + \frac{C}{\epsilon}\|D^j u\|^2 \\ &\le \epsilon\|D^{j+2}u\|^2 + \frac{C}{\epsilon}\Big\{\epsilon_1\|D^{j+1}u\|^2 + C_2(\epsilon_1)\|u\|^2\Big\}.\end{aligned}$$

If we choose here $\epsilon_1 = \epsilon/(2C)$, then the desired estimate for $\|D^{j+1}u\|$ follows. Similarly, (by (A.3.8))

$$\begin{aligned}|D^{j+1}u|_\infty^2 &\le \epsilon\|D^{j+2}u\|^2 + \frac{2}{\epsilon}\|D^{j+1}u\|^2 \\ &\le \epsilon\|D^{j+2}u\|^2 + \frac{2}{\epsilon}\Big\{\epsilon_1\|D^{j+2}u\|^2 + C_2(\epsilon_1)\|u\|^2\Big\},\end{aligned}$$

and the choice $\epsilon_1 = \epsilon^2$ finishes the induction w.r.t. j. The generalization from $k = 1$ to arbitrary k is straightforward.

In the previous lemma we did not keep track of the dependence of C_i on ϵ. However, since the constants are independent of $L \geq 1$, we can determine this dependence by scaling the x-variable. One obtains

Theorem A.3.11. *Let $j \geq 0$, $k \geq 1$ denote integers. There are constants $\bar{C}_i = \bar{C}_i(j,k)$, $i = 1,2$, independent of $0 < \epsilon \leq 1$, $L \geq 1$, and $u \in C^\infty[0,L]$ with*

$$|D^j u|_\infty^2 \leq \epsilon \|D^{j+k} u\|^2 + \bar{C}_1 \epsilon^{-(2j+1)/(2k-1)} \|u\|^2,$$

$$\|D^j u\|^2 \leq \epsilon \|D^{j+k} u\|^2 + \bar{C}_2 \epsilon^{-j/k} \|u\|^2.$$

Proof. Suppose that $\tilde{u}(\tilde{x})$, $0 \leq \tilde{x} \leq L$, is given; we define for $0 < h \leq 1$

$$u(x) := \tilde{u}(hx), \quad 0 \leq x \leq L/h.$$

If one observes that

$$D^j u(x) = h^j D^j \tilde{u}(\tilde{x}), \quad \tilde{x} = hx,$$

$$\|u\|^2 = \frac{1}{h} \|\tilde{u}\|^2$$

and applies the previous lemma to $u(x)$ with $\epsilon = 1$, then the result follows.

L_2-estimates for functions in a strip. Suppose that $u = u(x)$ is a C^∞-function defined in the strip

$$x = (x_1, x_-), \quad 0 \leq x_1 \leq 1, \quad x_- = (x_2, \ldots, x_s) \in \mathbf{R}^{s-1},$$

which is 1-periodic in $x_2, \ldots, x_s$. For each fixed x_1 one obtains, by Fourier expansion,

$$u(x) = \sum_{\omega_-} \hat{u}(x_1, \omega_-) e^{2\pi i \langle \omega_-, x_- \rangle}$$

where the sum extends over all vectors

$$\omega_- = (\omega_2, \ldots, \omega_s) \text{ with } \omega_j \in \mathbf{Z}$$

and

$$\hat{u}(x_1, \omega_-) = \int_0^1 \cdots \int_0^1 u(x_1, x_-) e^{-2\pi i \langle \omega_-, x_- \rangle} \, dx_2 \ldots dx_s.$$

Parseval's relation yields

$$\|u\|^2 = \sum_{\omega_-} \|\hat{u}(\cdot, \omega_-)\|^2$$

with

$$\|\hat{u}(\cdot, \omega_-)\|^2 = \int_0^1 |\hat{u}(x_1, \omega_-)|^2 \, dx_1.$$

We first want to show that we can estimate the L_2-norm of *any* derivative by "clean" derivatives. As in the periodic case, let

$$J_j^2(u) = \sum_{l=1}^{s} \|D_l^j u\|^2, \quad D_l = \frac{\partial}{\partial x_l}.$$

Lemma A.3.12. *There is a constant $C = C(j)$ independent of u with*

$$\|D^\nu u\|^2 \le C J_j^2(u), \quad D^\nu = D_1^{\nu_1} \dots D_s^{\nu_s}, \quad j = \nu_1 + \dots + \nu_s.$$

Proof. With Parseval's relation,

$$\|D^\nu u\|^2 = \sum_{\omega_-} \|D_1^{\nu_1} \hat{u}(\cdot, \omega_-)\|^2 (2\pi\omega_2)^{2\nu_2} \dots (2\pi\omega_s)^{2\nu_s}. \tag{A.3.11}$$

Case 1. $\nu_1 = 0$. From

$$\omega_2^{2\nu_2} \dots \omega_s^{2\nu_s} \le \omega_2^{2j} + \dots + \omega_s^{2j}$$

one obtains the estimate

$$\|D^\nu u\|^2 \le \|D_2^j u\|^2 + \dots + \|D_s^j u\|^2.$$

Case 2. $\nu_1 = j$, thus $\nu_2 = \dots = \nu_s = 0$. Here trivially $\|D^\nu u\|^2 = \|D_1^j u\|^2$.

Case 3. $1 \le \nu_1 \le j - 1$. An application of Theorem A.3.11 yields

$$\|D_1^{\nu_1} \hat{u}(\cdot, \omega_-)\|^2 \le \epsilon \|D_1^j \hat{u}(\cdot, \omega_-)\|^2 + C_0 \epsilon^{-\nu_1/|\nu_-|} \|\hat{u}(\cdot, \omega_-)\|^2,$$

$$|\nu_-| = \nu_2 + \dots + \nu_s.$$

We can assume that

$$\omega_2^{2\nu_2} \dots \omega_s^{2\nu_s} \neq 0$$

and set

$$\epsilon = \left\{(2\pi\omega_2)^{2\nu_2}\dots(2\pi\omega_s)^{2\nu_s}\right\}^{-1}.$$

Then we find from (A.3.11)

$$\begin{aligned}\|D^\nu u\|^2 &\le \sum_{\omega_-}\|D_1^j\hat{u}(\cdot,\omega_-)\|^2\\ &+ C_1\sum_{\omega_-}\|\hat{u}(\cdot,\omega_-)\|^2\left(\omega_2^{2\nu_2}\dots\omega_s^{2\nu_s}\right)^{1+\nu_1/|\nu_-|}.\end{aligned}$$

If we set

$$\alpha_l = \nu_l(1+\nu_1/|\nu_-|),$$

then

$$\left(\omega_2^{2\nu_2}\dots\omega_s^{2\nu_s}\right)^{1+\nu_1/|\nu_-|} = \omega_2^{2\alpha_2}\dots\omega_s^{2\alpha_s}$$

and

$$\alpha_2+\dots+\alpha_s = \nu_2+\dots+\nu_s+\nu_1 = j.$$

Therefore,

$$\omega_2^{2\alpha_2}\dots\omega_s^{2\alpha_s} \le \omega_2^{2j}+\dots+\omega_s^{2j},$$

and the result follows.

Next we generalize Lemma A.3.2 to functions defined in a strip.

Lemma A.3.13. *Let $j, k \ge 1$ denote integers. There is a constant $C = C(j,k)$ independent of $0 < \epsilon \le 1$ and u with*

$$J_j^2(u) \le \epsilon J_{j+k}^2(u) + C\epsilon^{-j/k}\|u\|^2.$$

Proof. We first estimate $D_1^j u$:

$$\|D_1^j u\|^2 = \sum_{\omega_-}\|D_1^j\hat{u}(\cdot,\omega_-)\|^2.$$

For each fixed ω_- we can apply Theorem A.3.11 to $\hat{u}(\cdot,\omega_-)$, and summation over ω_- gives the desired estimate of $D_1^j u$. For $l = 2,\dots,s$,

$$\|D_l^j u\|^2 = \sum_{\omega_-} \|\hat{u}(\cdot, \omega_-)\|^2 (2\pi\omega_l)^{2j},$$

and we can argue as in the proof on Lemma A.3.1.

Maximum norm estimates for functions in a strip. As before, let $u = u(x)$ denote a C^∞-function defined in the strip

$$x = (x_1, x_-), \quad 0 \le x_1 \le 1, \quad x_- = (x_2, \ldots, x_s) \in \mathbf{R}^{s-1},$$

which is 1-periodic in $x_2, \ldots, x_s$. The following estimate of $|u|_\infty$ is sometimes useful since it requires only one differentiation in the nonperiodic x_1-direction. Specializations of the result to 2- and 3-dimensional strips will be given below.

Theorem A.3.14. *Let β, γ denote integers with*

$$2\alpha := \beta + \gamma > s - 1 \text{ and } \gamma \ge \beta \ge 0.$$

There is a constant $C = C(s)$ independent of u and

$$0 < \epsilon, \epsilon', \epsilon'' \le 1$$

with

$$|u|_\infty^2 \le C\epsilon^{2\alpha-(s-1)} \Big\{ \epsilon' \tilde{J}_\beta^2(D_1 u) + \frac{1}{\epsilon'} \tilde{J}_\gamma^2(u) \Big\} + C\epsilon^{-(s-1)} \Big\{ \epsilon'' \|D_1 u\|^2 + \frac{1}{\epsilon''} \|u\|^2 \Big\}.$$

Here we have used the notation

$$\tilde{J}_j^2(v) = \|D_2^j v\|^2 + \ldots + \|D_s^j v\|^2.$$

Proof. From the Fourier representation of $u(x)$,

$$\begin{aligned} |u(x)|^2 &\le \Big\{ \sum_{\omega_-} |\hat{u}(x_1, \omega_-)| \Big\}^2 \\ &\le \Big\{ \sum_{\omega_-} |\hat{u}(x_1, \omega_-)|^2 \big(1 + \epsilon^{2\alpha} |\omega_-|^{2\alpha}\big) \Big\} \\ &\quad \cdot \Big\{ \sum_{\omega_-} \big(1 + \epsilon^{2\alpha} |\omega_-|^{2\alpha}\big)^{-1} \Big\} \\ &=: P_1 P_2. \end{aligned}$$

As in the proof of Theorem A.3.6,

$$P_2 \le C_1 \epsilon^{-(s-1)}.$$

Also, by (A.3.8),

$$|\hat{u}(x_1,\omega_-)|^2 \le \|\hat{u}(\cdot,\omega_-)\|^2 + 2\,\|\hat{u}(\cdot,\omega_-)\|\,\|D_1\hat{u}(\cdot,\omega_-)\|.$$

This yields

$$\sum_{\omega_-} |\hat{u}(x_1,\omega_-)|^2 \le \|u\|^2 + \epsilon''\|D_1 u\|^2 + \frac{1}{\epsilon''}\|u\|^2.$$

Furthermore,

$$\begin{aligned}|\hat{u}(x_1,\omega_-)|^2|\omega_-|^{2\alpha} &\le \|\hat{u}(\cdot,\omega_-)\|^2|\omega_-|^{2\alpha}\\ &\quad + 2\|\hat{u}(\cdot,\omega_-)\|\,|\omega_-|^{\gamma}\|D_1\hat{u}(\cdot,\omega_-)\|\,|\omega_-|^{\beta}\\ &\le \|\hat{u}(\cdot,\omega_-)\|^2|\omega_-|^{2\alpha}\\ &\quad + \epsilon'\|D_1\hat{u}(\cdot,\omega_-)\|^2|\omega_-|^{2\beta} + \frac{1}{\epsilon'}\|\hat{u}(\cdot,\omega_-)\|^2|\omega_-|^{2\gamma}.\end{aligned}$$

If we observe that $\gamma \ge \alpha$ and sum over ω_- , then the assertion follows.

A 2D strip. In the previous theorem we set

$$s = 2, \quad \alpha = \beta = \gamma = 1, \quad \epsilon' = \epsilon'' = \epsilon,$$

and obtain the estimate

$$\text{(A.3.12)} \quad |u|_\infty^2 \le C\Big\{\epsilon^2\|u_{xy}\|^2 + \|u_y\|^2 + \|u_x\|^2 + \frac{1}{\epsilon^2}\|u\|^2\Big\}, \quad 0 < \epsilon \le 1.$$

As an implication we note

Lemma A.3.15. *Consider a C^∞-function $u = u(x,y)$ defined in the strip*

$$0 \le x \le 1, \quad -\infty < y < \infty,$$

which is 1-periodic in y. Then

$$|u|_\infty^2 \le \text{const}\Big\{\|u\|\|u_{xy}\| + \|u_x\|^2 + \|u_y\|^2 + \|u\|^2\Big\}.$$

Proof. If

$$\epsilon^2 = \frac{\|u\|}{\|u_{xy}\|} \le 1,$$

then (A.3.12) yields the estimate. Otherwise $\|u_{xy}\| < \|u\|$, and we obtain the result from (A.3.12) upon choosing $\epsilon = 1$.

A 3D strip. In Theorem A.3.14 we set

$$s = 3, \quad \beta = 1, \quad \gamma = 2, \quad \epsilon' = 1, \quad \epsilon'' = \epsilon,$$

and obtain, for $0 < \epsilon \le 1$,

$$\text{(A.3.13)} \qquad \begin{aligned} |u|_\infty^2 \le C\epsilon\Big\{ \|u_{xy}\|^2 + \|u_{xz}\|^2 + \|u_{yy}\|^2 + \|u_{zz}\|^2 \Big\} \\ + \frac{C}{\epsilon}\|u_x\|^2 + \frac{C}{\epsilon^3}\|u\|^2. \end{aligned}$$

Similarly as before, we can derive from this an estimate by products.

Lemma A.3.16. *Consider a C^∞-function $u = u(x, y, z)$ defined in the strip*

$$0 \le x \le 1, \quad -\infty < y, z < \infty,$$

which is 1-periodic in y and in z. Then

$$|u|_\infty^2 \le \text{const}\Big\{ \|u_x\| H + \|u_x\|^2 + \|u\|^{1/2} H^{3/2} + \|u\|^2 \Big\}$$

with

$$H^2 = \|u_{xy}\|^2 + \|u_{xz}\|^2 + \|u_{yy}\|^2 + \|u_{zz}\|^2.$$

Proof. If

$$\epsilon = H^{-1}\Big\{ \|u_x\| + \|u\|^{1/2} H^{1/2} \Big\} \le 1$$

then (A.3.13) yields the estimate. Otherwise

$$H < 2\|u_x\| \quad \text{or} \quad H < 2\|u\|^{1/2} H^{1/2},$$

and we obtain the result from (A.3.13) upon choosing $\epsilon = 1$.

Discrete Sobolev inequalities for periodic functions. Let $m \in \{1, 2, \ldots\}$, $h = (2m+1)^{-1}$, and define a grid $x_\nu = \nu h$, $\nu = 0, \pm 1, \pm 2, \ldots$. Suppose that $u = u(x)$ is a 1-periodic gridfunction, $u(x) \equiv u(x+1)$, which is defined at all gridpoints $x = x_\nu$. We want to show that discrete analogs of the Sobolev inequalities hold with constants independent of h. For notations, see Section 3.2.2.

Lemma A.3.17. *For $\epsilon > 0$, j, $k \ge 1$ it holds that*

$$\|D_+^j u\|_h^2 \le \epsilon \|D_+^{j+k} u\|_h^2 + C_1(\epsilon)\|u\|_h^2,$$

$$|D_+^j u|_\infty^2 \le \epsilon \|D_+^{j+k} u\|_h^2 + C_2(\epsilon)\|u\|_h^2$$

with constants $C_{1,2}(\epsilon)$ independent of h.

Proof. Denote by $w = w(x)$ the Fourier interpolant of u. Since

$$D_+^j u_\nu = D_+^j w_\nu = D^j w(\xi_\nu)$$

for some point ξ_ν, we have $|D_+^j u|_\infty \le |D^j w|_\infty$. Now the second inequality follows from Lemma A.3.7 (with s=1) and Theorem A.2.2. Clearly, the first inequality follows from the second.

In the same way we can obtain discrete Sobolev inequalities for periodic functions in more space dimensions.

Nonperiodic discrete functions on an interval. Now we assume that $u_\nu = u(x_\nu)$, $x_\nu = \nu h$, $\nu = 0, 1, 2, \dots, N$, $Nh = 1$, is defined for gridpoints x_ν with $0 \le x_\nu \le 1$. As before, $D_+ u_\nu = (u_{\nu+1} - u_\nu)/h$ for $\nu = 0, \dots, N-1$, and we use the notation

$$(u, v)_{r,s} = h \sum_{\nu=r}^{s} \bar{u}_\nu v_\nu, \qquad \|u\|_{r,s}^2 = (u, u)_{r,s}.$$

We have

$$\begin{aligned}(u, D_+ v)_{r,s} &= \sum_{\nu=r}^{s} \bar{u}_\nu v_{\nu+1} - \sum_{\nu=r}^{s} \bar{u}_\nu v_\nu = \sum_{\nu=r+1}^{s+1} \bar{u}_{\nu-1} v_\nu - \sum_{\nu=r}^{s} \bar{u}_\nu v_\nu \\ &= -(D_- u, v)_{r+1,s+1} + \bar{u}v\Big|_r^{s+1},\end{aligned} \tag{A.3.14}$$

where

$$\bar{u}v\Big|_r^{s+1} = \bar{u}_{s+1} v_{s+1} - \bar{u}_r v_r.$$

Let

$$|u_k| = \min_\nu |u_\nu|, \quad |u_l| = \max_\nu |u_\nu|$$

and assume for definiteness that $l > k$. By (A.3.14),

$$\begin{aligned}|u_l|^2 - |u_k|^2 &= (u, D_+ u)_{k,l-1} + (D_- u, u)_{k+1,l} \\ &\le \epsilon \|D_+ u\|_{0,N-1}^2 + \frac{1}{\epsilon} \|u\|_{0,N}^2.\end{aligned}$$

Since $|u_k|^2 \le \|u\|_{0,N}^2$, we obtain the analog of (A.3.8):

$$\max_{0 \le \nu \le N} |u_\nu|^2 \le \epsilon \|D_+ u\|_{0,N-1}^2 + (\epsilon^{-1} + 1) \|u\|_{0,N}^2. \tag{A.3.15}$$

This yields, for any $p \geq 0$,

$$\max_{0 \leq \nu \leq N-p} |D_+^p u|^2 \leq \epsilon \, \|D_+^{p+1} u\|_{0,N-1-p}^2 \; + \; (\epsilon^{-1} + 1) \, \|D_+^p u\|_{0,N-p}^2.$$

If we apply (A.3.14) with $v = D_- u$, we obtain

$$\begin{aligned}(u, D_+ D_- u)_{1,N-1} &= -\|D_- u\|_{2,N}^2 + \bar{u} D_- u\Big|_1^N \\ &= -\|D_+ u\|_{1,N-1}^2 + \bar{u}_N D_+ u_{N-1} - \bar{u}_1 D_+ u_0 \\ &= -\|D_+ u\|_{0,N-1}^2 + \bar{u}_N D_+ u_{N-1} - \bar{u}_0 D_+ u_0.\end{aligned}$$

Therefore,

$$\|D_+ u\|_{0,N-1}^2 \leq \|u\|_{1,N-1} \|D_+^2 u\|_{0,N-2} \; + \; 2 \, \{\max_\nu |u_\nu|\} \, \{\max_\nu |D_+ u_\nu|\}.$$

Now we can proceed as in the proof of Lemma A.3.9 and obtain

$$\text{(A.3.16)} \qquad \|D_+ u\|_{0,N-1}^2 \leq \epsilon \, \|D_+^2 u\|_{0,N-2}^2 \; + \; \frac{C}{\epsilon} \, \|u\|_{0,N}^2.$$

As in the continuous case, we can use (A.3.15), (A.3.16) to prove

Lemma A.3.18. *Let $h = 1/N$ and let $u_\nu = u(x_\nu)$, $x_\nu = \nu h$, $\nu = 0, 1, 2, \ldots, N$, denote a gridfunction. Then the following discrete Sobolev inequalities hold:*

$$\|D_+^j u\|_{0,N-j}^2 \leq \epsilon \|D_+^{j+k} u\|_{0,N-j-k}^2 + C_1(\epsilon) \|u\|_{0,N}^2,$$

$$\max_{0 \leq \nu \leq N-j} |D_+^j u_\nu|^2 \leq \epsilon \|D_+^{j+k} u\|_{0,N-j-k}^2 + C_2(\epsilon) \|u\|_{0,N}^2, \quad k \geq 1,$$

with constants independent of h.

Discrete analogs of our estimates for functions defined in a strip are also valid; this can be shown by Fourier interpolation in the periodic directions.

APPENDIX

4 Application of the Arzela-Ascoli Theorem

We will use the following result proved in analysis. It is (a special case of) the Arzela-Ascoli Theorem.

Theorem A.4.1. *Let $\Omega \subset \mathbf{R}^s$ be a closed bounded set, and let $u_m : \Omega \to \mathbf{C}^n$ denote a sequence of functions with the following properties:*

(i) *For each $\epsilon > 0$, there is a $\delta > 0$ independent of m with*

$$|u_m(y) - u_m(y')| \leq \delta \ \text{ if } \ |y - y'| \leq \epsilon, \quad y, y' \in \Omega.$$

(ii) *There is a K independent of m with*

$$|u_m(y)| \leq K \text{ for all } \ y \in \Omega \ \text{ and all } m.$$

Then there is a continuous function $u : \Omega \to \mathbf{C}^n$, and a sequence of indices $m_j \to \infty$ with

$$\max_{y \in \Omega} |u_{m_j}(y) - u(y)| \to 0 \ \text{ as } \ m_j \to \infty.$$

To apply the result, assume that we have a sequence of functions

$$u_m(x, t), \quad 0 \leq x \leq 1, \quad 0 \leq t \leq T,$$

which is *uniformly smooth*; i.e., $u_m \in C^\infty$ for all m, and for all nonnegative integers p, q there is a constant $C(p,q)$ independent of m with

$$\max_{x,t} \left| \frac{\partial^{p+q}}{\partial x^p \partial t^q} u_m(x,t) \right| \le C(p,q).$$

We want to show:

Theorem A.4.2. *There is a function $u \in C^\infty$, and a sequence $m_j \to \infty$ with*

$$\max_{x,t} \left| \frac{\partial^{p+q}}{\partial x^p \partial t^q} u_{m_j}(x,t) - \frac{\partial^{p+q}}{\partial x^p \partial t^q} u(x,t) \right| \to 0 \quad as \quad m_j \to \infty.$$

In short, a subsequence of u_m converges along with all its derivatives to a C^∞-limit.

Proof. By the Arzela-Ascoli Theorem there is a continuous u and a sequence $m = m_j \to \infty$ with

$$|u_m - u|_\infty \to 0 \quad \text{as} \quad m = m_j \to \infty.$$

Now apply the Arzela-Ascoli result to the sequence

$$\partial u_{m_j} / \partial x,$$

and obtain the existence of a continuous v and a sequence $j_k \to \infty$ with

$$\left| \frac{\partial u_m}{\partial x} - v \right|_\infty \to 0 \quad \text{as} \quad m = m_{j_k} \to \infty.$$

From

$$u_m(x,t) - u_m(0,t) = \int_0^x \frac{\partial u_m}{\partial x}(\xi,t)\,d\xi$$

we conclude that for $m = m_{j_k} \to \infty$,

$$u(x,t) - u(0,t) = \int_0^x v(\xi,t)\,d\xi.$$

Hence u is differentiable w.r.t. x, and

(A.4.1) $$\frac{\partial u}{\partial x} = v.$$

Let us show that

(A.4.2) $$\left| \frac{\partial u_m}{\partial x} - v \right|_\infty \to 0 \text{ as } m_j \to \infty;$$

i.e., we have convergence as $m_j \to \infty$, not just for the subsequence m_{j_k}. This follows from a general argument which we formulate for a sequence of numbers only. The generalization to a sequence of functions is straightforward.

Lemma A.4.3. *Let b_j denote a sequence of numbers, and suppose that*

(i) *Each subsequence b_{j_k} has a convergent subsequence.*
(ii) *All convergent subsequences of b_j converge to the same limit b.*

Then $b_j \to b$ as $j \to \infty$.

Proof. If b_j does not converge to b, then there is an $\epsilon > 0$ and a sequence $j_k \to \infty$ with

$$|b_{j_k} - b| \geq \epsilon \quad \text{for all} \quad j_k.$$

However, according to (i), (ii) there is a subsequence of b_{j_k} converging to b, a contradiction.

Let us continue the proof of Theorem A.4.2. Since all possible limits of subsequences of $\partial u_{m_j}/\partial x$ equal $\partial u/\partial x$ (see (A.4.1)), we obtain the result (A.4.2). Clearly, the existence of $\partial u/\partial t$ and convergence

$$\left|\frac{\partial u_m}{\partial t} - \frac{\partial u}{\partial t}\right|_\infty \to 0 \quad \text{as} \quad m = m_j \to 0$$

follow in the same way. Also, we can apply the previous arguments to $\partial u_{m_j}/\partial x$, and differentiability of $\partial u/\partial x$ follows, etc. This proves the theorem.

References

AGMON, S. (1962). "Problème mixte pour les équations hyperboliques d'ordre supérieur". In: *Les équations aux dérivées partielles,* Centre Nat. de la Recherche Sci., Paris, pp. 1-6.

AGRANOVICH, M. S., and VISHIK, M. I. (1964). "Elliptic problems with a parameter and parabolic problems of a general form". *Uspehi Mat. Nauk* **19**:3, pp. 53-161. (Russian; English translation in *Russian Math. Surveys* **19**:3, pp. 53-157.)

AGRANOVICH, M. S. (1971). "Boundary value problems for systems with a parameter". *Mat. Sbornik* **84**, pp. 25-65.

AGRANOVICH, M. S. (1972). "Theorem of matrices depending on parameters and its application to hyperbolic systems". *Functional Anal. Appl.* **6**, pp. 85-93.

BRENNER, P. (1966). "The Cauchy problem for symmetric hyperbolic systems in L_p". *Math. Scand.* **19**, pp. 27-37.

CARRIER, G. F., and PEARSON, C. E. (1976). *Partial differential equations: Theory and Techniques*, Academic Press, New York.

CHORIN, A., and MARSDEN, J. E. (1979). *A mathematical introduction to fluid mechanics*, Springer-Verlag, New York.

COURANT, R., FRIEDRICHS, K. O., and Lewy, H. (1928). "Über die partiellen Differenzengleichungen der mathematischen Physik". *Math. Ann.* **100**, pp. 32-74.

COURANT, R., and FRIEDRICHS, K. O. (1948). *Supersonic flow and shock waves*, Interscience Publishers, New York.

COURANT, R., and HILBERT, D. (1962). *Methods of mathematical physics*, Vol. **II**, Interscience Publishers, New York.

DIPERNA, R. (1985). "Measure-valued solutions to conservation laws". *Arch. Rational Mech. Anal.* **80**, pp. 223-270.

EIDELMAN, S. D. (1964). *Parabolicheskie sistemy*, Nauka, Moscow. Translation: *Parabolic systems*, North-Holland, Amsterdam (1969).

ESKIN, G. I. (1973). *Boundary value problems for elliptic pseudo-differential equations*, Moscow; Transl. Math. Monographs **52**, Amer. Math. Soc., Providence, R.I. (1981).

ESKIN, G. I. (1983). "Initial-boundary value problem for hyperbolic equations". In: *Proceedings of the International Congress of Mathematicians*, Warszawa, pp. 1165-1173.

FRIEDLAND, S. J., ROBBIN, J. W., and SYLVESTER, J. H. (1984). "On the crossing rule". *Comm. Pure Appl. Math.* **37**, pp. 19-37.

FRIEDMAN, A. (1964). *Partial differential equations of parabolic type*, Prentice-Hall; reprint R.E. Krieger Publishing Co., Malabar, (1983).

FRIEDRICHS, K.O. (1954). "Symmetric hyperbolic linear differential equations". *Comm. Pure Appl. Math.* **7**, pp. 345-390.

FRIEDRICHS, K.O. (1958). "Symmetric positive linear differential equations". *Comm. Pure Appl. Math.* **11**, pp. 333-418.

GÅRDING, L. (1951). "Linear hyperbolic partial differential equations with constant coefficients". *Acta Math.* **85**, pp. 1-62.

GÅRDING, L. (1953). "Dirichlet's problem for linear elliptic partial differential equations". *Math. Scand.* **1**, pp. 55-73.

GARABEDIAN, P. R. (1964). *Partial differential equations*, Wiley, New York.

GILBARG, D. (1951). "The existence and limit behavior of the one-dimensional shock layer". *Amer. J. Math.* **73**, pp. 256-274.

GLIMM, J. (1965). "Solutions in the large for nonlinear hyperbolic systems of equations". *Comm. Pure Appl. Math.* **18**, pp. 695-715.

GLIMM, J., and LAX, P. (1970). *Decay of solutions of systems of nonlinear hyperbolic conservation laws*, Mem. Amer. Math. Soc. No. **101**, Amer. Math. Soc., Providence, RI

GURTIN, M., and MARTINS, L. (1976). "Cauchy's theorem in classical physics". *Arch. Rational Mech. Anal.* **60**, pp. 305-324.

GUSTAFSSON, B., KREISS, H. -O., and SUNDSTRÖM, A. (1972). "Stability theory for difference approximations of mixed initial boundary value problems II". *Math. Comp.* **26**, pp. 649-686.

GUSTAFSSON, B., and SUNDSTRÖM, A. (1978). "Incompletely parabolic problems in fluid dynamics". *SIAM J. Appl. Math.* **35**, pp. 343-357.

HADAMARD, J. (1921). *Lectures on Cauchy's problem in linear partial differential equations*, Yale.

HARABETIAN, E. (1986). "A convergent series expansion for hyperbolic systems of conservation laws". *Trans. Amer. Math. Soc.* **294**, pp. 383-424.

HENSHAW, W. D., KREISS, H. -O., and REYNA, L. G. (1988). *On the smallest scale for the incompressible Navier-Stokes equations*, ICASE Report No. 88-8.

HERNQUIST, C. D. (1988). "Smoothly symmetrizable hyperbolic systems of partial differential equations". To appear in *Math. Scand.*

HERSCH, R. (1963). "Mixed problems in several variables". *J. Math. Mech.* **12**, pp. 317-334.

HOPF, E. (1950). "The partial differential equation $u_t + uu_x = \mu u_{xx}$." *Comm. Pure Appl. Math.* **3**, pp. 201-230.

HUGHES, T. J. R., and MARSDEN, J. E. (1976). *A short course in fluid mechanics*, Publish or Perish, Wilmington, DE.

JOHN, F. (1974). "Formation of irregularities in one-dimensional nonlinear wave propagation". *Comm. Pure Appl. Math.* **27**, pp. 377-405.

JOHN, F. (1978). "Algebraic conditions for hyperbolicity of systems of partial differential equations". *Comm. Pure Appl. Math.* **31**, pp. 89-106.

KREISS, H. -O. (1959). "Über Matrizen die beschränkte Halbgruppen erzeugen". *Math. Scand.* **7**, pp. 71-80.

KREISS, H. -O. (1960). "Über die Lösung von Anfangsrandwertaufgaben für partielle Differentialgleichungen mit Hilfe von Differenzengleichungen". *Transactions of the Royal Institute of Technology, Stockholm, Sweden*, Nr. **166**.

KREISS, H. -O. (1963). "Über sachgemässe Cauchyprobleme". *Math. Scand.* **13**, pp. 109-128.

KREISS, H. -O. (1968). "Stability theory for difference approximations of mixed initial boundary value problems I". *Math. Comp.* **22**, pp. 703-714.

KREISS, H. -O. (1970). "Initial boundary value problems for hyperbolic systems". *Comm. Pure Appl. Math.* **23**, pp. 277-298.

KREISS, H. -O., and OLIGER, J. (1973). *Methods for the approximate solution of time dependent problems*, Garp Publications No. 10, Geneva.

KREISS, H. -O. (1985). *Problems with different time scales*, Academic Press, New York.

KREISS, H. -O., LORENZ, J., and NAUGHTON, M. (1988). "Convergence of the solutions of the compressible to the solutions of the incompressible Navier-Stokes equations". To appear.

LADYZHENSKAYA, O. A., SOLONNIKOV, V. A., and URALCEVA, N. N. (1968). *Linear and quasilinear equations of parabolic type*, Translation from the Russian, Amer. Math. Soc., Providence, RI

LADYZHENSKAYA, O. A. (1969). *The mathematical theory of viscous incompressible flow*, Translation from the Russian, Gordan and Breach, New York.

LADYZHENSKAYA, O. A. (1984). *The boundary value problems of mathematical physics*, Translation from the Russian, Springer-Verlag, New York.

LANDAU, L. D., and LIFSHITZ, E. M. (1959). *Fluid mechanics*, Translation from the Russian, Pergamon Press, Oxford.

LAX, P. D. (1964). "Development of singularities of solutions of nonlinear hyperbolic partial differential equations". *J. Math. Phys.* **5**, pp. 611-613.

LAX, P. D. (1957). "Hyperbolic systems of conservation laws II ". *Comm. Pure Appl. Math.* **10**, pp. 537-556.

LAX, P. D. (1972). *Hyperbolic systems of conservation laws and the mathematical theory of shock waves*, SIAM Regional Conference Series in Appl. Math. No. **11**.

LERAY, J. (1953). *Hyperbolic differential equations*, The Institute for Advanced Study, Princeton, N.J.

LIONS, J. L. (1961). *Équations différentielles operationelles et problèmes aux limites*, Springer-Verlag, Berlin.

LIU, T. P. (1983). *Quasilinear hyperbolic partial differential equations and the mathematical theory of shock waves*, Contemporary Mathematics Vol. **17**, Amer. Math. Soc., Providence, RI

MAJDA, A., and OSHER, S. (1975). "Initial-boundary value problems for hyperbolic equations with uniformly characteristic boundary". *Comm. Pure Appl. Math.* **28**, pp. 607-675.

MAJDA, A. (1984). *Compressible fluid flow and systems of conservation laws in several space variables*, Applied Mathematical Sciences Vol. **53**, Springer-Verlag, New York.

MATANO, J. (1982). "Nonincrease of the lap-number of a solution for a one-dimensional semilinear parabolic equation". *J. Fac. Sci., Univ. Tokyo*, Sect. I A, Math. **29**, No. 2, pp. 401-441.

MEYER, R. (1971). *Introduction to mathematical fluid dynamics*, Interscience Publishers, New York.

MICHELSON, D. (1980). *Initial-boundary value problems for hyperbolic equations and their difference approximation with uniformly characteristic boundary*, PhD Thesis, Dept. of Math. Sci., Tel-Aviv University.

MICHELSON, D. (1983). "Stability theory of difference approximations for multi-dimensional initial-boundary value problems". *Math. Comp.* **40**, pp. 1-45.

MICHELSON, D. (1987). "Convergence theorem for difference approximations of hyperbolic quasi-linear initial-boundary value problems". *Math. Comp.* **49**, pp. 445-459.

MICHELSON, D. (1988). *Initial-boundary value problems for incomplete singular perturbations of hyperbolic systems*, Report. Inst. of Math., Hebrew University, Jerusalem 91904, Israel.

MIZOHATA, S. (1956). "Le problème de Cauchy pour les équations paraboliques". *J. Math. Soc. Japan* **8**, pp. 269-299.

NIRENBERG, L. (1970). "Pseudo-differential operators". In: *Proc. Symp. Pure Math.* **16**, pp. 149-167.

NIRENBERG, L. (1973). *Lectures on linear partial differential equations*, Regional Conf. Series in Math. **20**, Amer. Math. Soc., Providence, RI

OLEINIK, O. A. (1957). "Discontinuous solutions of nonlinear differential equations". *Uspehi Mat. Nauk* **12**, No. 3(75), pp. 3-73. (Russian; English Translation in *Amer. Math. Soc. Transl.* (2) **26**, pp. 95-172.)

OLEINIK, O. A. (1959). "Uniqueness and stability of the generalized solutions of the Cauchy problem for a quasi-linear equation". *Uspehi Mat. Nauk* **14**, No. 2(86), pp. 165-170. (Russian; English Translation in *Amer. Math. Soc. Transl.* (2) **33**, pp. 285-290.)

OLIGER, J., and SUNDSTRÖM, A. (1978). "Theoretical and practical aspects of some initial boundary value problems in fluid dynamics". *SIAM J. Appl. Math.* **35**, pp. 419-446.

OSHER, S. (1972). "Stability of parabolic difference approximations to certain mixed initial boundary value problems". *Math. Comp.* **26**, pp. 13-39.

PETROVSKII, I. G. (1937). "Über das Cauchysche Problem für Systeme von partiellen Differentialgleichungen". *Mat. Sbornik. N. S.* **44**, pp. 814-868.

PETROVSKII, I.G. (1938). "Über das Cauchyproblem für ein System linearer partieller Differentialgleichungen im Gebiete der nichtanalytischen Funktionen". *Bull. Univ. Etat. Moscow,* Ser. Int. Sect. A Fasc. **7**, pp. 1-74.

PETROVSKII, I. G. (1954). *Lectures on partial differential equations*, Interscience Publishers, New York.

PROTTER, M. H., and WEINBERGER, H. F. (1967). *Maximum principles in differential equations*, Prentice-Hall, Englewood Cliffs, NJ

RALSTON, J. V. (1971). "Note on a paper of Kreiss". *Comm. Pure Appl. Math.* **24**, pp. 759-762.

RAUCH, J. (1972). "L_2 is a continuable condition for Kreiss' mixed problems". *Comm. Pure Appl. Math.* **25**, pp. 265-285.

RAUCH, J. (1972). "Energy and resolvent inequalities for hyperbolic mixed problems". *J. Differential Equations* **11**, pp. 528-450.

RAUCH, J. (1973). "General theory of hyperbolic mixed problems". In: *Proc. Symp. Pure Math.* **23**, pp. 161-166.

RICHTMEYER, R. D., and MORTON, K. W. (1967). *Difference methods for initial value problems*, Interscience Publishers, New York.

SAKAMOTO, R. (1970). "Mixed problems for hyperbolic equations, I, II ". *J. Math. Kyoto Univ.* **10**; *Matematika* **16**:1, pp. 62-99.

SAKAMOTO, R. (1982). *Hyperbolic boundary value problems*, Translated from the Japanese, Cambridge University Press, Cambridge.

SARASON, L. (1977). "Hyperbolic and other symmetrizable symbols in regions with corners and edges". *Indiana Univ. Math. J.* **26**, pp. 1-39.

SERRIN, J. (1959). "Mathematical principles of classical fluid mechanics". In: S. Flügge, *Handbuch der Physik*, Band **8**, Springer-Verlag, Berlin.

SJÖBERG, A. (1970). "On the Korteweg-de Vries equation: existence and uniqueness". *J. Math. Anal. Appl.* **29**, pp. 569-579.

SMOLLER, J. (1983). *Shock waves and reaction-diffusion equations*, Springer-Verlag, New York.

SOLONNIKOV, V. A. (1965). "On boundary-value problems for linear parabolic systems of differential equations of general form". *Trudy Mat. Inst. Steklov* **83**, pp. 3-163.

STRANG, W. G. (1966). "Necessary and insufficient conditions for well posed Cauchy problems". *J. Differential Equations* **2**, pp. 107-114.

STRIKWERDA, J. (1977) "Initial boundary value problems for incomplete parabolic systems". *Comm. Pure Appl. Math.* **30**, pp. 797-822.

TEMAM, R. (1983). *Navier-Stokes equations and nonlinear functional analysis*, CBMS-NSF Regional Conference Series in Appl. Math. No. **41**.

TEMAM, R. (1984). *Navier-Stokes equations*, North-Holland, Amsterdam.

VOLEVICH, L., and GINDINKIN, S. (1980). "The method of energy estimates in mixed problems". *Russian Math. Surveys* **35**:5, pp. 57-137.

VON WAHL, W. (1985). *The equations of Navier-Stokes and abstract parabolic equations*, Vieweg & Sohn, Braunschweig.

WHITHAM, G. B. (1974). *Linear and nonlinear waves*, Interscience Publishers, New York.

YAMAGUTI, M., and KASAHARA, K. (1959). "Sur le système hyperbolique à coefficients constants". *Proc. Japan Acad.* **35**, pp. 547-550.

Author Index

Subject Index

G

H

I

K

L

M

N

P

R

S

T

V

W

PURE AND APPLIED MATHEMATICS

Vol. 1 Arnold Sommerfeld, *Partial Differential Equations in Physics*

Vol. 2 Reinhold Baer, *Linear Algebra and Projective Geometry*

Vol. 3 Herbert Busemann and Paul Kelly, *Projective Geometry and Projective Metrics*

Vol. 4 Stefan Bergman and M. Schiffer, *Kernel Functions and Elliptic Differential Equations in Mathematical Physics*

Vol. 5 Ralph Philip Boas, Jr., *Entire Functions*

Vol. 6 Herbert Busemann, *The Geometry of Geodesics*

Vol. 7 Claude Chevalley, *Fundamental Concepts of Algebra*

Vol. 8 Sze-Tsen Hu, *Homotopy Theory*

Vol. 9 A. M. Ostrowski, *Solution of Equations in Euclidean and Banach Spaces, Third Edition of Solution of Equations and Systems of Equations*

Vol. 10 J. Dieudonné, *Treatise on Analysis: Volume I, Foundations of Modern Analysis; Volume II; Volume III; Volume IV; Volume V; Volume VI*

Vol. 11* S. I. Goldberg, *Curvature and Homology*

Vol. 12* Sigurdur Helgason, *Differential Geometry and Symmetric Spaces*

Vol. 13 T. H. Hildebrandt, *Introduction to the Theory of Integration*

Vol. 14 Shreeram Abhyankar *Local Analytic Geometry*

Vol. 15* Richard L. Bishop and Richard J. Grittenden, *Geometry of Manifolds*

Vol. 16* Steven A. Gaal, *Point Set Topology*

Vol. 17 Barry Mitchell, *Theory of Categories*

Vol. 18 Anthony P. Morse, *A Theory of Sets*

Vol. 19 Gustave Choquet, *Topology*

Vol. 20 Z. I. Borevich and I. R. Shafarevich, *Number Theory*

Vol. 21 José Luis Massera and Juan Jorge Schaffer, *Linear Differential Equations and Function Spaces*

Vol. 22 Richard D. Schafer, *An Introduction to Nonassociative Algebras*

Vol. 23* Martin Eichler, *Introduction to the Theory of Algebraic Numbers and Functions*

Vol. 24 Shreeram Abhyankar, *Resolution of Singularities of Embedded Algebraic Surfaces*

Vol. 25 François Treves, *Topological Vector Spaces, Distributions, and Kernels*

Vol. 26 Peter D. Lax and Ralph S. Phillips, *Scattering Theory*

Vol. 27 Oystein Ore, *The Four Color Problem*

Vol. 28* Maurice Heins, *Complex Function Theory*

*Presently out of print

Vol. 29 R. M. Blumenthal and R. K. Getoor, *Markov Processes and Potential Theory*
Vol. 30 L. J. Mordell, *Diophantine Equations*
Vol. 31 J. Barkley Rosser, *Simplified Independence Proofs: Boolean Valued Models of Set Theory*
Vol. 32 William F. Donoghue, Jr., *Distributions and Fourier Transforms*
Vol. 33 Marston Morse and Stewart S. Cairns, *Critical Point Theory in Global Analysis and Differential Topology*
Vol. 34* Edwin Weiss, *Cohomology of Groups*
Vol. 35 Hans Freudenthal and H. De Vries, *Linear Lie Groups*
Vol. 36 Laszlo Fuchs, *Infinite Abelian Groups*
Vol. 37 Keio Nagami, *Dimension Theory*
Vol. 38 Peter L. Duren, *Theory of H^p Spaces*
Vol. 39 Bodo Pareigis, *Categories and Functors*
Vol. 40* Paul L. Butzer and Rolf J. Nessel, *Fourier Analysis and Approximation: Volume I, One-Dimensional Theory*
Vol. 41* Eduard Prugovećki, *Quantum Mechanics in Hilbert Space*
Vol. 42 D. V. Widder, *An Introduction to Transform Theory*
Vol. 43 Max D. Larsen and Paul J. McCarthy, *Multiplicative Theory of Ideals*
Vol. 44 Ernst-August Behrens, *Ring Theory*
Vol. 45 Morris Newman, *Integral Matrices*
Vol. 46 Glen E. Bredon, *Introduction to Compact Transformation Groups*
Vol. 47 Werner Greub, Stephen Halperin, and Ray Vanstone, *Connections, Curvature, and Cohomology*:
Volume I, De Rham Cohomology of Manifolds and Vector Bundles
Volume II, Lie Groups, Principal Bundles, and Characteristic Classes
Volume III, Cohomology of Principal Bundles and Homogeneous Spaces
Vol. 48 Xia Dao-xing, *Measure and Integration Theory of Infinite-Dimensional Spaces: Abstract Harmonic Analysis*
Vol. 49 Ronald G. Douglas, *Banach Algebra Techniques in Operator Theory*
Vol. 50 Willard Miller, Jr., *Symmetry Groups and Theory Applications*
Vol. 51 Arthur A. Sagle and Ralph E. Walde, *Introduction to Lie Groups and Lie Algebras*
Vol. 52 T. Benny Rushing, *Topological Embeddings*
Vol. 53* James. W. Vick, *Homology Theory: An Introduction to Algebraic Topology*
Vol. 54 E. R. Kolchin, *Differential Algebra and Algebraic Groups*
Vol. 55 Gerald J. Janusz, *Algebraic Number Fields*
Vol. 56 A. S. B. Holland, *Introduction to the Theory of Entire Functions*

Vol. 57 Wayne Roberts and Dale Varberg, *Convex Functions*
Vol. 58 H. M. Edwards, *Riemann's Zeta Function*
Vol. 59 Samuel Eilenberg, *Automata, Languages, and Machines: Volume A, Volume B*
Vol. 60 Morris Hirsch and Stephen Smale, *Differential Equations, Dynamical Systems, and Linear Algebra*
Vol. 61 Wilhelm Magnus, *Noneuclidean Tesselations and Their Groups*
Vol. 62 François Treves, *Basic Linear Partial Differential Equations*
Vol. 63* William M. Boothby, *An Introduction to Differentiable Manifolds and Riemannian Geometry*
Vol. 64 Brayton Gray, *Homotopy Theory: An Introduction to Algebraic Topology*
Vol. 65 Robert A. Adams, *Sobolev Spaces*
Vol. 66 John J. Benedetto, *Spectral Synthesis*
Vol. 67 D. V. Widder, *The Heat Equation*
Vol. 68 Irving Ezra Segal, *Mathematical Cosmology and Extragalactic Astronomy*
Vol. 69 I. Martin Isaacs, *Character Theory of Finite Groups*
Vol. 70 James R. Brown, *Ergodic Theory and Topological Dynamics*
Vol. 71 C. Truesdell, *A First Course in Rational Continuum Mechanics:* Volume *1*, *General Concepts*
Vol. 72 K. D. Stroyan and W. A. J. Luxemburg, *Introduction to the Theory of Infinitesimals*
Vol. 73 B. M. Puttaswamaiah and John D. Dixon, *Modular Representations of Finite Groups*
Vol. 74 Melvyn Berger, *Nonlinearity and Functional Analysis: Lectures on Nonlinearity Problems in Mathematical Analysis*
Vol. 75 George Gratzer, *Lattice Theory*
Vol. 76 Charalambos D. Aliprantis and Owen Burkinshaw, *Locally Solid Riesz Spaces*
Vol. 77 Jan Mikusinski, *The Bochner Integral*
Vol. 78 Michiel Hazewinkel, *Formal Groups and Applications*
Vol. 79 Thomas Jech, *Set Theory*
Vol. 80 Sigurdur Helgason, *Differential Geometry, Lie Groups, and Symmetric Spaces*
Vol. 81 Carl L. DeVito, *Functional Analysis*
Vol. 82 Robert B. Burckel, *An Introduction to Classical Complex Analysis*
Vol. 83 C. Truesdell and R. G. Muncaster, *Fundamentals of Maxwell's Kinetic Theory of a Simple Monatomic Gas: Treated as a Branch of Rational Mechanics*
Vol. 84 Louis Halle Rowen, *Polynomial Identities in Ring Theory*
Vol. 85 Joseph J. Rotman, *An Introduction to Homological Algebra*
Vol. 86 Barry Simon, *Functional Integration and Quantum Physics*

Vol. 87 Dragos M. Cvetkovic, Michael Doob, and Horst Sachs, *Spectra of Graphs*
Vol. 88 David Kinderlehrer and Guido Stampacchia, *An Introduction to Variational Inequalities and Their Applications*
Vol. 89 Herbert Seifert, W. Threlfall, *A Textbook of Topology*
Vol. 90 Grzegorz Rozenberg and Arto Salomaa, *The Mathematical Theory of L Systems*
Vol. 91 Donald W. Kahn, *Introduction to Global Analysis*
Vol. 92 Eduard Prugovečki, *Quantum Mechanics in Hilbert Space, Second Edition*
Vol. 93 Robert M. Young, *An Introduction to Nonharmonic Fourier Series*
Vol. 94 M. C. Irwin, *Smooth Dynamical Systems*
Vol. 96 John B. Garnett, *Bounded Analytic Functions*
Vol. 97 Jean Dieudonné, *A Panorama of Pure Mathematics: As Seen by N. Bourbaki*
Vol. 98 Joseph G. Rosenstein, *Linear Orderings*
Vol. 99 M. Scott Osborne and Garth Warner, *The Theory of Eisenstein Systems*
Vol. 100 Richard V. Kadison and John R. Ringrose, *Fundamentals of the Theory of Operator Algebras:*
Volume 1, Elementary Theory; Volume 2, Advanced Theory
Vol. 101 Howard Osborn, *Vector Bundles: Volume 1, Foundations and Stiefel-Whitney Classes*
Vol. 102 Avraham Feintuch and Richard Saeks, *System Theory: A Hilbert Space Approach*
Vol. 103 Barrett O'Neill, *Semi-Riemannian Geometry: With Applications to Relativity*
Vol. 104 K. A. Zhevlakov, A. M. Slin'ko, I. P. Shestakov, and A. I. Shirshov, *Rings that Are Nearly Associative*
Vol. 105 Ulf Grenander, *Mathematical Experiments on the Computer*
Vol. 106 Edward B. Manoukian, *Renormalization*
Vol. 107 E. J. McShane, *Unified Integration*
Vol. 108 A. P. Morse, *A Theory of Sets, Revised and Enlarged Edition*
Vol. 109 K. P. S. Bhaskara-Rao and M. Bhaskara-Rao, *Theory of Charges: A Study of Finitely Additive Measures*
Vol. 110 Larry C. Grove, *Algebra*
Vol. 111 Steven Roman, *The Umbral Calculus*
Vol. 112 John W. Morgan and Hyman Bass, editors, *The Smith Conjecture*
Vol. 113 Sigurdur Helgason, *Groups and Geometric Analysis:* Integral Geometry, Invariant Differential Operators, and Spherical Functions
Vol. 114 E. R. Kolchin, *Differential Algebraic Groups*
Vol. 115 Isaac Chavel, *Eigenvalues in Riemannian Geometry*

Vol. 116 W. D. Curtis and F. R. Miller, *Differential Manifolds and Theoretical Physics*

Vol. 117 Jean Berstel and Dominique Perrin, *Theory of Codes*

Vol. 118 A. E. Hurd and P. A. Loeb, *An Introduction to Nonstandard Real Analysis*

Vol. 119 Charalambos D. Aliprantis and Owen Burkinshaw, *Positive Operators*

Vol. 120 William M. Boothby, *An Introduction to Differentiable Manifolds and Riemannian Geometry, Second Edition*

Vol. 121 Douglas C. Ravenel, *Complex Cobordism and Stable Homotopy Groups of Spheres*

Vol. 122 Sergio Albeverio, Jens Erik Fenstad, Raphael Høegh-Krohn, and Tom Lindstrøm, *Nonstandard Methods in Stochastic Analysis and Mathematical Physics*

Vol. 123 Alberto Torchinsky, *Real-Variable Methods in Harmonic Analysis*

Vol. 124 Robert J. Daverman, *Decomposition of Manifolds*

Vol. 125 J. M. G. Fell and R. S. Doran, *Representations of *-Algebras, Locally Compact Groups, and Banach *-Algebraic Bundles: Volume 1, Basic Representation Theory of Groups and Algebras*

Vol. 126 J. M. G. Fell and R. S. Doran, *Representations of *-Algebras, Locally Compact Groups, and Banach *-Algebraic Bundles: Volume 2, Induced Representations, the Imprimitivity Theorem, and the Generalized Mackey Analysis*

Vol. 127 Louis H. Rowen, *Ring Theory, Volume I*

Vol. 128 Louis H. Rowen, *Ring Theory, Volume II*

Vol. 129 Colin Bennett and Robert Sharpley, *Interpolation of Operators*

Vol. 130 Jürgen Pöschel and Eugene Trubowitz, *Inverse Spectral Theory*

Vol. 131 Jens Carsten Jantzen, *Representations of Algebraic Groups*

Vol. 132 Nolan R. Wallach, *Real Reductive Groups I*

Vol. 133 Michael Sharpe, *General Theory of Markov Processes*

Vol. 134 Igor Frenkel, James Lepowsky, and Arne Meurman, *Vertex Operator Algebras and the Monster*

Vol. 135 Donald Passman, *Infinite Crossed Products*